LAZARE CARNOT
SAVANT

Les talents écartent l'ennui, chassent le vice
et sement la vie des fleurs: puissance, richesse,
vous n'offrez point ces avantages!

Legend of a lithograph depicting the spirit of
enlightened industry and belonging to a
great-great granddaughter of Lazare Carnot

LAZARE CARNOT SAVANT

by Charles Coulston Gillispie

*A monograph treating Carnot's scientific work,
with facsimile reproduction of his unpublished writings
on mechanics and on the calculus,
and an essay concerning the latter by*
A. P. YOUSCHKEVITCH

Princeton University Press · Princeton · New Jersey · 1971

The silhouette on the title page, drawn by the brother
of Lazare Carnot, C.-M. Carnot de Feulint, is from
*Centenaire de Lazare Carnot, 1753–1823: Notes et
Documents Inédits*, Paris: La Sabretache, 1923.
The tailpieces, too, are from this work.

This book was composed in Linotype Baskerville
Printed in the United States of America
by Princeton University Press, Princeton, New Jersey

In Memoriam
Daniel Martin Sachs III
1938–1967

Preface

THIS BOOK derives from the author's long-standing interest in the interactions of science with politics in revolutionary and Napoleonic France. Lazare Carnot remains one of the very few men of science and of politics whose career in each domain deserves careful attention on its own merits. Serious scientific and political talents seldom go together, and it was curiosity that led to the present study, curiosity about whether his career was the exception in fact that it was in appearance, and whether his scientific writings merited him the epithet "savant" ceremonially accorded in the days of his prominence as a statesman.

An answer to that question, originally imagined as an article, has issued in the present monograph. To it my colleague, Professor A. P. Youschkevitch of the Institute for the History of Science and Technology in Moscow, and I are appending a facsimile printing of certain of Carnot's hitherto unknown early memoirs in mathematics and mechanics. That the book has developed in this fashion is owing in larger measure than is usual even in the most disinterested scholarly work to the assistance of associates and friends, including a number of Carnot's descendants.

Among them I should like to express the most profound gratitude to Monsieur Pierre-Sadi Carnot and his sister, Mademoiselle Lucie Carnot, who not only welcomed my interest in their ancestor, but also invited my wife and me to stay in the house where he was born in Nolay (Côte-d'Or) during periods in the summer of 1964 when I was beginning the research and consulting the books and many papers that he left. It was there, surrounded by the record of the drafts and writings of a lifetime, that I realized that Carnot's concerns with mechanics and mathematics were occupations sustained throughout his career and not the mere recreations of an engineer turned statesman. An apology is owing to my hosts of that summer in that personal and professional preoccupations have too long postponed my completing the monograph that has now at last grown out of their hospitality.

In some respects the delay has been fortunate, however, for it was in the interval that I learned of Professor Youschkevitch's interest in Carnot's treatise of 1797, *Réflexions sur la métaphysique du calcul infinitésimal*. A new Russian translation of this work was published in 1933 to which Professor Youschkevitch contributed a learned and illuminating historical introduction. It chances that among the papers at Nolay is an undated draft of a letter from Carnot to one Johann-Karl-Friedrich Hauff, who had undertaken the German translation of the *Réflexions*. From the correspondence I learned that Carnot had originally composed the essay in competition for a mathematical prize to be awarded in 1786 by the Académie royale des sciences, arts, et belles-lettres of Berlin on the subject "how it is that so many correct theorems have been deduced from a contradictory supposition"—i.e., that of infinitesimal magnitude. I was much intrigued, partly because Carnot observed at the outset of the *Réflexions* that, though published in 1797, the treatise had been composed much earlier, and even more because I had already seen in the archives of the Académie des sciences in Paris the manuscripts of two memoirs on the science of machines that Carnot had submitted successively to the academy in a contest proposed for 1779 and reopened for 1781 on the theory of friction in "simple machines" and naval cordage. It was immediately evident that these memoirs constituted early drafts of the first treatise Carnot ever published, the *Essai sur les machines en général* of 1783, and it was most interesting thus to discover that in both the branches of science to which he contributed, his earliest efforts were elicited by competitions of the kind regularly set by eighteenth-century scientific academies and learned societies.

It was in May 1968, Professor Youschkevitch and I happening to meet then in Paris, that I learned of his much earlier work on Carnot through mentioning my own, and asked him whether in his studies of the *Réflexions* he had ever come on any evidence to confirm Carnot's remarks about its origin. He did not recall that he had, but was much interested and offered to investigate during a future visit to Berlin in order to determine whether the manuscript of Carnot's prize memoir might still be extant in the archives of the Academy. He found that it is indeed, and tells of it in the essay he has contributed to the present volume comparing the text of that "Dissertation sur la théorie de l'infini mathématique" to the later, published development that Carnot gave to his views on the justification of the calculus. I am most grateful to him for writing this essay, which handles the background and origin of Carnot's mathematical thought with a mastery I could not have achieved, and which he entrusted to me for translation from the French. It seemed to both of us that the coincidences that had thus led to the discovery of these early versions of Carnot's work were such happy ones that it would be agreeable to publish the documents in question. We are pleased that Princeton University Press has made this possible. In consequence all of the important scientific writings that Carnot left in finished form will now be readily available to scholars.

In the case of the two memoirs on the science of machines, it is the central, theoretical portions that present sufficient interest to warrant publication. I am most grateful to Monsieur Louis de Broglie, *Secrétaire Perpétuel* of the *Académie des sciences, Institut de France*, for conveying permission thus to print these documents, and to Madame Pierre Gauja of the Archives of the Academy for arranging to provide photocopies and more largely for the graciousness with which she and her staff accommodate the research of those who interest themselves in the history of French science.

Professor Youschkevitch and I are equally grateful to the Secretariat of the *Deutsche Akademie der Wissenschaften zu Berlin* for permission to publish the complete text of Carnot's "Dissertation" and in particular to Dr. Christa Kirsten, its director, who generously arranged to provide us with photocopies. We should also like to acknowl-edge the kindness of Dr. Kurt-R. Biermann, who was of great assistance in facilitating Professor Youschkevitch's study of the various documents in the archives concerning the competition of 1786 sponsored by the then Prussian Academy.

The purpose of this monograph being the relatively narrow one of explicating Carnot's scientific work, I have not attempted a full biography. To have done so would in any case have been superfluous because there has appeared in Paris a com-plete and authoritative life by Marcel Reinhard, whose book is based upon a thorough mastery of the sources, published and unpublished. I have drawn very largely upon the narrative of the two volumes of his *Le Grand Carnot* for my opening chapter. Its limited objective is to recount Carnot's personal and public life in such a way as to situate within it the incidence of his scientific work, with which Professor Reinhard did not concern himself. I have indicated the few junctures where my reading of the technical materials has led me to differ somewhat from Professor Reinhard's inter-pretations, and I should like to acknowledge the obligation that I, in common with all students of the French Revolution, must feel to his book, the first truly scholarly appraisal of Carnot's public career by one of the accomplished masters of the his-toriography of the period.

An older work remains indispensable: the *Mémoires sur Carnot* that Carnot's younger son, Hippolyte, born in 1801, composed many years after his father's death for the edification of his own children (one of whom became President of the French Republic) and published in two volumes in 1861 and 1863. The elder son, Sadi, of thermodynamics fame, having died young and a bachelor, it is from Hippolyte that later generations of the family are descended, and through his memoir that its pub-lished tradition of their ancestor was conserved. That book is to be treated by the historian with the combination of respect and skepticism appropriate to an admirable work of family piety: skepticism about Hippolyte's ability to remember accurately in later life events of his boyhood or things his father had told him about the Revolu-tion, respect for the regard in which he held the father whose exile he shared from his fifteenth year, who died when the son was twenty-two, and whose inwardness and magnanimity he knew as the scholar of later times may not, except through the con-fidence that in these personal matters Hippolyte Carnot inspires.

One such scholar is tempted to conclude from the encouragement and cordiality extended by others of Carnot's descendants that magnanimity is a hereditary charac-teristic. I should like to record my own deep personal gratitude to Monsieur Lazare Carnot, to two granddaughters of President Carnot, Madame René Giscard d'Estaing and Madame Roger Crépy-Carnot, to great friends of the next generation, Monsieur Philippe Giscard d'Estaing and Monsieur Pierre Vignial together with their wives and families, to Madame C. de Gaudart d'Allaines, and finally to two fine friends, related to the Carnot connection by marriage rather than descent, Monsieur and Madame Jean Darde, through whose kindness I was first introduced to Monsieur and Mademoiselle Carnot, the proprietors of the Nolay property, and who over the years have opened to my wife and me countless windows on the French civilization of which we enjoy the actuality while studying scientific aspects of its history. My wife joins in these expressions of our gratitude, and for my part I wish to thank her for the assistance, careful reading, and criticism she has given to this as to my other writings.

That this is becoming a long preface to a short book will not deter acknowledgment of invaluable criticism from colleagues who have read the text in rougher form than thanks to them it now is. Dr. J. R. Ravetz of the University of Leeds and Monsieur René Taton, Director of the *Alexandre Koyré Centre d'Histoire des Sciences et des Techniques, Ecole Pratique des Hautes Études,* have read the entire text and returned general and specific criticism and correction. Dr. Stephen Fox of the University of Leicester has given it the benefit of his expert knowledge of Sadi Carnot. Professor Robert C. McKeon of Tufts University has been particularly helpful with regard to the subjects treated in Chapter IV. My greatest obligation, however, is to my colleague and former student, Professor Stewart Gillmor of Wesleyan University, who has been studying work more famous than Carnot's scientifically, though in many ways it was complementary, that of Coulomb. Professor Gillmor and I have been talking Coulomb, Carnot, and early engineering science ever since the seminar that started these interests, and he has pointed out to me many things I did not see and corrected numerous errors I should otherwise have printed. I should be glad if this book were to be read as a companion to his on Coulomb, which is about to issue from the press.

My book is dedicated to the memory of a splendid friend who, though not a bookish man, took a great interest in its composition during the most difficult period in his life. His concern for those near him fortified them then in the belief that what they were doing was important. He had in common with the subject of this study, not science, but uncommon courage and the capacity, though not the chance, for statesmanship. His qualities of personal power, forbearance, intelligence, and grace were altogether extraordinary and should by rights be at the service of another Republic, our own, which sorely needs their like in this, its time of trouble.

Princeton, New Jersey
June, 1970

Contents

Preface   vii

CHAPTER I.   Biographical Sketch   3

CHAPTER II.   The Science of Machines   31

     A.   Summary of *Essai sur les machines en général*   31

     B.   Geometric Motions   40

     C.   Moment-of-Momentum   45

     D.   Moment-of-Activity—The Concept of Work   51

     E.   Practical Conclusions   58

CHAPTER III.   The Development of Carnot's Mechanics   62

     A.   Argument of the 1778 Memoir on the Theory of Machines   62

     B.   Argument of the 1780 Memoir   72

     C.   Argument of the *Principes fondamentaux de l'équilibre et du mouvement* of 1803   81

     D.   Comparing the Work of Sadi Carnot   90

CHAPTER IV.   The Carnot Approach and the Mechanics of Work and Power, 1803-1829   101

CHAPTER V.   An Engineering Justification of Algebra and the Calculus   121

     A.   Geometric Analysis and the Problem of Negative Quantity   123

     B.   The Compensation of Error in Infinitesimal Analysis   134

     C.   Analysis and Synthesis   143

ESSAY BY A. P. YOUSCHKEVITCH: Lazare Carnot and the Competition of the Berlin Academy in 1786 on the Mathematical Theory of the Infinite   149

APPENDIX A.   Carnot's "Dissertation sur la théorie de l'infini mathématique" with notes by A. P. Youschkevitch   169

APPENDIX B.   Text of the theoretical sections of the 1778 memoir on the theory of machines   269

APPENDIX C.   Text of the theoretical sections of the 1780 memoir on the theory of machines   297

     Transcribed Footnotes   341

     Bibliographical Note   347

     Index   355

LAZARE CARNOT

SAVANT

NOLAY – Maison Natale de Carnot.   D'après un dessin du Président Carnot - 1856

CHAPTER I

Biographical Sketch

IN 1797 a small book called *Réflexions sur la métaphysique du calcul infinitésimal* was published in Paris. The author, Lazare Carnot, was the leading member of the Directory of the French Republic, in which office he, together with four not very congenial colleagues, exercised the executive power of the state in the regime that from 1795 to 1799 intervened between the demise of the revolutionary Convention and the Consulate of Napoleon Bonaparte. A little later in 1797 a coup d'état named for the revolutionary month of Fructidor displaced Carnot from government. Prudence led him to take refuge from the consequences for the next two years in Switzerland and Germany. Back from exile in 1800, he was named Minister of War by Napoleon and served in that post for some months until a fundamental incompatibility with the Bonapartist order became clear to him. After his resignation Carnot published further treatises of a scientific nature, most notably *Principes fondamentaux de l'équilibre et du mouvement* and *Géométrie de position*, both in 1803. Alone among the members of the Napoleonic Tribune, an honorific and juridical council of civic leaders, Carnot refused in 1804 to support naming Napoleon first consul for life, the step preparatory to converting the Consulate into the Empire. Nevertheless, he rallied to the Emperor in 1815 during the extremity of the Hundred Days, served him as his last Minister of the Interior, and in consequence ended his own days in Magdeburg in 1823, once again an exile.

Prior to the Directory, Carnot, a captain of engineers at the outset of the Revolution, had been one of the three most powerful members of the Committee of Public Safety (Robespierre and Saint-Just were the others) who, governing France throughout the early years of warfare against monarchical Europe and under the Terror, converted anarchy into revolutionary authority and defeat into victory. Carnot's own responsibility was that of dictatorial war minister in time of national emergency. In textbooks of French history he is "L'Organisateur de la Victoire." Many towns in France have an avenue, a street, or a square called Carnot, and people now have usually forgotten whether it is after Lazare himself, "le grand Carnot," after his son Sadi, a founder of the science of thermodynamics, or after his grandson, the latter's nephew, who was also called Sadi and who was President of the French Republic from 1887 until his assassination in 1894.

Among active statesmen, Lazare Carnot must surely have been unique in occupying periods of eclipse by preparing treatises on mechanics and mathematics. For his life reversed the usual order, wherein scientists often become public figures in consequence of their attainments. Carnot was unknown scientifically until after he had organized victories. Reading in his writings, however, soon establishes the problem of his career to be one of more than merely incidental significance in the history of science.

Most engineers will have heard of a "Principle of Carnot" in the science of mechanics. Unlike French schoolchildren, who can usually say what Carnot accomplished in the Revolution, physicists and engineers, when asked what he did, normally look vague for a few moments, hesitate between the concept of a reversible process and the second law of thermodynamics, and tentatively propose the expression $\frac{T_2 - T_1}{T_1}$, which symbolizes that the efficiency of a heat engine is a function of the difference in temperature on the absolute scale at the commencement and completion of each cycle of operation. They are quite right in a way, for that expression did indeed evolve out of the argument advanced by Carnot's son Sadi in his famous and his only published memoir: the *Réflexions sur la puissance motrice du feu* in 1824. When reminded, they recall another, apparently more axiomatic principle concerning machines of every sort, to the effect that a condition of maximum efficiency is the elimination of percussion or turbulence (in the case of hydraulic machines) in the transmission of power—the very phrase "waste motion" derives from considerations started by Carnot. That principle, sometimes promoted to the dignity of a theorem or a law and stated in various ways, was known by the name of Carnot until sometime after the middle of the nineteenth century. Only rarely is it still so called. Why its parentage should have been gradually obscured, is difficult to say. Perhaps it was so by virtue of its very generality, for it follows directly from conservation of energy (of which sensed but unstated law it was, in a way, one of many early partial instances).

Perhaps, also, Lazare Carnot's reputation in science has suffered by comparison to his son's from a certain glamor that has attached to thermodynamics at the expense of its own parent science of mechanics, an appeal heightened by the pathos attaching to Sadi's short life. It may further be that Lazare Carnot's scientific career, even though made possible by his political and military achievements, has also been overshadowed by them. At all events it was Lazare Carnot who, while a young engineering officer, enunciated the principle of continuity in the transmission of power in a small treatise called *Essai sur les machines en général* first printed in 1783 in Dijon, provincial capital of his native Burgundy.

The essay contains much matter worthy of remark. Its very conception, the application of the science of motion to the determination of the general principles underlying the operation of machines, pertains to the development of mechanics as an engineering science rather than at one extreme a mode of mathematical analysis or at the other a merely practical subject. From among the elementary principles of mechanics, Carnot selected the conservation of live force (*vis viva* or in later parlance kinetic energy) as the basis from which propositions in the science of machines might most naturally be derived. In doing so, he ignored the metaphysical disputes that had discredited it earlier in the century, brought it forward from the status it retained of an auxiliary principle used mainly in hydrodynamics and occasionally in celestial mechanics, and accelerated one of the sequences that half a century later issued in the law of conservation of energy. In the course of deriving from the principle of live force his own principle of continuity, Carnot recognized its equivalent, the product of the dimensions of force times distance, to be the physical quantity that the science of mechanics might employ to estimate the efficacy of machines. Carnot called that quantity "moment-of-activity." The name that has lasted is "work," proposed by Coriolis in 1829. Two further features of his method of reasoning, while they are not to be identified so explicitly as contributions or as findings, did point the way toward later developments in mechanical analysis of problems involving work and power. Carnot employed trigonometric expressions for representing the projections of a directed force or velocity upon a second direction, and for combining analytically rather than by geometric resolution what later were distinguished as vector quantities. Even more interesting, however, is the analytical device that he called "geometric motions," which also belongs in the pre-history of vector analysis, and which further suggested the idea of a reversible and cyclic mechanical process later applied to thermodynamics by his son Sadi.

Given the fundamental and elementary character of these findings, it may seem surprising that, despite a second printing of Carnot's *Essai* in Paris in 1786, his treatise should have been ignored by the scientific community until after he had become great and famous in the Revolution and a member of the *Institut de France*. Thereafter, he published further writings, on the calculus, on mechanics, and on geometry, all of which repay attention. The study of these works is interesting, not simply to re-establish as a matter of simple justice the parentage of important principles of mechanics, for although Carnot had them it is doubtful that mechanics had to have them from him or not at all, nor because of the pathetic spectacle of genius unappreciated and finally coming into its own, for as will appear from analysis there was in them more of naïveté than of genius. What is interesting is the subject matter forcing itself into expression, so to say, through the mind of an ambitious young engineer, unsubtle and unafraid, and the access which that spectacle gives to the interaction at a critical juncture in the history of science proper between developments in mechanics, physics, and mathematics, developments not to be understood except in virtue of the institutional, the social, and even the political development of the profession of engineering. Before going into the requisite detail of Carnot's subject matter, therefore, it will be well to review the circumstances wherein his works were written.

Lazare-Nicolas-Marguerite Carnot was born on 13 May 1753 into a Burgundian bourgeois family that occupied a leading position in its locality. His father was a lawyer and a notary in the town of Nolay, situated on the rocky reverse slope of the ridge along the eastward side of which march the vineyards producing the great burgundies of the Côte-d'Or. Beaune is the nearest city of any consequence. The manor house where Carnot was born still stands, the most considerable private dwelling in the town, and although the rooms fronting over a decayed terrace onto the modest square are no longer used, it is still in the possession of the Carnot family.

Carnot's father, Claude, provided for the education of his three older sons who were then expected to make their return as tutors and preceptors to the three younger. Two of the six showed an early aptitude for mathematics and technical subjects, Lazare, the second son, and his younger brother, Claude-Marie, usually called Carnot de Feulint, who followed him into the Royal Corps of Engineers. Feulint also entered into the politics of the Revolution, wherein his career paralleled and in the early days seemed likely to outshine his elder brother's.

In the eighteenth century a French education would often begin with a clerical tutor or family preceptor at home, followed by a strict curriculum of classical studies or "humanities" in a boarding school or "college," after which paths diverged to lead to the university for law, medicine, or divinity, or else to service schools for military or naval training. In the army the Royal Corps of Engineers and to a lesser degree the artillery required technical proficiency, and indeed were called the "armes savantes." Carnot learned his humanities in the College of Autun, formerly Jesuit and then Oratorian, the same in which Napoleon Bonaparte was placed ten years later by his father. Carnot finished at the usual age of sixteen. The memoir by Hippolyte recounts a family story: in defending his thesis he refused the customary presence at his side of a professor in case his Latin should need prompting and stood up alone to questioning.[1]

Thereafter, the usual course would have been to enroll in one of the schools that gave intensive cramming in mathematics and mechanics looking to the entrance examinations for the service schools of the "learned arms."[2] Instead, Carnot tried preparing for the engineering school at Mézières on his own, wishing to save his family the expense. He failed badly in the examinations given in 1769. Thereupon, his father determined to send him to Paris for the requisite schooling, bespoke the protection of a patron, the Duc d'Aumont, who by virtue of one of his titles held the seigneurie of Nolay, and enrolled Lazare in the most highly reputed of the several such establish-

[1] Hippolyte Carnot, *Mémoires sur Carnot*, 2 vols. (Paris, 1861–63), I, 82.

[2] An outline of the technical possibilities of military and naval education is given in two chapters contributed by Roger Hahn to the symposium *Enseignement et diffusion des sciences en France au XVIIIe siècle*, ed. René Taton, vol. XI of the series "Histoire de la pensée" published by the *Ecole pratique des hautes études* (Paris, 1964), pp. 511-558. The same volume contains a chapter by René Taton on "L'Ecole royale du génie de Mézières," pp. 559-615.

ments in the capital, the school kept by one Louis-Siméon de Longpré in the rue Culture-Sainte Catherine in the Marais.[3]

Already in his school days Carnot was encountering two of those who marked his way of apprehending the problems of mechanics, Bossut and d'Alembert. The abbé Charles Bossut conducted the examinations for Mézières and in 1763 published the first edition of a widely used textbook of mechanics.[4] Longpré could claim his friendship, and that strengthened the drawing power of the establishment. It was quite consistent with an elderly d'Alembert's penchant for taking up and paternalizing gifted young men that he, too, should have been a friend of Longpré and a frequent visitor, gathering the pupils round his chair and trying them on mathematical puzzles. In old age Carnot recalled having been singled out by the great man to answer thorny questions and encouraged to feel equal to a successful career.[5]

He began by succeeding in the next round of examinations, and entered at Mézières in 1771 for the two years of training that completed the formal education of the engineering cadets. It was a requirement of the regime under which the school operated that candidates for its examinations should furnish prior proof of noble ancestry, or if that was impossible, of deriving from a family "vivant noblement," that is to say one engaged in no degrading occupation. In fact the great majority of cadets in engineering were of bourgeois background, and the Carnot family made no difficulty about securing the necessary certificates, which even provided Lazare with certain putative if distant uncles and cousins distinguished in the profession of arms. The matter seemed less shocking then than it does according to democratic canons of social and professional propriety.

It is natural, however, that the revolutionary generation and later historians judging by those canons should have been impressed by a famous indignity inflicted upon the greatest talent ever associated with the school at Mézières, that of Gaspard Monge. Monge came from circumstances in Beaune that were very humble indeed. Refused admission as a cadet for that reason, he was taken on instead as an assistant in the shop.

---

[3] In Paris, there were three such establishments, those of Bouffet, Berthaud, and Longpré. (Taton, op. cit., p. 593n; Marcel Reinhard, Le grand Carnot, 2 vols. [Paris, 1950–52], I, 23-24.) Only the latter two were mentioned by Hippolyte Carnot, who gives a bit of detail on the regime of the Longpré establishment, which became increasingly fashionable and prosperous in the later 1770's and 1780's. It may be interesting to note in passing that Hippolyte (I, 85), writing from notes or recollections of his father's reminiscences,

ascribed to these schools the role more often imputed to Mézières—i.e., they "tenaient la place si gloirieusement prise plus tard par l'Ecole polytechnique. Elles ont, l'une et l'autre, formé des ingénieurs éminents."

[4] Traité élémentaire de mechanique et de dinamique appliqué principalement aux mouvemens des machines (Charleville, 1763). There were many later editions and companion volumes on algebra, geometry, and arithmetic. See Taton, op. cit., p. 584, n. 2.

[5] H. Carnot, op. cit., I, 85-86.

His ability in drafting and in pedagogy there appearing, he was made professor first of mathematics and then also of physics to the cadets from whose ranks he had been excluded. This situation, too, will look less startling when it is appreciated that only in the classroom did the professor in a military school exercise authority. In the perspective of the larger mission, his post was one of service like that of a preceptor in a collective household—Monge's predecessors in physics and mathematics, Nollet and Bossut, both were clerics.[6]

Generalizing from the figure cut by Monge in the part of a geometrical Figaro, certain historians, Reinhard among them, have tended to make the entire Corps of Engineers who studied under him into a band of frustrated and semi-humiliated technicians, an intelligentsia within the armed forces that would have its due in the Revolution.[7] Nor is there any doubt of Monge's own participation and that of his close entourage in the politics of the Jacobin left. Of that group, however, only Meusnier was a former student, and among other graduates of Mézières an insignificant number made themselves at all prominent in the Revolution. There were Carnot, his brother Feulint, Prieur de la Côte-d'Or, to a less degree Bénézech de Saint-Honoré and Bureaux de Pusy. It is difficult to be confident that the activities of this handful testify to greater affinity for the Revolution among the engineers than among any representative body of able men of the Third Estate, and Reinhard may have glimpsed a somewhat oversensitized vision of the significance to Carnot's political career of the social milieu of engineering school. There is no reason to look more closely there than elsewhere in the regime of eighteenth-century France for an explanation either of his ardent republicanism or of his fidelity to the tradition of personal solidity and magisterial probity in which he was brought up.

At any rate Hippolyte says little of Mézières, and the further possibility suggests itself that Monge and Carnot were not actually close—despite the apparent filiation of their mathematics.[8] The first suspicion that they were not arises from an absence of evidence. There was little or no mention of the one by the other, although Monge's disciples were given to rather fulsome tributes. Nor is it recorded that Monge took the slightest interest in Carnot's writings in mechanics or mathematics or did anything to secure them a hearing. Some further inference may be permitted from their differences in temperament: Monge's that of a mathematician and pedagogue who, when he involved himself in public affairs, proved to be theoretical, emotional, incapable in decision, deficient in judgment, and inattentive to detail; Carnot's that of an engineer, eminently practical, able in conception and in execution, and capable of the

[6] For the clerical aspect of Bossut's career, see Thomas F. Mulcrone, "A note on the mathematician abbé Charles Bossut," *Bulletin of the American Association of Jesuit Scientists*, 42 (1965), 16-19.

[7] Reinhard, *op. cit.*, I, 61-66, 148-160.

[8] On the career of Monge, see the standard study, René Taton, *L'oeuvre scientifique de Monge* (Paris, 1951).

deeper consistency that inheres in judging of circumstance and practicality rather than in asserting moral absolutes.

Carnot did, it is true, lose his judgment in one episode, but that was a love affair and no affair of state. Frequently on leave from garrison duties in the 1780's, he was often in Dijon and much in the company there of the family of a certain chevalier de Bouillet, with whom his father had business relations of long standing, and who was the prouder of his title and particle for having acquired them shortly before. Carnot fell in love with his daughter, Ursule. She became his mistress. Almost surely his intention was that they marry when his circumstances should permit. Her father had grander views. Unknown to Carnot he arranged her marriage to one chevalier de Duesme. Carnot learned the news only from his brother, Joseph, who had it from the publication of banns a week before the intended wedding. Carnot hastened up from Nolay, arrived in Dijon the day before the scheduled ceremony, was told by Ursule that matters had advanced too far to rearrange, upon which rebuff, furious, he forced himself upon Duesme, apprised him of the true nature of his relations with Ursule, and documented his conquest by showing her letters. The scandal did break up the marriage. It also landed Carnot in prison by letter of cachet, for it is not surprising that Bouillet should have charged him before his superiors with conduct unbecoming an officer and a gentleman. Thus it happened that Carnot was in confinement in Béthune—for two months only, it is true—when the Revolution began its course in April 1789 with the meeting of the States-General in Versailles. He was released on 29 May by order of the chief of engineers.[9]

His professional life in the sixteen years prior to that affair had not on the whole been stimulating. Graduated from Mézières on 1 January 1773 and commissioned a lieutenant, Carnot entered upon a sequence of garrison duties first at Calais and then, successively, at Le Havre, Béthune, Arras, and Aire. Only in Arras did he find cultivated company. There in the 1780's he was admitted to the predominantly literary society of Rosati, a group of beaux esprits famous rather for having counted Carnot and Robespierre among the membership (according to Hippolyte, they were never congenial) than for any lasting contribution to letters. Carnot himself had a minor gift for versifying that he exercised more often, perhaps, than the quality of his inspiration warranted. Not that he lacked employment for his training, for during his tour at Calais the fortifications of the port were being rehabilitated, and from Le Havre he was detached for three years of duty on the harbor under construction at Cherbourg, the most considerable and elaborate engineering works undertaken by the French military establishment in the latter part of the century. In these assignments he made the reputation of a sound, reliable, and enterprising young officer. His personnel file contains a series of commendations by his superiors.

[9] For this episode, see Reinhard, I, 66-70, 137-147.

Nevertheless, the life was not one to engage the full talent or fulfill the ambition of an able and educated man. In a memoir of 1776, Coulomb, a fellow engineer and in science a far more famous one, took occasion to testify to its limitations: "After graduating from Mézières, a studious young man who would withstand the boredom and monotony of his work had no choice except to cultivate some branch of science or literature that was entirely independent of his professional duty."[10] Later on Carnot described himself as having been "neglected, lonely, absent-minded, preoccupied, what was called a kind of 'philosophe' or in other words sort of an odd type,"[11] during the years between 1773, when at the age of twenty he received his commission, and 1784, when he won a competition set by the Academy of Dijon for an essay on Vauban. In the perspective of his later military and political career, the *Eloge de Vauban* appears, naturally enough given the subject and the discussion that ensued, to be its opening episode.[12]

In the course of the present monograph it will appear that, although no detailed account of his studies during that interval is possible, the individuality of his scientific work is such that he must have been occupying it with a self-education in mechanics and mathematics that carried him far beyond the curriculum of Mézières. That training, completed when the cadets were twenty on the average, amounted to no more than the equivalent of the first two years of engineering school in later times. The scientific work required in 1772 consisted of Camus' four-volume textbook of arithmetic, geometry, and statics; Bossut's treatises of dynamics and hydrodynamics; and one-year courses in engineering drawing, perspective or descriptive geometry, and experimental physics.[13] It must probably have been in the isolation of garrison life that Carnot set himself to the reading and re-reading of d'Alembert, Bossut, Bélidor, the mechanics of Euler, and the hydrodynamics of Daniel Bernoulli. His early work bespeaks familiarity, but an autodidactic familiarity. Unlike two fellow engineers who were making important reputations in science, Coulomb and Meusnier, and who spent long intervals in Paris, Carnot had no access to the scientific milieu of the capital. He passed his lengthy and frequent leaves at home in Nolay and in Dijon.

That he accompanied these first ten years of his career with private mechanistic

---

[10] Quoted by Reinhard, *op. cit.*, I, 44, from a document in the *Archives historiques du ministère de la guerre.* The memoir from which this passage is quoted is being printed in full as Appendix B of C. Stewart Gillmor's forthcoming monograph, *Coulomb and the Evolution of Physics and Engineering in Eighteenth-Century France*, to be published by Princeton University Press. Coulomb wrote it while stationed at La Hougue, near Cherbourg, and dated it 1 September 1776.

[11] H. Carnot, I, 93.

[12] The prize was first offered in 1783, but, in the absence of sufficiently worthy entries, the competition was adjourned until the following year. See Reinhard, I, 76-86; H. Carnot, I, 99-117.

[13] For details of the curriculum and full bibliographical identification of the successive editions of these elementary courses, see Taton, "L'Ecole royale du génie de Mézières," *loc. cit.*, esp. pp. 575, n. 2; 584, n. 2; 593-594.

study is more than plausible conjecture. We do in fact know that he was doing so. It was while he was just out of school and still at Calais that his father sent the younger brother, Feulint, to reside with him and to be tutored by Lazare in mathematics and mechanics for the entrance examinations to Mézières.[14] (Feulint succeeded in late 1773 on the first try.) What is more substantial, Carnot's earliest published piece of writing was not the *Eloge de Vauban* of 1784, but the *Essai sur les machines en général*, of 1783, which is analyzed at length in the next chapter, and which contains in principle all that he actually contributed to mechanics, while also marking the starting point of the French engineering tradition in the science of machines. From documents in the Archives of the Academy of Sciences in Paris, in those of the German Academy of Sciences in Berlin, and in those of the Carnot family in their house at Nolay, it is possible to fix with precision the occasion for Carnot's earliest formal compositions in the literature of mechanics and mathematics.

Like the *Eloge de Vauban* itself, those writings were elicited by competitions set by learned societies. He composed two early drafts of the *Essai sur les machines*, the first completed in March 1778, while at Cherbourg and the second in July 1780 at Béthune, in response to a competition first announced by the Academy in Paris in April 1777 for the year 1779, and later adjourned to 1781 since the judges found none of the initial entries to be worthy. The prize on the second round went to a memoir submitted by Coulomb, and Carnot's was awarded honorable mention. The contents and their conformity to the Academy's desiderata, are topics best reserved for discussion following analysis of the published *Essai sur les machines* itself.

As for Carnot's complementary interests in the foundations of mathematics, the *Réflexions sur la métaphysique du calcul infinitésimal* grew out of the memoir that Carnot submitted to the Royal Prussian Academy of Sciences in 1785: it is printed below as Appendix A. Carnot's entrance into that competition followed his one success among these early efforts, the *Eloge de Vauban*. It may well have been the presence of Prince Henry of Prussia at the public meeting of the Academy of Dijon in which he read his *Eloge* that turned Carnot's thoughts to Berlin. The occasion was a ceremonial one presided over by the Prince de Condé. Afterwards (according to Hippolyte's account) Prince Henry asked to have the author presented, and in the course of conversation offered him a commission in the Prussian army. However that may have been, there is no doubt that throughout his life Carnot displayed a marked sympathy for things German.

Carnot made one further attempt just prior to his recognition for the Vauban *Eloge* to win the approbation of the Academy of Sciences. On 17 January 1784 he sent in a "Lettre sur les Aerostats" in response to its invitation to interested parties to communicate reflections inspired by the first human flight. On 5 June 1783 the brothers

---

[14] H. Carnot, I, 94-95.

Montgolfier had created a sensation by sending a balloon filled with heated air soaring to a height of 6,000 feet above the town of Annonay. Immediately their feat raised the engineering problems of locomotion and stability of flight. The studies evoked were by no means all sterile. A memoir of Meusnier contains the principle that much later governed adjustment of the specific gravity of the submarine.[15] Carnot addressed himself rather to the problem of the "dirigeable" and proposed a scheme for a propeller to be powered by a motion of systole and diastole created in the balloon itself by the dissipation of heat, a sort of jellyfish effect. The notion was visionary enough. What excites interest in retrospect is the discussion of heat, fluids, and the potentiality awaiting the steam engine. "Notice by the way, gentlemen," he wrote, "how much labor will be saved in factories when the mechanism of heat is better understood." And again: "The engine powered by heat provides a very powerful motive principle, and the principle can as easily be adapted to moving blades and wheels as beams and pistons."[16] Throughout his life, indeed, Carnot maintained a lively interest in the actual work of mechanical invention—rather in point of principle, criticism, and appreciation, however, than actual participation: he does not appear to have been gifted with practical inventiveness himself.

Only with his military writings did Carnot win attention. Biographers and historians interested in Carnot, wartime leader and member of the Committee of Public Safety, have naturally looked in these early compositions for clues to what their man would become.[17] In good philosophic vein Carnot invoked the example of Vauban to convey his own eighteenth-century sentiments about natural equality in men of merit and inequities in the social order. The central thrust of the essay was strategic, however. It celebrated, in the declamatory style thought appropriate to ceremonial occasions, the complements of geometry and humanitarianism that impelled Vauban, the founder of professional military engineering, to reconceive warfare in the civic spirit of a defense of civilization instead of the barbarous one of destruction of an enemy.

Historians have sometimes waxed a little wry at Carnot's expense for having advocated, when a young Turk, the humane warfare of position, and then, when actually in charge of a war, having put into effect in the Revolution the contrary strategy of mobility, mass, offensive, and conquest. What must be remembered is the difference

[15] Jean-Baptiste Meusnier de La Place, "Mémoire sur l'équilibre des machines aérostatiques . . ." (3 December 1783). This paper is republished in *Mémoires et travaux de Meusnier relatifs à l'aérostation*, ed. Gaston Darboux (Paris, 1910). Darboux's preface gives an admirable account of the early history of flight.

[16] H. Carnot, I, 125.

[17] The Academy of Dijon, embarrassed by its experience with its most famous laureate, Jean-Jacques Rousseau, thereafter took care to edit the texts it crowned before publication. Hence it came about that the published version of Carnot's *Eloge* was a toned down version of what he actually had delivered. Its original tenor has been plausibly reconstructed by G. Duthuron, "Un éloge de Vauban," *Annales historiques de la révolution française*, 17 (mai-juin, 1940), 152-165.

in circumstances. In the 1780's Carnot was a young engineering officer writing out of the specialist tradition of his corps and upholding its mission in the design and construction of formal works of siegecraft and defensive fortification. Doing so required him to contend against the emphasis of the combat arms on gallantry, movement, and command under fire, for chivalric pretentions were the line officers' part in the aristocratic resurgence. In power in the 1790's, however, Carnot disposed, not of the disciplined, technically trained, professional, and careful forces that such a conservative strategy presupposed, but of untrained levies, some under arms because of patriotism and some because of conscription. Soldiers of the one sort furnished a commander with dash and those of the other with mass, and both sorts were altogether different from the armies of Louis XIV and the eighteenth century.

Some rather forced attempts have been made to draw analogies between Carnot's military-political thought and his approach to the science of mechanics, largely through license taken from Arago's expansiveness.[18] Thus Hippolyte, the tendency of whose memoir is to tone down the revolutionary sentiments of his father, based himself on a "most ingenious" remark of Arago about "the analogy that can almost always be pointed out between scientific theories and the rules of conduct of their authors. The Carnot who labored to demonstrate the drawbacks of sudden changes in mechanical processes was certainly the same Carnot who disliked violent perturbations in politics and who would have recourse to revolutionary methods only when the path of reform seemed to him completely unfeasible."[19] It is odd that Reinhard, whose quite opposite purpose it was to exhibit Carnot in the guise of the revolutionary man of talent rather than the non-political technician of victory, should have chosen to cite the same analogy. His book does so in evidence of an alleged tendency on the part of Carnot to "move from the general to the concrete in applying mechanical laws to society," and tells us that this bent to abstraction prevented him from discovering the

[18] François Arago, "Carnot," in *Oeuvres complètes* (Paris, 1854), vol. I, "Notices biographiques," pp. 511-638. The *Eloge* was read before the academy on 21 August 1837.

[19] H. Carnot, I, 120. "Dans ma jeunesse, encouragé par la bienveillance, par l'amitié dont Carnot voulait bien m'honorer, je prenais quelquefois la liberté de reporter ses souvenirs sur ces grandes époques de nos annales révolutionnaires où les partis, dans leurs convulsions frénétiques, furent anéantis, vaincus, ou seulement apaisés par des mesures brusques, violentes, par de véritables coups d'Etat. Je demandais alors à notre confrère comment, seul entre tous, il avait constamment espéré d'arriver au but sans secousses, et sans porter atteinte aux lois; sa réponse, toujours la même, s'était profondément gravée dans ma mémoire; mais quelle ne fut ma surprise lorsque, sortant un jour du cercle d'études qu'un jeune astronome doit toujours s'imposer, je retrouvai textuellement la réponse constante dont il vient d'être question dans l'énoncé d'un théorème de mécanique; lorsque je vis que notre confrère m'avait toujours entretenu de l'organisation politique de la société, précisément comme dans son ouvrage il parle d'une machine où des changements brusques entraînent nécessairement de grandes déperditions de force, et tôt ou tard amènent la dislocation complète du système!" Arago, *Oeuvres complètes*, "Carnot," I, 539-540.

turbine in extension of his principle of continuity and from enlarging his political involvements into true statesmanship. In the few paragraphs that Reinhard devotes to Carnot's mechanics, he takes the criticism of the principles of the science in the *Essai sur les machines* to be symptomatic of intellectual pride and an instance of Carnot's desiring to prove everyone wrong, not only the vulgar in their prejudices, but also Descartes and even Archimedes in their mechanics.[20]

In fact, as will appear, Carnot's mechanics offers no basis for any of these judgments, nor is it through such literal or metaphorical analogies that the serious interactions of science and politics ever do become manifest. All one can reasonably infer from the *Essai sur les machines* is that Carnot was a still youngish officer who must have been disappointed in the reception of scientific essays before he managed, in the five years remaining prior to the Revolution, to break out of routine, isolation, and obscurity by means of his writings on military doctrine and strategy. The success of his *Eloge de Vauban* drew him into the midst of the skirmish of the books then being fought amongst elements of the French armed forces, wherein the political interests of the several arms and services were inextricably mingled with conflicting theories of warfare. He was immediately attacked by a young officer of artillery whose own extra-military writings were of a sort quite different from Carnot's mechanics (though not perhaps irrelevant to his private life): Choderlos de Laclos, author of *Les liaisons dangereuses*,[21] who addressed a "Lettre à MM. de l'Académie française" ridiculing the style, the argument, and the scholarship of Carnot's *Eloge*.[22] Almost simultaneously appeared a more damaging assault, a pirated reprinting of the *Eloge* accompanied by an invidious array of insulting annotations calculated to destroy the work entirely. This latter was the work of one of the protagonists in the main strategic debate, General the Marquis de Montalembert, whose protégé and disciple Choderlos de Laclos then was.

The story is a very complicated one, in itself hardly worth the space it takes to clarify the misunderstandings through which it developed. Montalembert, formerly of the cavalry, was the author of a novel system of fortification that he called perpendicular.[23] In effect, his notion was to replace Vauban's complex fortification in depth by vertical redoubts studded with casements to permit the massing of cannon and saturation of attacking forces by firepower. He had been authorized to carry out experimental demonstrations of his construction in tests conducted in the isle of Aix in 1780, and had

[20] Reinhard, I, 103.

[21] Four vols. (Amsterdam and Paris, 1782).

[22] Lettre à MM. de l'Académie française sur l'Eloge de M. le Maréchal de Vauban (Amsterdam and Paris, 1786).

[23] *La Fortification perpendiculaire ou l'art défensif supérieur à l'art offensif*, 5 vols. (Paris, 1776–1784). Five supplementary volumes and many ancillary writings appeared in the next ten years. On these issues, see Robert S. Quimby, *The Background of Napoleonic Warfare: The Theory of Military Tactics in 18th-Century France* (New York: Columbia University Press, 1941).

been severely criticized, not only by advocates of offensive warfare, but by the higher officers of the Corps of Engineering, whose defensive doctrine he shared but whose designs and methods he would supplant. Smarting under the strictures of the latter, Montalembert took the *Eloge de Vauban* for an essay by a young engineering officer inspired by his superiors to discredit his proposals. In fact, his conclusion was gratuitous. Carnot had never read Montalembert's works when he composed the *Eloge*. That gap in his own information he admitted in a letter published in a collection of essays on perpendicular fortification acknowledging the interest and importance of Montalembert's ideas for their common belief in defensive warfare. And, in fact, Laclos and Montalembert were later associated with Carnot in the military effort of the Year II, though rather in the development of new weapons and tactics than in the perfection of defense.

To Carnot's superiors, however, his open-mindedness in this discussion and subsequent alliance with Montalembert appeared to be less a piece of military scholarship than a betrayal, for they, and particularly Fourcroy, the chief of engineers (not to be confused with the chemist Antoine Fourcroy), had rallied to his defense against the onslaught of Laclos.[24] It is true that the enemy Fourcroy had in view was in the other camp altogether, the Comte de Guibert, proponent of the offense, disciple of Frederick the Great, and apostle of the modern concept of total war. Guibert held the Corps of Engineering in contempt. He reminded it that the name derived etymologically from engine and not from genius,[25] and campaigned to demolish the eighteenth-century system of fortification and base the order of battle on a mobile army modelled after the Prussian. As in any entrenched military establishment, there were, indeed, many superfluous installations. In order to save those that were essential, Fourcroy admitted as much in a publication that he inspired purporting to derive from Vauban himself.[26]

Not content with having seemed to repudiate Fourcroy's support against Laclos and the artillery, Carnot now turned around to attack this set of essays for the fabrication that it was. His memoir is a passionate defense of fortification in general.[27] In retrospect his conduct does appear to have been guided by that impracticality that became the higher practicality of the revolutionary mentality. It is not surprising that his brief imprisonment, coinciding as it did with the last weeks of the old regime, should often have been attributed to the professional disfavor of his superiors rather

[24] C.-F. Fourcroy de Ramecourt, *Mémoire sur la fortification perpendiculaire, par plusieurs officiers du Corps Royal du Génie . . .* (Nyon and Paris, 1786).

[25] H. Carnot, I, 128. Guibert's most important treatise was *Essai général de tactique*, 2 vols. (London and Paris, 1772). On Guibert see Robert R. Palmer, "Frederick the Great, Guibert, Bulow," *Makers of Modern Strategy*, ed. Edward Mead Earle (Princeton, 1943), pp. 49-74.

[26] *Recueil de quelques mémoires sur la trop grande quantité des places fortes qui subsistent en France* (Paris, 1788).

[27] L. Carnot, *Mémoire à présenter au conseil de la guerre au sujet des places fortes qui doivent être démolies ou abandonnées* (Paris, 1789). Composed in August, 1788.

than to the personal imbroglio of his love affair. Such, indeed, was the account of the matter that he put about himself, one that in the natural service of self-respect he probably found it easy to believe.[28]

The unexpectedness of what happened to persons swept along in the ineluctability of what happened to society always enhances the historical drama in the French Revolution. There is no need to follow Carnot's political career in detail: the tale has been well told by Reinhard in two full volumes. It will be enough to evoke it as the context, the quite disproportionately important context, in which recognition for scientific originality together with great political authority came the way of one who entered the Revolution a no longer young engineering officer betraying personal and professional frustrations in increasingly headstrong and angry writings. The earliest phase, the period of the Constituent Assembly, saw that pattern relieved. For Carnot the relief came mainly through the new facility of expressing grievances and advocating change.

At the outset he did not think to go beyond professional matters. The spirit, however, was that of revolution. That the army must be reformed to secure promotion in accordance with merit rather than birth; that equity and dignity must be assured to the Corps of Engineering; that the engineers and the artillery would best be amalgamated into a single, scientifically-trained corps headed by an elective council, rather than an authoritarian commander; that warfare was itself to be limited and humanized by adoption of Montalembert's combination of firepower with fortification—these were ideas shared by many of Carnot's colleagues. His advocacy was distinguished mainly in its imperative tone and further in that like many countrymen, though few professional colleagues, he went over the heads of his superiors and chose to address his reclamations not to the War Office but directly to the National Assembly.[29]

The way to politics itself was opened by his reassignment to duty in the North, specifically at Aire in the region of Calais, where he and his brother Feulint had put down roots among the interests and sympathies of the locality. There for a time he followed in the way smoothed by his younger brother's readier congeniality. Lazare did so personally and married into the same family as Feulint. He did so politically and presented himself alongside Feulint as candidate for deputy to the Legislative Assembly from the Pas-de-Calais in September 1791. Having taken part in various local councils and deputations to the Constituent Assembly, Feulint was the better known of the brothers regionally and was elected to the first seat on the second ballot. Lazare had joined the Society of Friends of the Constitution in his garrison town of Aire and

[28] Reinhard, I, 155-157.
[29] *Réclamation adressée à l'Assemblée nationale contre le régime oppressif sous lequel est gouverné le Corps Royal du Génie . . . A* manuscript copy is preserved in the Carnot family archives at Nolay. On these matters, see Reinhard, Chapters 11 and 12 (I, 148-176).

had been its president for a time. That was his only formal political activity prior to his election, which he won for the ninth seat and then on the second ballot.

That so limited a political experience as Feulint's, and even more so Lazare's, should suffice for election as representative of the people will be less surprising when it is recalled that the outgoing Constituent Assembly had stipulated, unwisely in the judgment of most historians, that none of its members might be reelected. The French people having had no voice in their own affairs, there were in 1791 no communities in which political experience more extensive than that of the brothers Carnot might naturally have been attained. In Reinhard's view, Lazare, now married into a substantial family, entered upon his legislative duties in a much calmer spirit than he had upon the Revolution itself, which already in its early stage had struck down aristocracy and the impediments to promotion and dignity that had engendered the angers of his youth. Like the majority of his fellow deputies, he supposed the Revolution to be behind and looked forward to participating in the constitutional regime that would succeed it. He can as yet have had no thought of wielding power himself in posts of the highest responsibility.

He came to do so because a new regime in which men of his kind, the qualified and able bourgeoisie of France, could feel at ease was most unstably founded. In reality the Revolution had only begun when in October 1791 the Legislative Assembly convened. In its further course Carnot's professional military competence in harness with his engineer's ability to improvise arrangements and organize procedures proved to be great advantages, both to him and to the Republic. The truly critical faults in the political structure were precisely those that made his qualities relevant. Throughout the first six months of his legislative career the central untrustworthiness of the king himself and of many of his ministers dissipated the credibility of combining monarchy with constitutionalism and confirmed Carnot in a kind of latent republicanism. Thereafter, following the declaration of war in April, instigated by the Girondist faction with a view to exhibiting the duplicity of the crown, the widespread untrustworthiness of the leadership of the army and the misfortunes of the opening campaigns put at a premium the services of any patriotic deputy who could pretend to military competence.

Carnot's actual political interventions in the Assembly were maladroit. He was never a notable orator. Yet, knowing how he conducted himself in the pressures soon to be upon him, the historian can understand how it was that his traits apparently made themselves felt in the heightened sensibility of those intense and febrile days when the French monarchy was proving its caducity, and no one dared plan openly for what might take its place. Nothing was then observed more narrowly than character. Men took one another's measure, and Carnot's began to come through: his contemporaries felt the presence of the intelligence, the probity, the sense of reality that proved adequate to judging of the rightness of republicanism (for he was never a democrat and never in rapport with the idolators of humanity and the demagogues); the instinct for concentrating upon the job at hand and the fund of toughness about what he could not help (the Terror, later, and the guillotine) so long as it did not touch him in his own pride; the selflessness about material interests in a man alert in defense of his dignity; the passion to prevail with the right answer; the unwillingness to compromise his principles combined with some insensitivity to those of others that might possibly conflict.

Reinhard alludes to a mathematical dogmatism in his spirit.[30] It may not be necessary to read his temperament that way. On the whole, he was not a notable mathematician, and his scientific work was at the opposite pole from dogmatism. It might be preferable to attribute the rigidity in him to a fundamental decency, manifest in the political stands he took, in which he appears in retrospect the more sympathetic for a certain maladroitness, his fervor not always suited to his matter but well suited to the times. His brother's reluctant behavior in contrast gradually revealed a man of the old dispensation insufficiently tempered by an occasional opportunism. Judging of events with a surer and deeper instinct, Lazare Carnot moved towards the left and the Republic. Originally the less popular of the two brothers, only he was returned from the Pas-de-Calais to the Convention in September 1792, following the overthrow of the monarchy and the dissolution of the Legislative Assembly at the hands of the city of Paris in its rising of 10 August.

Among the hundreds of deputies to the Convention, it was Carnot who most effectively improvised the function of the representative-on-mission by means of which the new sovereignty of the people was converted into specific actions throughout the mass of the nation. The substitution of the first French Republic for the monarchy had been the act in the first instance of the city of Paris. Other municipalities, provincial administrations, civil servants, officers of the army and navy—in short, the agencies and agents of government all remained in the inertial grip of the routine procedures of the old regime when they were not, as many of them were and especially so in the armed forces, actively or potentially disposed to counter-revolution at home or to treating with the enemy abroad or to both these manifestations of an opposition now becoming treason. Into such situations of anarchy, confusion, or active disaffection in the winter of 1792 and into the summer of 1793 would come sometimes one, though more usually two or three, of the elected deputies to the Convention, Representatives of the People, dressed in blue coats with brass buttons, red, white, and blue sashes, and felt hats decorated with tricolored feathers. To them was delegated carte-blanche the sovereign authority of the people, and in them was incarnated for the time being its revolutionary will. They might suspend local officials and replace them with others;

[30] Reinhard, I, 187.

they might set up special tribunals and sit as judges; they might take over property, fix prices, or requisition whatever transport, enterprises, livestock, foodstuffs, or supplies they required for civil or military purposes; if on mission to the armies, they might supersede commanders and overrule or countermand orders. They might do whatever they thought necessary.

Thus, to cite the example that was most influential in establishing the practice, the National Assembly on the morrow of 10 August 1792 designated commissioners to apprise the several armies that the Republic had supplanted the monarchy and to secure their allegiance to the new regime. Carnot, Prieur de la Côte-d'Or, and Coustard were dispatched to the Army of the Rhine. Carnot was chief of the mission. They arrived in Phalsbourg on the 14th, summoned the acting commander, the former duc de Biron and de Lauzun (the titular commander, Lückner, was an elderly German anachronism), together with his staff and immediate subordinates, and required them on the spot to reply one after the other to the question: "Do you purely and simply accept the decrees of the National Assembly, YES or NO?" All accepted but one, whom they immediately destituted. Thereupon they moved among the rank and file from one unit to the next explaining the portent of events in Paris, staged a triumphal entry into Strasbourg as embodiments of the sovereign republic, outmaneuvered Dietrich, the feuillant or constitutional-monarchist mayor of the city, purged the departmental administration of persons of his persuasion, replaced a number of suspect municipal officials throughout Alsace, entered into diplomatic negotiations with the Swiss on problems of the scrambled border, and returned triumphantly to Paris on 4 September to assure the Assembly that the Army of the Rhine was trusty and the eastern departments loyally affected to the Revolution.[31]

Between Carnot's election to the Convention a few weeks later and his entry into the Committee of Public Safety in August 1793, he was much absent from Paris in the discharge of further such missions—to the Army of the Pyrenees, when Spain entered into the war, to the Pas-de-Calais and the North when the initial conquest of Belgium turned into defeat and confusion, to the Army of the North itself when its commander, Dumouriez, deserted to the Austrians taking captive with him the Minister of War and several of Carnot's colleagues. At that humiliating nadir in military misfortune, Carnot overrode the authority of Dumouriez' demoralized successor and organized the defense to his own prescription.

In politics itself his views were no less firm. He voted the execution of Louis XVI; he moved the decree to annex Monaco and Belgium; he proposed conscription for all men between 20 and 25 ("every citizen is born a soldier"); his acts were consistent with a principle he proposed for the basis of a declaration of rights: that the "safety of the

[31] On this episode, see Reinhard, Chapter 18 (I, 254-287), and on the government of the Committee of Public Safety, Robert R. Palmer, *Twelve Who Ruled* (Princeton, 1941).

people is the supreme law," and again, "That every political measure is legitimate if it is required for the safety of the state."[32] He was never an ideologist, a demagogue, or a zealot, never really a democrat. His was a tough republican will and his reputation that of a reliable, not to say a ruthless, patriot when in the late summer of alarm, defeat, civil war, and treason, he was called by the more politically minded men already constituting the great Committee of Public Safety to serve in its membership together with his younger countryman and fellow engineer, Prieur de la Côte-d'Or.

The historical vision of the French Revolution tends to be dominated by the climactic events of the Year II of the Republic, which is to say the autumn, winter, spring, and early summer of 1793-94 prior to the overthrow of Robespierre on 9 Thermidor (27 July 1794) and the termination of the Jacobin Terror associated as an instrument of government with his influence and spirit. Those months were dramatic ones from any standpoint, and not least Carnot's. Barely two years previously he had been a captain of engineers relieved that the way to promotion was finally opened, settling into family life, and resigned to the near approach of middle age. Now he found himself called to membership in the body of twelve men that was ruling France and determining the destiny of the Revolution.

The Committee of Public Safety was like no other government that had ever been. Originating in early 1793 to exercise the Convention's general supervision, first over the War Office and then over all the ministries, it not so gradually came to supersede them and to take executive power into its own hands. Everything about it was anomalous, except its energy. All powerful in action, it existed only at the will of the Convention, which renewed its mandate monthly and only for a month. Implementing its decrees and decisions through the network of Jacobin clubs throughout France and in Paris through that of the central Jacobin club, it both controlled and depended upon what was in effect the agency of a one-party political consciousness. Complicating its life on the left was a similar reciprocity of power through, and subjection to, the will of the Paris sections, running sometimes parallel to that of the Jacobins and sometimes contrary. By its own definition this government was merely provisional, "revolutionary until the peace." With no precedents it evolved its own procedures. Its responsibility was collective. Any or several members signed and validated decrees for the whole, working round the clock behind the closed doors of their council room in the Tuileries, sometimes two or three of them there at a time, sometimes five or six, never the entire twelve. Some were always off on missions to the armies, to the navy, or to some point where counter-revolution threatened or revolution needed firm guidance. Their creation was the levée-en-masse, which mobilized the French people into the first incarnation of a modern nation in arms beating back its enemies, overrunning its frontiers, and destroying the assent of history to the old order of Europe.

[32] Reinhard, II, 27, 34.

At home they destroyed dissent. Their instruments were the revolutionary tribunal and its executioner, the guillotine. No man involved in the application of that Terror could ever after escape either the inner necessity to shift the responsibility from his conscience elsewhere, or the external necessity to justify himself when public revulsion set in after Thermidor. The purposes of the present monograph do not require reaching a moral judgment on Carnot's part in the Terror, whether it was one of application or acquiescence. Perhaps, however, there are several things that ought to be said. For, on the one hand, his ruthlessness is not to be gainsaid. He encouraged Turreau, deputy on mission in the Vendée, to put down counter-revolution and "exterminate the brigands down to the last man." He signed the decree enjoining that the city of Lyons should be brought into line, "torch in hand and bayonets fixed." Toulon (also in rebellion) was to be burned to the ground by bombardments of red-hot cannon balls.[33]

On the other hand, there is justice in the view, originating with Hippolyte if not with Carnot himself, that dissociates him from the zealotry of Robespierre, Saint-Just, and Couthon, who would have made of the Revolution a moral crusade for the regeneration of humanity, and woe in the Republic of Virtue to whoever faltered or fell short. He is to be distinguished, too, from those other colleagues who appear rather to have been adventurers and malcontents than crusaders in a great cause: Collot d'Herbois, a paranoid and unsuccessful actor; Herault de Séchelles, a renegade aristocrat; Billaud-Varenne, an unscrupulous drifter and would-be lawyer; Barère, the politician who could always please. It was not in the company of such as these, fanatics, sadists, or demagogues, that Carnot was with his own kind, but in that of the two Prieurs, one a lawyer and the other an engineer though no relation to each other; Jeanbon Saint-André, a Protestant and sometime pastor and ship's captain; and Robert Lindet, careful, industrious, and the senior member in point of age.[34]

Carnot concentrated on supreme direction of the war (with Prieur's assistance in the matter of ordnance, supply, and the revolutionary production of gunpowder and materiel). To do so was to exclude neither politics, since no war in European history had ever been so political, nor yet the men of politics, than whom none have ever been more bellicose or intruded themselves more decisively into the management of campaigns: witness the youthful and bloodthirsty Saint-Just. Nevertheless, the reorganization of the armies and their command in the autumn and winter of 1793 was primarily of Carnot's designing, and so also was the strategy of the three-pronged offensive into Belgium in the spring of 1794. Thereby the French armies passed over into the attack that, under one regime after another, they sustained against all of Europe for the better part of twenty years. As time went on, Robespierre, who had no head for military

[33] *Ibid.*, II, 86-87.
[34] H. Carnot, I, 510-537. For the personalities and factions on the Committee, see Palmer, *op. cit.* n. 31, *passim*.

affairs, avoided the responsibility of signing the Committee's decrees concerning the prosecution of the war, just as Carnot tended to eschew those concerning domestic matters of police and revolutionary justice.[35] War and Terror were intimately related, however. Military security was invoked at the time, and has often been accepted by historians since, as justification for the strong hand of Terror. What is more to the present point, it was certainly military success in the spring of 1794 that rendered the Terror unnecessary and finally intolerable to a people no longer frightened of foreign invasion and in a position instead to export despotic interference in ordinary life from its own territory to that occupied by its armies.

Therein, no doubt, lay the reason that Carnot alone among his erstwhile colleagues continued in office (though not uninterruptedly) through the immediate reaction following 9 Thermidor and into the regime of the Directory that followed. "Carnot organized the victory," cried the anonymous voice that first phrased the judgment of history from the floor of the Convention when at one tumultuous session of that expiring body, that of 20 May 1795, he was under attack as an erstwhile terrorist.[36] Fearful of despotism, the framers of the Constitution of the Directory adopted later that year placed the executive power of the Republic in the collective hands of five men to be elected by the upper house of the legislature upon nomination of the lower. In November 1795 Carnot was elected a Director, and during the nearly two years he held the office, he was the leading figure in the government of that ill-fated regime, paralyzed in the absence of a stable, sturdy center by the irreconcilability of surviving radicalism on the left and reviving royalism on the right.

In the early days the threat to the Republic from the left seemed the more dangerous to Carnot, and he took charge of putting down with a very heavy hand indeed the so-called Conspiracy of the Equals led by an obscure journalist called Babeuf, in which episode connoisseurs of revolution have seen the germ of the proletarian revolution struggling to be born from the social bankruptcy of its bourgeois predecessor. Regrouping the armies after the disarray of Thermidor, and directing them across the Rhine and the Alps, Carnot seemed at first to handle military administration with his old sure touch. The sword he chose had two edges, however. It was almost surely his decision that confided the Army of Italy to the youthful General Bonaparte, with whom he conducted a correspondence that for a short while only bore the appearance of exchanges between patron and protégé, mentor and pupil, Aristotle and Alexander. This time the success of military strategy doomed the Republic instead of saving it.

Two years before that dénouement, Carnot was driven from office. In the spring of 1797 the legislative elections resulted in a royalist resurgence. Unwilling to accept the undoing of the Revolution, the factions of the left prepared to reject the verdict of the polls and to purge the two assemblies of their reactionary members. The coup

[35] Reinhard, II, 89.   [36] Reinhard, II, 158.

occurred on the 18 Fructidor (4 September). Carnot was persuaded that the Constitution could not be preserved by violating its provisions and refused to lend himself to the maneuver. Warned that his fellow directors were preparing his arrest, he escaped from his quarters in the Luxembourg while his brother detained the would-be captors in conversation, went into hiding for a few weeks in Paris, and then fled into Switzerland. So ended the decisive portion of the political career in which, though never holding authority alone, Carnot shared in the supreme power for a longer period—August 1793 until September 1797—than any revolutionary statesman prior to Bonaparte (except for the time-serving Barras, whose tenure as Director lasted the whole four years of that regime), and longer, too, than any major minister of state had done in the entire reign of Louis XVI. At no time in modern French history would four almost uninterrupted years of power seem short.

Thereupon, turning fifty years of age, Carnot resumed the studies in mechanics and in mathematics that he had begun with the ambitions of a young engineer in the Old Regime. In Hippolyte's account his father reverted to the quiet life of the scientist in his study with a sense of relief. It was the "interior asylum where he took refuge in order to breathe pure air and turn his eyes away from a painful spectacle."[37] Such an attitude would suit the mythology about the abstraction from public affairs proper to a true scientist. A sense of public duty selflessly overcoming reluctance to get into the battle hardly seems, however, to offer plausible motivation for Carnot's conduct in the Revolution. His first publication in exile was a passionate defense of his own course prior to Fructidor.[38] More interesting than the pious stereotype is the evidence that, after the strenuous life of a national leader throughout one of the great crises of all history, Carnot was able to reengage his interest in professional studies, one will not say with undiminished but rather with every appearance of enhanced enthusiasm. That he could do so bespeaks a magnanimity of temperament (of which his life offers other testimony) for he never recorded even the hint of a reproach at having remained unappreciated by the scientific community when young and obscure.

More to the general point, however, is the reflection that although the contents of mechanics and politics had nothing in common, nevertheless, given the historical juncture, the two aspects of his life were not different in the qualities required of him. The activism animating the nascent profession of engineering was congruent with, and indeed a manifestation of, the activism animating the large political affirmations of the Revolution. In the great world of politics men acted on the belief that by taking

[37] H. Carnot, II, 86.

[38] *Réponse de L.-N.-M. Carnot, citoyen français, l'un des fondateurs de la République et membre constitutionnel du Directoire exécutif . . . au rapport fait . . . par J-Ch. Bailleul.* This pamphlet, perhaps the most natural and eloquent thing Carnot ever wrote, was composed in May 1798 in the village of Lutzelburg near Augsburg, where Carnot had taken refuge. Published also in German (Nürnberg, 1799) and English (London, 1799).

charge of their own affairs they could add to their stature. In the special world of engineering, the practices on which men acted in building and working contained problems that were brought into the house of science where they alimented analytical mechanics, and helped turn it into modern physical science.

To analyze the actual content and significance of Carnot's scientific work in sufficient detail to exemplify that assertion forms the purpose of the present monograph. This introductory chapter is intended merely to sketch the background and circumstances. Both according to Hippolyte and according to Carnot himself in the preface to the *Réflexions sur la métaphysique du calcul infinitésimal*, that work, though published in 1797 during his tenure as Director, had been written some years previously.[39] That must almost surely have been prior to Carnot's entry into the Committee of Public Safety, and very possibly prior to the Revolution itself. As Professor Youschkevitch makes clear in his essay in the present volume, the published text differs in certain significant respects from that of the memoir Carnot had sent to the Berlin contest in 1785 (Appendix A below). After Napoleon's seizure of power in December 1799 (18 Brumaire) Carnot returned to France in accordance with a general amnesty for the victims of Fructidor. It is not quite true, as legend has it, that Napoleon sent for him forthwith to make him Minister of War in the Consulate. In fact he was given a trivial post at first, was appointed to the Ministry of War on 2 April, and resigned after five months of cross-purposes with a regime in which he could not be master in his own department, and of which in any case he sensed the inimicality to the Republic. His remaining technical books followed very rapidly upon his retirement from government: later in the same year a *Lettre au citoyen Bossut contenant des vues nouvelles sur la trigonométrie* (1800); in 1801 *De la corrélation des figures de géométrie*; in 1803 both the revision of his *Essai sur les machines* under the title *Principes de l'équilibre et du mouvement* and the book that he regarded as his major work, *Géométrie de position*; in 1806 *Mémoire sur la rélation que existe entre les distances respectives de cinq points quelconques pris dans l'espace, suivi d'un essai sur la théorie des transversales*.

His sons, Sadi and Hippolyte, were born in 1796 and 1801 respectively. Hippolyte's memoir attributes to Carnot and to his younger countryman, political associate, and fellow graduate of Mézières, Prieur de la Côte-d'Or, the initial conception of the most famous technical creation of the Revolution, *Ecole polytechnique* (or as it was first called *Ecole centrale des travaux publics*) founded by the expiring Convention in 1795 in one of its final measures. Of Monge, usually considered the founder, Hippolyte, relying on reminiscences of Prieur, said that he "took the idea over with his habitual impetuosity."[40] That Carnot should have been sympathetic to the foundation of *Polytechnique* is plausible enough: the leading notion of the school was to combine mili-

---

[39] H. Carnot, II, 86.　　　　　　　　[40] H. Carnot, I, 568.

tary and civil engineering into a single profession, to carry the momentum imparted to war production by the Committee of Public Safety into civil engineering, and to open widely the possibility of the kind of training that for Carnot, Prieur, and their colleagues had been restricted to military purposes, and rather menial military purposes at that. Nevertheless, the assertion that he had an actual part in the first organization has been disputed, and though Sadi graduated in the class of 1812, there is in fact no convincing evidence that Carnot himself took a leading part in its direction.[41] The year following Sadi's graduation he was, it is true, in company with Berthollet and Laplace elected by their colleagues of the *Institut de France* to serve on the *Conseil de perfectionnement* of *Polytechnique*.[42] But by then all three were elders.

It was rather through his membership in the Institute itself that Carnot participated in the new scientific order that emerged from the Revolution. On 8 August 1793, in the opening phase of what was to become the Terror, the Academy of Sciences, together with the other learned academies of the Old Regime, had been abolished as privileged bodies unsuited to the spirit of a Republic. Carnot entered the Committee of Public Safety only on 7 August and cannot be charged with greater responsibility for that measure than any other member of the Jacobin Convention. Nor, however disappointed he may have been in his hopes for scientific recognition, had he ever taken part in the campaign of vilification of the Academy that other unsuccessful aspirants for its favor joined in the months prior to its downfall. All the more interesting is it, therefore, that he was early elected a member of the Institute. The function of the new body in effect continued that of the former Academy although its regime and principle were intended to be different: national rather than privileged; unifying arts, letters, and science, all the branches of a single culture, there at the learned apex of the Republic.

The Constitution adopted in 1795 establishing the Directory laid down the provisions founding the Institute in fundamental law itself. In Carnot's capacity of Director he did not wish to be a charter member, and awaited a vacancy to present his candidacy. On 1 August 1796, he was elected to one created in the section of mechanics by the death of Vandermonde.[43] It cannot be said that his unsuccessful competitors, Bréguet or Janvier, have left greater names in mechanistic science. Yet neither is it easy to imagine that Carnot's existing contribution to the literature of science, consisting of his thirteen-year-old *Essai sur les machines*, which had made no noticeable impression, would in the Old Regime have entitled him to election to the Academy. The part that prominence of another sort might play in fixing scientific attention even

[41] See documents printed in James Guillaume, ed., *Procès-verbaux du comité d'instruction publique de la Convention* (V, 1904), p. 653.

[42] Institut de France, Académie des sciences,

Procès-verbaux des séances . . ., 23 August 1813, V, p. 241. The same three were reelected for a further term on 17 October 1814. See p. 407.

[43] *Ibid.*, 21 Prairial an 4, I, p. 61.

in a Republic, or perhaps especially in a Republic, will be suggested yet more forcibly by reflecting on the identity of the person who took Carnot's place when, in the year following his election, he was eliminated from the Institute in consequence of Fructidor: his successor in the section of mechanics was General Napoleon Bonaparte, who often used thereafter to specify below his signature, "Membre de l'Institut."

Returning from exile in 1800, Carnot was reelected to membership on 26 March, and the Institute, together with his family, became his chief interest. He was named Vice-President of the First Class, i.e., the scientific branch, in 1804 (30 Ventôse an 11) and succeeded to its presidency in the following year.[44] Like the Academy before it, the First Class of the Institute fulfilled two main functions. On the one hand, it saw to the advancement of science by publishing its memoirs and by endorsing publication of worthy treatises submitted to its judgment. On the other hand, it advised government departments and the public on technological matters by examining the design and merit of new machines, industrial processes, or agricultural methods submitted by inventors and entrepreneurs in hopes of a patent, premium, or other subsidy or favor. Throughout the eighteenth century in the great days of the old Academy its leading members esteemed the former of those functions and regarded the latter with ill-concealed impatience. Something of that differential is no doubt inherent in the value structure of science, and it would oversimplify and exaggerate matters to say that the utilitarian and humanitarian emphasis of the Revolution transvalued the assessments. Nevertheless, the Institute did devote a larger share of its attention even in scientific publications to subjects of applied science and engineering. There was a displacement towards engineering; or perhaps it would be more exact to say that engineering problems occupied a larger share of the attention of scientists, or better yet of their respect. Such was certainly Carnot's preference. He came into the Institute from that world and was at ease in it as perhaps one may say he would not have been in the company of the Academy, for the effect of the Revolution was to legitimate and accelerate this shift though not to cause it.[45] Once his own treatises just mentioned were published (which had no immediately observable effect in science proper whatever their significance for its later course), Carnot devoted his time, energy, and interest to many commissions charged with examining the merits of many of the numerous mechanical inventions that attest to the fertility of French technical imagination in the early nineteenth century.

These tasks were no mere desultory hobbies of retirement. In an average year Carnot would sit on twelve or fifteen such commissions, most often with Bossut, Charles, Coulomb, and Prony when the subject was machine technology, with Guyton de

[44] *Procès-verbaux*, II, p. 638; III, p. 3.
[45] For a discussion of the emergence of engineering science in the somewhat earlier instance of Coulomb's work, see Gillmor, *op. cit.*, Chapters I and II.

Morveau when it was military technology, and with Bossut and Legendre and sometimes Laplace, Lagrange, and Monge when it was mathematics. Frequently it would be Carnot who was responsible for writing the commission's findings. Many of these reports amounted to memoirs reviewing not merely the project but the state of a whole subject.[46] Such was the report on a notational scheme for descriptive algebra by one Samson Michel, for which Carnot summarized the development of what is now called topology, *analysis situ,* from its creation by Leibniz through the contributions of d'Alembert and Euler to his own applications of geometry to algebra.[47] On several occasions he participated in evaluating early contributions of young men soon to become famous—Poinsot, for example, and Cauchy, the most notable instance, whose theory of limits in the 1820's rendered Carnot's metaphysics of the calculus irrelevant to the foundations of mathematics.[48] It says much of the direction in which all physical science was moving that a commission on which Carnot served in 1807 with Laplace, Lagrange, Legendre, and Lacroix should have chosen for the subject of a mathematics prize the topic of a mathematical theory of double refraction to be verified by experiment.[49] Normally, however, Carnot scrutinized inventions of a more mundane sort—a canal boat fitted out for mowing the banks (hardly worth the trouble), a new form of knitting frame (interesting), a more efficient force-pump (wholehearted approval), a motor powered by the combustion of hydrogen (entirely impractical), a twenty-seven-foot submarine capable of carrying nine persons (the problem of air supply was not solved).

Among many such inventions, dreams, and realizations, a few stand out that hold a more permanent interest for the history of technology, notably in the realm of engines and motors. For example, Carnot was a member of the Commission that Robert Fulton invited to watch his steamboat mount the Seine.[50] Even more intriguing because less familiar was the model of an internal combustion motor called the Pyréolophore by its authors, the brothers Niepce, Claude and Joseph-Nicéphore, of whom the latter was known later for his invention of the tintype and association with Daguerre in the early development of photography. The Niepce engine, said Carnot in an enthusiastic

---

[46] Volumes III through V of the *Procès-verbaux,* which record the actions of the First Class of the Institute from 1804 through 1815, show Carnot's employment on its work to have been steady and full. The tabulation below summarizes the number of reports for which Carnot had full or partial responsibility in the first column and in the second the number of commissions on which he served:

Vol. III (1804–1807) 15 37
Vol. IV (1808–1811) 16 38
Vol. V (1812–1815) 17 39

[47] *Procès-verbaux,* 22 Nivôse an 11, II, pp. 611-613.

[48] *Procès-verbaux,* 16 October 1809 (IV, pp. 263-266), Report on Poinsot's memoir on polygons and polyhedrons; 12 April 1813 (V, pp. 198-199), Report on Cauchy's memoir on symmetrical functions.

[49] *Procès-verbaux,* 21 December 1807, III, p. 632. This prize led to the discovery of polarization of light by Malus in 1809.

[50] *Procès-verbaux,* 20 Thermidor an 11, II, p. 690.

report, was the first device ever imagined for drawing motive power from the expansive force of heated air, or (to employ his own usage) of air combined with caloric. In contrast to the steam engine, which consumed a large quantity of caloric in heating and vaporizing the water before harnessing the expansibility of steam, an air engine presented the advantage in principle that all the fuel was employed to produce the expansion that went into motive force. As will appear in the detailed analysis of Carnot's theory of machines in the next chapter, considerations of this type were what he brought to the subject as a whole, and of course it was the special question of heat engines that Sadi took up after him, a lad of ten when in 1806 his father was studying the operation of the Niepce engine.[51]

In 1809 Carnot reported to the Institute on another model heat engine imagined and designed by a prolific inventor called Cagniard de Latour.[52] Recently it has been argued, and convincingly so, that the Cagniard engine must have been peculiarly suggestive to Sadi if he considered it in later years.[53] As in the Niepce engine, the expansible fluid was air but employed in a much subtler manner. The fundamental notion was to introduce atmospherically-cool air to the bottom of a bath of heated water and to draw power from its thermal expansion on rising. For that purpose, Cagniard devised a reversed Archimedean screw that captured bubbles of air at the surface of the water and led them down to the bottom of the tank into a pipe that conducted the air into a tub of hot water where it was released. The bubbles, expanding as they rose, were caught in the inverted cups of a paddlewheel in a way to make it revolve: their expansion under heat enabled the wheel to deliver five times the power required to turn the screw, and thus supplied motive force.[54] The features that the Niepce and

[51] *Procès-verbaux*, 15 December 1806, III, pp. 465-467. Berthollet was the other member of the commission. The Niepce engine was essentially a one-cylinder motor. It consisted of a copper receptacle fitted with a cylinder and piston and a narrow pipe through which air might be injected by a bellows. The pipe contained two openings between the bellows and the chamber. The flame from a pilot light burned under the aperture closer to the chamber. Fuel was injected through the further intake. The best fuel was lycopode powder, but that being extremely expensive, the brothers Niepce employed a mixture of powdered coal with a small amount of resin. At each stroke of the bellows, the fuel was blown across the flame and carried into the receptacle where it burned with an explosive force sufficient to drive the piston. The great problem was to clear the chamber and renew the

air, which object the Niepces carried out by a system of valves and a diaphragm activated by the piston that swept the chamber clear on each stroke. The Niepce engine was no mere table-top model. The capacity of the receptacle was twenty-one cubic inches and the surface of the piston three square inches. It developed a pressure sufficient to balance a weight of 114 pounds. Mounted on a boat displacing nine quintals, the engine was able to make headway against the current of the Saône. In that experiment the machine ran at a rate of twelve or thirteen revolutions and burned 120 grains of fuel per minute.

[52] *Procès-verbaux*, 8 May 1809, IV, pp. 200-202.

[53] Thomas S. Kuhn, "Sadi Carnot and the Cagnard Engine," *Isis*, 52 (1961), 567-574.

[54] Professor Kuhn points out that the Cagniard engine (the name is usually spelled

Cagniard engines presented in common with that later imagined by Sadi—i.e., that they drew motive power from heat and that they depended on the property of expansibility common to all gases including air rather than on steam alone—render these devices more interesting as possible intermediaries between the work of father and son than certain others, perhaps equally ingenious, on which Lazare Carnot also framed reports, notably a force pump operating by air compression invented by a certain Lingois;[55] a number of hydraulic machines constructed and operated by an industrialist of some eminence, Mannoury d'Ectot;[56] and a memoir of Dupin of 1814 on the stability of floating bodies.[57]

Such were Carnot's occupations during the decade from 1804 to 1814 when he was in his fifties and his sons were boys. Hippolyte recalled that their circle was a largely technical one. He did not mention the lawgivers of the Institute, however: Laplace or Berthollet or the aging Monge, but rather Prieur, two elderly survivors of the Old Regime, the explorer Bougainville and the economist Dupont de Nemours, the engineer Girard, the chemist Adet, the mathematics teacher and writer Lacroix, the botanists Palissot de Beauvois and the two Thouins, father and son. When certain financial speculations turned out badly, Carnot accepted a pension of Napoleon's favor. He never abandoned his interests in military strategy nor his role as a defender of the Revolution. In 1810 he published a systematic treatise on his early topic of defensive warfare and the employment of fortifications.[58]

Amid the crumbling of the Napoleonic system, he wished to help defend the country and offered his services to Napoleon at the moment when the retreat from Moscow reached the Rhine. In those desperate circumstances Napoleon appointed him Governor of Antwerp. Carnot commanded the defense. It was the only actual command in his career, and he maintained French control even after Napoleon's first abdication. He then accommodated himself to the initial return of the Bourbons, and even addressed a memoir to Louis XVIII counselling him about the aspects of the previous

---

with an "i") exhibits certain characteristics later idealized by Sadi in the *Réflexions sur la puissance motrice du feu*, namely (1) its production of motion entirely from the transfer of heat from a high temperature reservoir, (2) its parallelism between flow of water and flow of heat, and (3) a coupling of two engines working in reverse directions permitted by the excess of work produced by one of them (although in fact this notion of a coupled pair of engines was not in Sadi's memoir, but has been imputed to it in the literature). Kuhn observes that the Cagniard engine and Sadi's were different in that the latter included the

operation of an idealized cylinder and piston. The one element in Sadi's analysis not presaged in the Cagniard engine, in Professor Kuhn's opinion, was reversibility. As will appear in the following chapters, that emerged from an adaptation of Lazare Carnot's construction of geometric motions.

[55] *Procès-verbaux*, 13 July 1812, V, pp. 72-75.

[56] *Ibid.*, 28 December 1812, V, pp. 133-137, and 21 June 1813, V, pp. 221-224.

[57] *Ibid.*, 30 August 1814, V, 392-395.

[58] *De la défense des places fortes* (Paris, 1810).

twenty-five years that the king must needs respect in order to stabilize his throne.[59] Nevertheless, Carnot rallied to Napoleon upon the return from Elba and took the Ministry of the Interior in the government of the Hundred Days.

That act rather than the mere fact of having voted death to Louis XVI established the consistency of the true old revolutionary and was not forgiven by a monarchy that had had to be restored twice, the less so as in the Commission of government that bridged the interregnum after Waterloo, Carnot sought forlornly enough to secure the accession of Napoleon II. His wife having died and Sadi having graduated from *Polytechnique* into a career, Carnot took his younger son Hippolyte into exile. They travelled by way of Brussels, Munich, Vienna, and Cracow to Warsaw. In that gay capital Carnot thought to stay, but prices were rising and the political barometer falling under the Grand Duke Constantine. Forbidden the Rhineland because of its proximity to France, he and Hippolyte settled in Magdeburg. There Carnot corresponded with German colleagues, published a few volumes of verse, and brought up his younger son, who became a Saint-Simonian, and there he died on 2 August 1823.

[59] A distorted version of what was meant to be a private communication was pirated to the publishers apparently with the purpose of discrediting Carnot and the surviving Jacobins. *Mémoire adressé à S.M. Louis XVIII, roi de France* (Bruxelles, 1814).

CHAPTER II

The Science of Machines

THE great events that Carnot touched and shaped never caused him to lose sight of the adaptation of the science of mechanics to the science of machines that he thought to initiate when still an obscure young engineer. Fortunately for the historian, the documentation is adequate to permit following his conceptions from their genesis in the entry he prepared for the prize contest set by the Academy of Science in April 1777 right through their development during his lifetime into the subject matter of a new branch of science in the 1820's. Since his own ideas were expressed in their most individual and unadorned form in his first publication, the *Essai sur les machines en général* of 1783, it will be best to present them through the medium of the detailed analysis of that work that occupies the current chapter. Thereupon it will be informative to look first back and then forward. The stages through which Carnot formulated his approach may be observed in the successive memoirs he submitted to the Academy in 1778 and 1780. Twenty years later, an ostensibly retired statesman, he extended and developed the subject in *Principes fondamentaux de l'équilibre et du mouvement*. Beyond this, it is one of the purposes of the present monograph to exhibit that Sadi Carnot's *Réflexions sur la puissance motrice du feu*, published in 1824, the year after his father's death, may properly be read not only as the foundation of thermodynamics, but also as the culmination of a methodologically and conceptually coherent series of Carnot essays on the science of machines. After surveying the literature of that subject a little more at large, it will be appropriate finally to consider the relevance of Carnot's mathematical writings to his mechanics and his work in science.

A. SUMMARY OF *ESSAI SUR LES MACHINES EN GÉNÉRAL*

There is no difficulty in understanding why the scientific community should have ignored Carnot's *Essai sur les machines en général* in the 1780's. His book does not read like the rational mechanics of the eighteenth century. It had long since become normal to compose treatises of mechanics addressed to a professional public in the language of mathematical analysis; though Carnot reasoned no less rigorously than did contemporary mathematical argument, he conducted the discussion verbally, conceived the mathematical expressions he did employ in a geometric or trigonometric rather than algebraic spirit, and usually went on to explain in words what the formulas contained. The genre was apparently of an altogether lower order than that of d'Alembert and Lagrange or Euler and the Bernoulli family. Judging by the style alone, prolix and naïve, a contemporary reader might easily have supposed the book to be among the many negligible writings that retailed merely elementary mechanics under one pretext or another.

Yet, the essay, despite its title, could never have served the purpose of a practical manual for designing or employing actual machinery. Carnot proceeded on the basis of a highly abstract definition of a machine: it was an intermediary body serving to transmit motion between two or more primary bodies that do not act directly one on another. Carnot lodged his complaint against the existing mode in mechanics right at the outset. Since its problems were normally limited to analyzing the interactions among primary bodies, the practice had tacitly arisen of abstracting from the mass of the machine itself as if it were inertia free. While simplifying problems in mechanics, however, that method of treatment, carried over from the geometrization of simple machines in statics, complicated a dynamical study of machines. It was standard procedure to deduce from the laws of mechanics the particular rules of equilibrium and motion in each class of simple machines, i.e., cords, the lever, the crank, the pulley, the wedge, the screw, and the inclined plane. No general principles existed, however, that were capable of containing the conditions of motion and equilibrium in all machines. In order to establish such principles, the mechanist needed to treat machines like other bodies and account for their mass in his analysis. Only then would the science of machines in general be feasible, and then it would come down to resolution of the following problem:

> Given the virtual motion of any system of bodies (i.e., that which each of the bodies would describe if it were free), find the real motion it will assume in the next instant in consequence of the mutual interaction of the bodies considered as they exist in nature, i.e., endowed with the inertia that is common to all the parts of matter.[1]

Already it will appear that, verbally expressed though the argument was, the *Essai sur les machines* was beyond the comprehension of readers without formal education. It presupposed the competence of scientifically literate persons, but was not couched in language that would attract their interest. The essay would have been accessible to trained engineers, in other words, and since the author himself was of a first generation of them, the impediment was that it could have been appreciated only in a profession that barely existed.

A difficulty about the content complements that about the form. The *Essai sur les machines* mingled novelty of approach with the most elementary aspects of mechanics in such an unassuming fashion that it is not easy to distinguish in retrospect what was new from what was obvious. Indeed, to treat Carnot's scientific career in point of discovery would be relatively unprofitable. Innovations there were, to be sure, and far

---

[1] Lazare Carnot, *Essai sur les machines*, § X, p. 21.

from negligible. But they did not occur on the frontiers of unsolved problems in the science of mechanics. They occurred behind the front lines, so to say, and rather than consider Carnot to have been a seeker and finder in the conventional scientific sense, it will be better to approach his work as the critic might that of a writer who should introduce a new idiom into a literature in order to bring existing, largely unnoticed resources to bear on different purposes. What Carnot added to mechanics accrued largely in virtue of what he thought to do with it.

Before entering into his reasoning, a reader needs to be reminded of certain contemporaneous physical conventions about the structure of matter. He will not get beyond the first few pages of the *Essai sur les machines* without encountering the distinction between hard bodies and elastic bodies that entered mechanics in the early stages of collision theory in the seventeenth century and disappeared into the theory of elasticity in the nineteenth. Carnot's work came into that development somewhat past its midpoint and inherited as assumptions the positions adopted by Maupertuis.[2] Perfectly hard bodies were held to be indeformable and perfectly elastic bodies to contain forces capable of restoring their initial shape and volume after compression or shock of impact. For completeness there should have been the third category of soft bodies, deformable and incapable of self-restitution. In the seventeenth century, John Wallis had addressed one of the classic papers inaugurating these studies to the problem of the behavior of inelastic bodies in collision.[3] He wrote of hard bodies but, paradoxically enough, soft bodies would behave in the same way in collision since they would not rebound. Resilience or its absence being the significant alternatives, elasticity and hardness sufficed for the idealized conditions. The polarity was unsymmetrical, however, because hardness did not admit of degree whereas the degrees of elasticity could be attributed to a complement of softness.

In modeling nature itself, hard bodies were taken to be fundamental, their properties a manifestation of the impenetrability of matter inhering in the ultimate corpuscles. In actual bodies, solid matter was imagined to consist in such corpuscles connected to each other by rods or shafts—rigid, inextensible, and incompressible rods in hard bodies, springs in elastic bodies. In order to follow Carnot's thought, it will further be important to appreciate the way in which other physical states related to the model, for that was not in the usual notion of progression from solid through liquid to gas. Instead, liquids were fluids congruent with hardness in mechanical properties in that they were incompressible though deformable, and gases were fluids mechanically congruent with elasticity in deformability and resilience.

[2] For a helpful account, see Wilson L. Scott, "The significance of 'hard bodies' in the history of scientific thought," *Isis*, 50 (1959), 199-210.

[3] John Wallis, "A Summary Account of the General Laws of Motion," *Philosophical Transactions*, 26 November 1668, III, 864-866.

These distinctions created dynamical difficulties for Carnot that might arouse impatience in his modern reader. Historically, however, it would be misleading to consider them false problems, for working through them was precisely what led Carnot in the direction of physics of work, power, and ultimately energy, wherein they did indeed become obsolete. The embarrassment grew out of the famous issue in eighteenth-century dynamics about whether momentum (MV) or "live force" (MV²) should be admitted to be the fundamental quantity conserved in interchanges of motion. Ignoring (as is legitimate for present purposes) metaphysical aspects, it was a limitation upon the employment of the principle of live force that although seventeenth-century collision theory conserved its quantity in the interaction of perfectly elastic bodies, in the supposedly more fundamental case of hard bodies, MV² was conserved only when motion was communicated smoothly ("by insensible degrees" is the usual phrase), and not in impact or collision.

Looking back at Carnot's writing, one can see that a treatise in which the quantity now called work (force·distance) was to be designated as the measure of what machines accomplish was in some sense bound to presuppose its equivalence to kinetic energy ($\frac{1}{2}$ MV²). How Carnot saw in advance that he must move towards such a convertibility can only be conjectured. Dimensionally it derived from the equivalence in the law of fall of "*mgh*" to "$\frac{1}{2}$ *mv²*," known since Galileo as the means of equating the velocity that a body would generate in falling a certain distance with the force required to carry it back. Beyond that, the relation had to do with his object in thinking of machines as massive bodies in action and not mere extensions of rigidity serving the geometrical transfer of inertial motion, or in simpler words with his having been an engineer. Certainly, too, his ideas carried over the role of live-force conservation from hydrodynamics into what he would have regarded as the deeper problem of hard-body interactions.

Following the publication of Daniel Bernoulli's famous treatise in 1738,[4] hydrodynamics was the subject in which the principle of live force was basic and indispensable in the solution of engineering problems. There is thematic evidence running through all Carnot's writings on mechanics that the hydraulic application of principles and findings, although ostensibly a special case, actually held a major if not a primary place in his thinking.[5] It would have been likely enough at the outset, though not strictly correct, to think of motion being communicated by liquids in accordance with the principle of continuity and to suppose from the common incompressibility of liquids and hard bodies that the conservation of live force might profitably be

[4] Daniel Bernoulli, *Hydrodynamica* (Strasbourg, 1738).

[5] Scholars have often cited the analogy between fluid flow and heat flow in discussing the influence of Lazare's work upon Sadi's, though, as will appear, the present monograph makes a much wider claim for the continuity between the work of father and son. Below, pp. 61, 90-100, 144-145.

taken to be the principle governing the latter types of interaction, in nature the most fundamental.

At any rate, whether for those reasons or others, the strategy of Carnot's *Essai sur les machines* was decided by the necessity to outflank the discrepancy between continuous and discontinuous change of motion in hard-body interaction and to define the sense in which the principle of live force might equally well be employed in continuous or discontinuous interchange of motion between elastic or inelastic bodies. To that end Carnot tacitly made use of distinctions that, although he never saw so far, later developed into those between potential and kinetic energy, work and power, input and output, scalar and vector quantity, reversible and irreversible process.

In his avowed point of view Carnot robustly admitted to the engineer's inaptitude for metaphysics. There had always been, he acknowledged, two distinct approaches to the science of mechanics, the experiential and the rational. The former, to which he adhered, took its departure from those basic notions that we draw from our gross experience of nature and to which we give the names body, power, equilibrium, and motion. These ideas neither could nor needed to be defined. They were primary, and other conceptions were derivative, such as velocity and the various types of force in terms of which laws of motion were framed. The second approach was one that would make mechanics a purely rational science. Its adherents began with hypotheses, deduced the laws that bodies in motion would exhibit if the hypotheses were correct, and then compared their conclusions with phenomena. What they thus gained in elegance was at the cost of incomparably greater difficulty, for nothing was more embarrassing in exact science, and especially in mechanics, than an effort to formulate definitions entirely free of ambiguity. That was something he would not and need not attempt.[6]

Carnot's reluctance with respect to definitions makes the *Essai sur les machines* hard to follow for persons schooled in nineteenth- and twentieth-century terminology. Only by the use he made of terms is it borne in on the modern reader what he meant by them. Perhaps Carnot's ambiguity may have been felt at the time, for in reworking the material twenty years later for the *Principes de l'équilibre et du mouvement*,[7] he

---

[6] Carnot, *Essai sur les machines*, pp. 105-106. Carnot did not pose the choice between the two schools of mechanics as explicitly as Jouguet implies—i.e., that what distinguishes them is that one takes mass to be fundamental and force for a derivative notion, and the other takes force to be fundamental and the laws of motion and impact for derivative. Carnot's own distinction may indeed come down to that, but he did not say so, and Jouguet like other commentators on his mechan-

ics confuses his *Essai sur les machines* of 1783 with the *Principes* of 1803. See Jouguet, *Lectures de la mécanique*, 2 vols. (Paris, 1924), II, 72.

[7] Perhaps it will be useful to reproduce the dimensional glossary from Carnot's *Principes fondamentaux de l'équilibre et du mouvement*, § 18, p. 13. Denoting mass by m, linear distance by e (espace), and time by t, then

"Any quantity of the form or reducible to the form $\frac{e}{t}$ is called velocity.

gave a glossary. Even those terms that are dimensionally what might be expected did not carry the full meaning of later usage. For example, the simplest, "vitesse," might better be rendered speed than velocity, for Carnot disposed of no convention for combining intensity and direction of quantity in a single expression, but always specified the sense in which a velocity or force was to be reckoned. The present discussion will follow Carnot's own usage, translating "vitesse" usually as velocity in order to avoid awkwardness, and understanding that by "force" he usually meant quantity-of-motion (momentum) in moving bodies. Given his point of view, only secondarily did he need to allude to Newtonian force, the product of mass times acceleration. When he did, he normally employed the phrase widely used in the eighteenth century, "force-motrice," or "motive-force," dimensionally identical with "dead force" or statical pressure.

From among the basic principles of mechanics proper, Carnot chose two for an introductory criticism that also established his own standpoint. The first stated for the general law of equilibrium in weight-driven machines the condition that the center of gravity of the system be at the lowest point possible. The second, which Carnot called "Descartes' famous law of equilibrium," will be unfamiliar under that designation both to the student of modern mechanics and the student of Descartes, who in fact made no such statement. The principle was that two forces in equilibrium were in inverse ratio to their "vitesses" (by which Carnot meant the motions they produce) at the instant when one prevailed infinitesimally over the other, thus initiating a "small" motion.

Carnot preferred to rely on the center of gravity principle. True, it was open to certain objections. For one, its applicability appeared to be limited to machines driven by weight. That defect, however, Carnot dismissed as merely apparent. It was always possible to reduce the operation of other forces to that of gravity by replacing their agency in principle with that of a weight acting over a pulley. This imaginary transformation of the system may at first annoy the modern reader as a somewhat sopho-

---

"Any quantity of this form, $m\frac{e}{t}$, is called quantity of motion.

"Any quantity of this form, $\frac{e}{t^2}$, is called accelerating or retarding force.

"Any quantity of this form, $m\frac{e}{t^2}$, is called motive force.

"Any quantity of this form, $m\frac{e}{t}$, or of this, $m\frac{e}{t^2}$, is called simply force or power. [The ambiguity here reflects the historical situation. Although the concept of power (puissance) as the rate of work or energy change with respect to time was implicit in Carnot's analysis, he could also use the word as a synonym for force, often with the implication of latent force as the ability to exert effort.]

"Any quantity of this form, $m\frac{e^2}{t^2}$, is named live force, moment of motive force, or moment-of-activity.

"Any quantity of this form, $m\frac{e^2}{t}$, is named moment of the quantity of motion, or quantity of action."

The significance for Carnot of the last two quantities will appear in what follows.

moric evasion of the difficulty, but it would be better to see in it the engineer's way of reducing an abstract problem to his own terms. Anyone who has been to engineering school will have resorted to similar devices, although it would be more germane to recall for a moment the analytic role of the experience and manipulation of weight in the mechanics of an Archimedes, a Stevin, or a Galileo. Even Lagrange was not above it.

A further objection was more serious. There were exceptions to the statement that equilibrium required that the center of gravity be at the lowest possible point, as indeed there were to all assertions involving maxima and minima. There might be provisional or temporary equilibrium in other configurations of the system. In order to obviate that difficulty, Carnot employed a type of argument that his reader soon comes to recognize for characteristic. He restated the principle by appealing to the operation of an ideal machine. The form is arbitrary. No forces are applied other than weights. No motion has been imparted. In the quiescent state, the sum of the resistances of the fixed supports estimated vertically must equal the weight of the system. Suppose now that a "small" motion begins. A portion of the weight must have gone into producing movement, and only with the remainder are the fixtures loaded. The difference between weight of the system and load on the fixed supports will be the force depressing the center of gravity at a rate [here Carnot used the word "vitesse" where we should say acceleration, a term he rarely employed] equal to that difference divided by the mass of the system. It follows that if the center of gravity did not descend, there would be equilibrium:

> To demonstrate that several weights applied to any machine whatever are in equilibrium, it suffices to prove that if the machine be left to itself, the center of gravity of the system will not descend.[8]

What is noteworthy is the mode of reasoning, which combined the ideal with the operational and did so in a negative and restrictive way. In a treatise purporting to rest upon experience, a reader might expect that a demonstration would consist in a generalization of experiments. Not at all—it was based upon the exclusion in an idealized, unrealizable thought experiment of what we recognize from our general experience would not occur in the behavior of real objects. Nothing was new about the logic: Stevin, in obtaining the law of the inclined plane from the exclusion of perpetual motion, had adapted the geometric proof from the absurd to physics. The assertion was that a proposition was correct if its contrary entailed consequences that were physically unthinkable. Nothing was new about the model: the motions of systems of mass-points had long been the object of an analysis that had idealized bodies and

---

[8] *Essai sur les machines*, § II, p. 14.

turned rational mechanics into a form of mathematics. But, as in other aspects of Carnot's work, what made the difference was the use he repeatedly drew out of such arguments until they became the distinctive idiom of the science of machine motion.

As for the so-called principle of Descartes, Carnot found in it disqualifying flaws. It was less general than the center-of-gravity principle thus transformed, from which it could be deduced by conversion of its forces into weights acting over pulleys. It applied only to systems in which no more than two forces were at work. More seriously, it envisaged only the relative amounts of the forces in equilibrium, whereas in requiring their vertical projections the center-of-gravity principle specified also the direction of those forces. (It is just in these passages that one may begin to appreciate how Carnot's attempts to analyze the manner in which forces transmitted by shafts, cords, and pulleys would constrain and move points within systems composed of rigid members, clumsy though these constructs seem, nevertheless belong to the pre-history of vector analysis in exhibiting awareness that the quantity of a force comprises direction as well as intensity.)

Nothing in the principle of Descartes even required that two forces in order to be in equilibrium must act in opposite senses, and it could not, therefore, specify in what their opposition consisted. In raising that problem, Carnot anticipated one of the clarifications of which he was always proudest, the distinction between what he called impelling forces ("forces-sollicitantes") and resisting forces ("forces-resistantes"). Not by a metaphysical differentiation between cause and effect would we tell which is which, but merely by the geometry of the system. As shown in the sketch, a force that made an acute angle with the direction wherein the motion occurred would be called impelling; a force that made an obtuse angle would be called resisting.

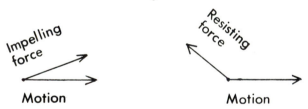

Finally, Carnot lodged another criticism against the principle of Descartes together with all those that invoked a "small"—i.e., infinitesimal—motion arising in the system. He had in view here the principle of virtual velocities, of course, although he did not identify it by name. This approach never specified what determined that infinitesimal motion. If it was necessary to invoke a new mechanical principle for that, then the original one was inadequate. If, on the other hand, the geometry of the system sufficed to determine the nascent motion, how did it do so? The identical objection lodged against analyses that considered two machines in two states infinitely close to each other. In specifying how the machine must move in passing from one to the other, either it was necessary to invoke a new principle, or else the determination was present in the geometry of the system, and in the latter case it was a defect of the principle that it failed to make evident the geometric conditions of the motion. In fact, however, such motions were subject to certain conditions, in definition of which Carnot was going to propose for the whole class the designation of *geometric motions*. As will appear, his geometric motions differed from virtual velocities in being finite, and amounted to possible or actual displacements in which no internal work was performed or energy consumed within the system.

Both of the foregoing principles applied only to equilibrium, and though Carnot alluded to d'Alembert's principle as justification for extending equilibrium principles to motion, he made no actual use of it, preferring in practice to found the reasoning on the conservation of live force. From among laws of motion and equilibrium themselves, Carnot chose two as axiomatic. The first was the equality in opposite senses of action and reaction, which he did not identify as Newton's third law of motion. It applied to all bodies without exception. The second applied only to hard bodies and asserted that in their interactions, whether by impact or pressure, their relative velocity in the next instant was zero. From these two laws, two corollaries followed: first, that the intensity of interactions between two bodies depended only on their motion relative to each other, and second, that the force they exerted on each other in impact was always along a perpendicular to their common surface at the point of contact.

Carnot began the argument itself by deducing from these fundamental principles an equation stating that in the motion of a system of hard bodies the net effect of mutual interactions among the corpuscles constituting the system was zero. Applied to generalized systems of hard bodies, this equation reduced to an expression equivalent to the conservation of live force. Just there, however, Carnot encountered the inapplicability of that principle to hard-body collision. It was mainly in order to turn this difficulty that Carnot defined for purposes of analysis the class of geometric motions, i.e., displacements depending for their possibility on the geometry of a system and not on rules of dynamics based in physics. For such motions Carnot might ignore the supposed loss of live force in inelastic collision. Invoking them permitted him to transform his fundamental equation into an indeterminate of general application into which arbitrary values determining the motions might be introduced in the solution of particular cases.

In later terms, this analysis amounted to a derivation of conservation of moment-of-momentum or torque from conservation of energy or work. Since what Carnot required for a theory of machines was precisely the license to disregard the internal constraints and interactions of systems of bodies and forces, he made conservation of moment-of-momentum ("moment of the quantity-of-motion," in his words) the fundamental principle in mechanics generally. From it he deduced the Principle of Least Action and a restatement of conservation of live force in an altered form expressing that what machines transmit is power rather than motion. Indeed, the central thrust of Carnot's contribution to mechanics was tacitly to transform the analysis of motion into the analysis of power.

The second part of the *Essai sur les machines* applied Carnot's adaptation of the principles of mechanics to the operation of machines themselves. The most important definition was of the quantity now called work. Carnot called it "Moment-of-Activity" and identified it clearly as the basis for measuring input against output in machine processes. Recurring to the principle of moment-of-momentum, Carnot deduced from it the most original of his propositions: that, in the case of a machine transmitting motion smoothly, the work done by the impelling forces equals the work done by the resisting forces. The principle, since named after Carnot, follows directly: characteristically, he thought of it and stated it restrictively to the effect that the work done by a system of moving parts equals the work done on it only if percussion or turbulence be eliminated in the transmission of motion (or power). A concluding scholium inveighed against the chimaera of perpetual motion, discussed the factors of mechanical advantage, developed the conditions for efficient design of machinery with particular emphasis on hydraulic power, and made the distinction already outlined between the experiential and rational approaches to the science of mechanics.

Such were the main heads and findings of Carnot's first publication. Readers interested largely in a qualitative summary of what he did may find this outline sufficient and prefer to pass on directly to Chapter III, which discusses the background of the *Essai sur les machines* and the ensuing development of the science of machines. Readers more immediately concerned with the inwardness of eighteenth-century mechanics itself may wish to follow the detailed précis and analysis of the argument that occupies the remainder of the present chapter.

B. GEOMETRIC MOTIONS

Imagine, Carnot charged his reader, any system of hard bodies in which the virtual motion is modified into some other motion in consequence of the internal constraints and interactions among the bodies. The example that he meant but did not specify was in consequence of their being assembled into a machine. The general problem of mechanics was to find that other motion, and Carnot began his resolution by summing the interaction of contiguous corpuscles, m' and m''. From the two basic axioms (equality of action and reaction and zero relative motion following collision) and designating by

F′ the action of m'' on m', i.e., the "force" (quantity of motion) that m'' imparts to m',

F″ the reaction of m' against m'',

V′ and V″ the respective velocities immediately after impact,

q' and q'' the angles between the directions respectively of V' and F' and those of V'' and F'',

Carnot obtained the expression[9]

$$\int F'V' \cos q' + \int F''V'' \cos q'' = 0$$

for the interaction of neighboring corpuscles taken two by two; which according to his wont he also stated verbally:

That is to say, that the sum of the products of the quantities of motion impressed on each other by the corpuscles separated by each of these little inextensible wires or incompressible rods . . . multiplied by the velocity of the corpuscle on which it is impressed, evaluated in the direction of that force, is equal to zero.[10]

Carnot next considered the system in motion as a whole in order to analyze the internal interactions of the constituent parts, and adopted designations that he thereafter used quite consistently for the several quantities. He called

The mass of each corpuscle	m
Its virtual velocity	W
Its actual velocity	V
The velocity that it *loses*,[11] in the sense that W is the resultant of V and of this velocity which is	U

[9] Perhaps it will be helpful to give here Carnot's explanation of this method of representing the projection of a force or any directed quantity upon another in terms of the elementary trigonometry of a right triangle.

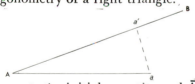

$\overline{Aa}$ represents the initial quantity and $\overline{AB}$ the direction in which it is to be "estimated." Thus, the projection $\overline{Aa'}$ of the force $\overline{Aa}$ estimated in the direction $\overline{AB}$ is given by $\overline{Aa} \cos \angle aAB$. The explanation is from the *Principes fondamentaux de l'équilibre et du mouvement* of 1803, § 26, p. 16. In the *Essai sur les machines* he did not stop to explain these elementary devices but simply employed them. (The interest lies mainly in the example they afford of his geometric and trigonometric way of visualizing relations. This comes out initially in the 1778 memoir for the *Académie des sciences*. See Appendix B below, esp. § 52-60.) In the expression given in the text above,

the velocity of m' imparted by the force F' estimated in the direction of F' is $V' \cos q'$. Since F' and F'' denote quantity of motion, what this expression says in the notation of a modern primer would be

$$\Sigma M' V'^2 + \Sigma M'' V''^2 = 0.$$

Carnot in this book used the sign Σ indifferently for integration and summation.

[10] *Essai sur les machines*, § XV, p. 26.

[11] It was one of Carnot's central conventions that when a system is set in motion, the difference between virtual and actual velocity in any component part is "lost to" or as the sign may be "gained from" the mutual interactions or constraints within the system. The usage comes from d'Alembert's mechanics. Carnot used the terms in the *Essai sur les machines* of 1783 and explained them in the *Principes* of 1803 (§ 41-43, pp. 24-26, and Pl. I, fig. 4), and did so in the following manner. If a body M tends to move with the virtual velocity $\overline{MW}$, but is constrained instead to take on the velocity $\overline{MV}$, then if a parallelogram of velocities is constructed, the velocity $\overline{MW}$ may be re-

The force (quantity of motion) that each contiguous particle imparts to m and
from which derives all the motion it receives from the system   F

The angle between the directions of W and V   X

The angle between the directions of W and U   Y

The angle between the directions of V and U   Z

The angle between the directions of V and F q

Given these designations, Carnot by means of simple algebra obtained from the previous equation an expression that he called his first fundamental equation (E), as follows:

$$\int m\text{VU} \cos Z = 0. \tag{E}$$

Only later in the argument did he point out that this expression was formally identical with conservation of live force. (Since $U \cos Z = V$, the expression reduces to $\int m\text{V}^2 = 0$.) He could not simply invoke that principle directly, however, given his assumption about the inapplicability of the principle to hard-body collision. He was bound by the generality in which he had set himself the problem to convert the expression into one applicable to interactions of elastic body and of hard body (the more so as the latter were generally taken for the term of comparison in nature) whether motion was communicated by impact or by insensible degrees.

To that end, Carnot introduced at this, the critical juncture of his argument, the notion that he always regarded as his most significant contribution, to mathematics as well as to mechanics: the idea of geometric motions, i.e., displacements that depended for their possibility only on the geometry of a system quite independently of the science of dynamics. Carnot's geometric motions became virtual displacements in later mechanics. The idea was an interesting one, therefore; it played a distinctive part in his own analysis; more signally than any other element it exhibits that a kind of operational economy in Carnot's reasoning was what differentiated it from the classical mode of analytic mechanics.

Imagine, Carnot asked in regard to a generalized system of hard bodies, that just as the system is struck, its actual motion be stilled and it be made instead to describe two successive movements arbitrary in character but subject to the condition that they be equal in velocity and opposite in direction. The configuration of the system being

---

solved into two components, of which one, $\overline{\text{MV}}$, will be the actual or remaining velocity and the other $\overline{\text{MU}}$ will be called the velocity "lost" by the body M. Prolonging $\overline{\text{MW}}$ in the opposite direction will give us the complementary case, a construction for velocity "gained," $\overline{\text{MU}'}$, in configurations in which the constraints augment initial virtual velocity.

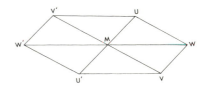

given, it was clear that such an effect could be accomplished in infinitely many ways, and (what was essential) by purely geometric operations.

In the *Essai sur les machines* Carnot did not at once achieve a definition of this class of motions that was both simple and adequate to the use he needed to make of them (although about the use itself, as will appear, he was perfectly clear). Thus, he laid down first that motions would be geometric if they involved the constituent bodies of a system in no displacement relative to each other. The converse would not be true, however, for motions that did involve such displacements might be geometric. For example, in the case of a machine composed of two weights suspended in equilibrium over a wheel and axle, a motion consisting of the descent of one from a height equal to the circumference of the wheel compensated by the raising of the other through a length equaling the circumference of the axle would be geometric since the equal and opposite movement could occur. Moreover, Carnot's purpose required him in a later passage to extend the concept to include motions of a system that meet the condition of reversibility in virtue of eliminating from consideration the components that (in later terminology) do no work. Perhaps, therefore, it will be legitimate for the sake of clarity to give the definition he did achieve in reworking the material in 1803 for the *Principes de l'équilibre et du mouvement*, wherein he suppressed allusions to these imaginary and reversible displacements, and laid down:

> Any motion will be called geometric if, when it is impressed upon a system of bodies, it has no effect on the intensity of the actions that they do or can exert on each other when any other motion is impressed upon them.[12]

In the *Essai sur les machines*, however, Carnot was thinking primarily in terms of bodies and displacements when it was a question of internal interactions in a system and only derivatively in terms immediately translatable as work and energy. Applying

---

[12] *Principes*, § 136, p. 108. It will help clarify both stages in his thought to give a few examples.

  a) From the *Essai sur les machines* of 1783:

    1) Two globes in contact. An impulse displacing both in the same direction along the line of centers would produce a geometric motion; an impulse separating them along the line of centers would not.

    2) Several bodies attached by flexible but inextensible wires to a common center. Any motion in which all remain equidistant from the center is geometric even if they shift among themselves; any motion altering the length of a radius is not.

    3) A body moving on a curved surface. A

movement tangential to the surface would be geometric; any departure from the tangent would not.

In each instance, the justification is that the equal and opposite motion is possible in the first case and not in the second (pp. 28-30n).

  b) From the *Principes* of 1803:

    1) Two bodies, A and B, are fixed to either extremity of a lever arm; the arm is rotated around a fixed point C. Each body assumes an angular velocity proportional to AC and BC respectively; since neither of these velocities influences the action of one body on the other whether by gravity or in any other manner, the rotation of the lever is a geometric motion (§ 139, p. 111).

the concept of geometric motion to a generalized system of hard bodies, he noted that by definition the relative velocity of neighboring corpuscles would be zero during the initial instant of such a motion. Designating by

u the absolute velocity of the corpuscle m in the initial instant of a geometric motion,

U as before, the velocity "lost" to the internal interactions,

z the angle between u and U,

then, since the corpuscles of the system would not tend to displace relative to each other in consequence of the velocity "u" alone, the mutual interactions within the system would be the same whether it was considered that "m" was animated solely by the velocity lost "U" or by the combined velocities "u" and "U." But if all the corpuscles were animated by the velocity "U" alone, equilibrium would necessarily obtain. Therefore, the real velocity after interaction would be "u," and by reasoning similar to that which yielded the first fundamental equation,

$$\int m V U \cos Z = 0, \tag{E}$$

Carnot had his second fundamental equation,

$$\int m u U \cos z = 0, \tag{F}$$

in which he has replaced an actual, physical velocity "V" with an idealized, geometric velocity "u."

Let us summarize, then, more explicitly than Carnot in the toils of his argument did himself, what precisely he thought to have gained by transposing the problem into terms of geometric motions. In the first place, that step justified him in extending the principle of conservation of live force to hard-body interactions, whether sudden or smooth, and that is how he interpreted Equation (F). As he observed later in the *Principes de l'équilibre*, all bodies are susceptible of geometric motion, whether hard, soft, or elastic, whether solid or fluid, for the reason that by definition these motions had no bearing on the internal interactions of bodies and were entirely independent of these dynamical distinctions.[13] Hence, Equation (F) extended to hard-body inter-action, and, since it contained as a special case Equation (E), which was itself a statement of conservation of live force, it authorized Carnot to take that principle as fundamental in the realm of geometric motion. And that, in the second place, is all he needed, since among the geometric motions of which a system was capable would be the real motion that it assumed upon any actual impulse. In the third place, therefore, Equation (F) represented the solution to the problem originally posed—i.e., given the virtual motion of any system of hard bodies, to find the actual motion of the system upon communication of external forces.

[13] *Principes*, § 144, pp. 114-115.

It will be well to specify what Carnot meant by a solution. Since Equation (F) was indeterminate, it would hold whatever the value of "u." Provided that the motion be geometric, particular values and arbitary directions might be attributed to the indeterminate according to the conditions of the specific problem, and it would then always be possible to formulate equations between the unknowns. For an example, Carnot produced the problem of a system of bodies unchanging in relative positions and containing no attachments to external fixtures. The solution could be drawn from Equation (F) by supposing arbitrarily that all points of the system were subjected to a geometric motion such that their velocities "u" were parallel to a given right line. Since "u" was then constant, Equation (F) became

$$\int m \mathrm{U} \cos z = 0,$$

which stated that the sum of the components of the forces "lost" to the constraints in the arbitrary direction of "u" was zero, and hence the resultant force was the same as if each body had been free. This was a "well-known principle," observed Carnot, not identifying it.

A second example was that of the same system made to rotate around a given axis, so that each of the points described a circle in a plane perpendicular to the axis. The movement being geometric, "R" being the radius of "m," and "A" a constant for all points, it was clear that

$$u = \mathrm{AR}$$

and that Equation (F) became

$$\int m\mathrm{RU} \cos z = 0.$$

That is to say, the sum of the moments relative to any axis of the forces "lost" to the mutual interactions was zero—another "well-known principle," said Carnot, not identifying it in that place for conservation of angular momentum or moment-of-momentum.[14] Of that, to which he recurred in a fundamental way, more needs to be said in the next section. What is characteristic to notice here is that such results were what Carnot meant by solutions to the problem. Contrary to what might have been expected of a young engineer, there were no numerical problems and solutions in the *Essai sur les machines*, just as there were no formulations not applicable generally to any type of machine.

C. MOMENT-OF-MOMENTUM

Having obtained a general solution for his fundamental problem of mechanics, Carnot now needed to state it in a form suited to drawing out the consequences for the theory of machines. To that end he turned to demonstrating its equivalence to the

---

[14] *Essai sur les machines*, § XVII, p. 33.

law called in later mechanics that of the moment-of-momentum, which involved much more for him than the example just cited concerning moments of forces relative to different axes of rotation. Carnot named the principle moment of the quantity-of-motion. Adducing it in terms of his geometric motions, he employed it for the same purpose that (according to Truesdell) physicists usually have done, in order "to obviate the need to specify the mutual forces among the particles of a rigid or deformable body."[15]

Referring still to a generalized dynamical system of bodies, wherein "m" is the mass of each body and "V" its velocity, suppose there be impressed upon it a geometric motion of velocity "u," which in its direction makes the angle "y" with the direction of the velocity "V." The quantity

$$mu\mathrm{V}\cos y$$

would then be the moment of the quantity of motion "$m\mathrm{V}$" with respect to the geometric motion "u," and its sum, $\Sigma\, mu\mathrm{V}\cos y$, would be the moment of the quantity-of-motion of the system with respect to the geometric motion imparted to it. Retaining the notation of the basic problem and expressing by

$\int mu\mathrm{W}\cos x$ the moment of the quantity-of-motion of the system before impact,

$\int mu\mathrm{V}\cos y$ the moment of the quantity-of-motion of the system after impact,

and

$\int mu\mathrm{U}\cos z$ the moment of the quantity-of-motion "lost" in impact,

Carnot showed by a simple trigonometric argument that

$$\int mu\mathrm{U}\cos z = 0$$

and hence

$$\int mu\mathrm{W}\cos x = \int mu\mathrm{V}\cos y$$

so that he could state a fundamental theorem:

In the impact of hard bodies, whether the impact be immediate or be transmitted by means of any springless machine [machine quelconque sans ressort], it is always true that with respect to any geometric motion—

1) The moment of the quantity-of-motion lost by the system as a whole is equal to zero.

2) The moment of the quantity-of-motion lost by any portion of the bodies of the system is equal to the moment of the quantity-of-motion gained by the remaining portion.

3) The moment of the quantity of real motion of the general system immediately

[15] Clifford A. Truesdell, "Whence the law of moment of momentum?," *Essays in the History of Mechanics* (New York, 1968), p. 243.

after impact is equal to the moment of the quantity-of-motion of the same system immediately before impact.[16]

These three propositions were identical at bottom, being simply interchangeable ways of stating the solution contained in Carnot's fundamental Equation (F). Nevertheless, the third was the most important to Carnot, for it was from that way of looking at the problem that he later in the argument drew out his own injunction to continuity in the transmission of power. Already in this passage, however, the principle appeared to him to be the most valuable in all the science of mechanics. For here was a quantity that did indeed remain unaltered in any impact, whether direct or indirect. The quantity was not what Descartes had thought it to be, the sum of the quantities of motion. That conservation holds only for particular directions and only when the system is free. Neither was it the sum of the live forces. That quantity is conserved only in the gradual transmission of motion. Here was this further quantity, however, that could not be diminished by obstacles interposed in the way of the motions of the system, nor yet by the machines that transmitted the motions, nor finally by percussions that might intervene; and that was the moment of the quantity-of-motion of the system in general with respect to any geometric motion it could perform. That principle (Carnot went on) contained all the laws of equilibrium and motion of hard bodies. He would next show that it might equally well be extended to other bodies whatever their nature or degree of elasticity.

In the inquiry just cited, Truesdell remarks that specialists in classical mechanics consider their subject to be one based on three fundamental laws: "the conservation or balance of *force, torque,* and *work,* or, in other terms, of *linear momentum, moment of momentum,* and *energy.*"[17] It would falsify the way it looked to Carnot himself to say that he saw the science so categorically. Yet it would not falsify what was actually in his reasonings to say that he found the first of little use for his purposes, felt that he could assume the third while requiring to give it more prominence than it was currently receiving, and while not claiming to originate the second, did believe that he had given an original argument for a principle the generality and profundity of which was then far from understood, and which he required in order to draw consequences that were his own. Even so it was not his originality on which he insisted but the conviction the argument gave to a rationality of mechanical practice that he felt should have been possible at any time. For Carnot was proud of his subject itself rather than vain of what he individually contributed to it.

Carnot justified his strong statement of conservation of moment of the quantity-of-motion with a series of corollaries and remarks that deduced from it principles he had been assuming and expressed them in a form applicable to the motion of machines. In the first he demonstrated that among the geometric motions of which a system was

---

[16] *Essai sur les machines,* § XXII, pp. 43-44.   [17] Truesdell, *op. cit.,* pp. 241-242.

capable, the one that would actually occur would give a minimum value for the sum of the products of each of the masses by the square of the velocity lost—though what he really meant, he hastened to point out was that the differential would be zero,

$$d \int m\mathrm{U}^2 = 0,$$

which could also be true of a maximum or, under certain conditions, of a value that was neither a maximum nor a minimum. (No doubt this possibility was what he had had in mind at the outset in noting the exceptions to which the conventional statement of the lowest center of gravity principle was liable.)[18] Carnot explicitly recognized the analogy of this proposition with Maupertuis' Principle of Least Action. The remark is revealing of the background of his own thinking, for, excepting d'Alembert, Carnot in the *Essai sur les machines* mentioned no other writer on mechanics more recent than Descartes. Departing from d'Alembert's point of view that the fundamental phenomena are those of inertial motion, the quantity of which constitutes force in his thinking, he directed the analysis to determining what forces could do, their action —though for Carnot this was an engineering concept and never a metaphysical one.

In the second corollary, Carnot turned to live force and stated that in hard-body impact the sum of live forces before collision equals the sum after collision plus the sum of the live forces that would obtain if the velocity remaining to each moving part were equal to what it had lost. The proof consists in a simple and obvious trigonometric transformation of Carnot's first fundamental Equation (E)

$$\int m\mathrm{VU} \cos \mathrm{Z} = 0$$

into the expression

$$\int m\mathrm{W}^2 = \int m\mathrm{V}^2 + \int m\mathrm{U}^2.$$

The analogy, Carnot here pointed out, between his Equation (E) and the conservation of live force thus amounted virtually to a demonstration. So also did the analogy of that equation with the same conservation in systems of hard bodies of which the motion changed insensibly, for then, "U" becoming infinitesimal, "U²" became an infinitesimal of the second order. He developed and established these points in the remaining two corollaries.

In the third corollary, Carnot turned attention to a subject that he had not yet discussed and that may be called Newtonian force only if it be disclaimed immediately that he saw it in such guise. For Carnot, the case was the one he had just mentioned, that of a system of hard bodies in which motion was communicated by insensible degrees. It was "motive force" that would have that effect, "force-motrice," and since he had not previously considered it, he required additional notation according to which

The mass of each body is	m
Its velocity is	V

[18] *Essai sur les machines*, § II, p. 12.

Its motive force is $\qquad$ p

The angle between V and p is $\qquad$ R

Its velocity in any geometric motion of the system is $\quad$ u

The angle between u and p is $\qquad$ r

The angle between V and u is $\qquad$ y

The element of time is $\qquad$ dt

Then the Equations (E) and (F) were shown to take the respective forms:[19]

$$\int m\mathrm{V}p\, dt \cos \mathrm{R} - \int m\mathrm{V}\, d\mathrm{V} = 0$$
$$\int mup\, dt \cos r - \int mu\, d(\mathrm{V} \cos y) = 0.$$

Considering, then, a body freely describing a trajectory in uniformly-accelerated motion, he integrated the former of those equations over time and space and obtained the conservation of live force in the form

$$\int m\mathrm{V}^2 = \int m\mathrm{K}^2 + \int m\mathrm{V}'^2$$

where "V" was final velocity, "K" initial velocity, and "V'" velocity acquired under a constant motive force in an undetermined time.

Finally, Carnot in his fourth corollary derived from the conservation of moment of the quantity-of-motion a proof that his fundamental indeterminate Equation (F)

$$\int mu\mathrm{U} \cos z = 0$$

governed motion and equilibrium not only in the case of hard body (that he had already proved in deriving it), but that it was a general law holding good for all bodies whatsoever. The proof being a digression from the theory of machines itself, it will not be necessary to follow it in detail. It turned on pointing out that in hard-body problems, Equation (E) became one of the determinates of the indeterminate (F) since by definition in that case, $u = \mathrm{V}$ (i.e., the real motion is geometric). For other types of body, expressions drawn from their nature would be needed in order to establish determinate values. For example, in respect of the force F that bodies exerted on each other, if in other bodies it was n times unity, then it would be possible to put U/n in place of U, and express Equation (E) in the form

$$\frac{n}{1-n} \int m\mathrm{V}\mathrm{U} \cos \mathrm{Z} = \int m\mathrm{U}^2.$$

In the case of perfect elasticity, in which convention would put $n = 2$, this expression reduced to

$$\int m\mathrm{W}^2 = \int m\mathrm{V}^2,$$

[19] The terms $p\, dt \cos \mathrm{R} - d\mathrm{V}$ represented the velocity lost by m in the direction of V in consequence of the interactions of the bodies, and were substituted for U cos Z in Equation (E). Similarly $p\, dt \cos r - d\, (\mathrm{V} \cos y)$ was the velocity lost by m in the direction u, and substituted for U cos z in Equation (F).

which, of course, was conservation of live force. That conservation was for elastic collision what the Equation (E) was for hard bodies, i.e., the characteristic determinant, which was just what he had undertaken to demonstrate at the outset.

Three remarks completed Carnot's formulation of the principles of equilibrium and motion themselves. The first referred the expressions he had derived to Cartesian coordinates, in which they appeared much simpler. Indicating by single, double, and triple primes the components of the quantities referred to three mutually perpendicular axes, and considering the conditions of equilibrium, Equation (E) reduced to $0 = 0$; the fundamental Equation (F) became

$$\int mu'W' + \int mu''W'' + \int mu'''W''' = 0;$$

and the general indeterminate for equilibrium in the case of motive forces became

$$\int mu'p' + \int mu''p'' + \int mu'''p''' = 0.$$

A second remark was more significant of Carnot's own style of thought. Traditionally mechanics had not only abstracted from the massiness of machines, but had treated obstacles and fixtures like the fulcrum of a lever either as stationary or as outside the system. Strictly speaking, however, no object in nature could be considered stationary, and no body was without mass. Rather than thus limiting the scope of the analysis, Carnot proposed a different mode of visualizing, one involving approximations to the material reality. He would let the fixtures and obstacles affecting the motions of a system be considered bodies of infinite mass and density though mobile in principle and hence susceptible of geometric motions. He would let bodies that merely communicated motion without offering any resistance to changes in the state of the system be considered bodies of infinitesimal mass and density. Of the two notions, the former seems the more significant, not for any practical difference it made in mechanics, but because this mode of accountability for mass bears a striking ancestral resemblance to that of the accountability for heat in Sadi Carnot's analysis of a heat engine drawing upon and discharging into reservoirs and sinks of heat of infinite capacity. By means of the latter notion, on the other hand, the mathematical theory of each of the classic types of machine might be derived from Equation (F).

Finally, Carnot began turning the discussion toward the object ultimately in view, the application of the principles of equilibrium and motion in the form just given them to the theory of machines. For that purpose, he found the word "power" (puissance) more natural than "force," and it would appear that Carnot's discussion marks an early stage of the process in which that word began making its transition from the usages of ordinary life into the service of exact science and engineering. Nothing could be more fundamental, of course, than the existence of bodies acting on and modifying the motions of other bodies in accordance with first principles. Cases would arise, however, in which it would prove convenient to abstract from the mass of bodies for the sake of considering only the effort being made by them—the pull of the cord against the load, the thrust of the wind against the sail, the drag of the current upon the boat, the resistance of the body to impulsion, the friction between the bearing and the moving part. When it was a question of effect, the size or nature of the agent made no difference, whether the source of power were man, beast, or machine; wind, water current, or weight. As for machines, they were considered to be assemblages of immaterial obstacles and moving parts capable of transmitting action without internal reaction. Carnot took them to be, in other words, bodies of infinitesimal density and mass, to which were applied forces external to the system—powers in a word—but Carnot was consistent with his own realism. Machines in fact possessed mass, and he would not neglect it. He would simply consider its effect as he did the other forces or powers exerted by agents external to the system. Any force was to be taken as a quantity of motion "lost" by the agent that exerted it, whatever that agent might be.

If, then, that force was designated "F," its quantity was evidently the same as that expressed by the product "$m\mathrm{U}$" in all the foregoing discussions; and if "Z" was the angle between the direction of that force and any geometric velocity "u" imparted to the system, then the fundamental Equation (F) might be expressed as follows:

$$\int Fu \cos Z = 0. \tag{AA}$$

That was the form in which Carnot employed it in the discussion of machines in general that ensued in Part II, wherein when he wrote of force he was thinking of what he also called power, and tacitly taking its quantity over time.

It is clear, therefore, that we have reached the crucial turn that Carnot gave to the arguments of classical mechanics in what was said of machines. Classically machines were taken to be agents for the transmission of motion. The quantity of motion was an expression of force and involved mass. Abstracting from the mass of machines in order to study motions, the formulations of mechanics took no account of the motions of machines themselves. Substituting the notion of power for that of force, Carnot was able to transpose the notion of machines into that of agents for the transmission of power. Instead of neglecting the mass of the machines, he considered it to be one of the powers affecting motions. The problem became, therefore, adaptation of the laws of motion to laws of power and specification of the optimum conditions for transmission of power, and it was in order to prepare the ground that Carnot had brought the concept of motions "lost" and "gained" into the purview of the classical conservation principles and models of the structure of matter.

D. MOMENT-OF-ACTIVITY—THE CONCEPT OF WORK

In all the foregoing, which constituted Part I of the *Essai sur les machines*, Carnot had not proposed to add to the science of mechanics but to state its principles in a form and arrange them in a sequence applicable to the science of "machines properly speaking" that he next inaugurated in Part II. There the writing went more directly to the point, the tone became more businesslike, and the yield of the successive propositions was more evident in the analysis they afforded of the transmission of power.

The analysis began with Carnot's distinction between forces applied to a machine in motion according to whether their direction made an acute or an obtuse angle with the direction of motion of the point of application. The former he called "moving" or "impelling" forces, the second being the adjective usually employed. The latter he called "resisting" forces. Elsewhere in the *Essai sur les machines* Carnot observed mildly that the definition would avert dispute about whether forces were to be considered causes or effects.[20] Here it sufficed to notice that impelling forces could become resisting and vice versa if the direction of the motion changed, and that in any power system each of the forces would be impelling or resisting with respect to any given geometric motion imparted to the system according to the angle between the respective directions.

Carnot needed those metaphysically indifferent definitions for themselves. He also needed the distinction between the two classes of force for its auxiliary importance in his discussion of what in retrospect appears the most significant recognition in all his writing on mechanics and on mathematics, that of the importance to the science of power of the quantity that is now called "work" and that he called "moment-of-activity." At first glance it might appear that extending as he did the distinction between impelling and resisting force into one between moments-of-activity "consumed" and "produced" by forces applied to a system arrested him on the threshold instead of in the presence of a unified concept indistinguishable in substance from that of work. Such was not the case, however—though a little complex in the phraseology, the distinction was simply that between work done on or by a system.

Let a force "P," he laid down, be moving with a velocity "u" (for, though he did not say so, such a movement could only be geometric), and let the angle between "P" and "u" be "z," then the quantity $P \cos z\, u\, dt$ (where dt is the differential of time) would be what Carnot defined as the *Moment-of-Activity* consumed by the force "P" during the time dt. Characteristically enough, Carnot was thinking about the gravitational instance for his example. Let a weight fall the distance "H" in time "t." Then, $dH = u \cos z\, dt$, and the moment-of-activity consumed during time dt would be $P \int dH$ or $P\,H$—the product of force by distance: foot-pounds in the Anglo-American dimensions of elementary physics.

Now, it is very difficult to establish categorical priorities in the history of mechanics. It is equally difficult, however, to find specialists in mechanics prior to Lazare Carnot who explicitly singled out that quantity to be the measure of what forces and powers accomplish and a unit in terms of which the fundamental conservations were to be

[20] *Essai sur les machines*, p. iv.

stated. It is significant, moreover, that he stated it as the sum of the moment-of-activity consumed in successive instants—i.e., as what was accomplished by a process occurring in time. He was thinking about power, and not merely its dimensional equivalent, the product of displacement by motive force.

For it is clear that he did know what he was doing. Considering an entire system of forces applied to a machine in motion, the moment-of-activity consumed by the totality of the forces would be the difference between the sum of moments-of-activity consumed by the impelling forces and the sum of moments-of-activity consumed by the resisting forces. Since impelling forces made an acute angle and resisting forces an obtuse angle with motions, the cosines would be positive in the former case and negative in the latter, and the sign of moments-of-activity of impelling forces would thus be positive and of resisting forces negative.

Turning now to the reciprocal point of view, the moment-of-activity *produced* by a force was to be understood as the product of its magnitude evaluated in a direction *contrary* to its velocity multiplied by the distance that the point of application traversed in an element of time. Obviously, then, the moments-of-activity *consumed* and *produced* were equal quantities of opposite signs, and the difference between them was identical in the terminology of power to that in mechanics proper between moments-of-the-quantity-of-motion gained or lost with respect to a given geometric motion. In a final definition, Carnot combined his two classes of force and with them his two classes of moment-of-activity into a single designation. He called the moment-of-activity *exerted* by a force that which was *consumed* by an impelling force or *produced* by a resisting force. Since he did not know which was which, he would in practice always take the cosine of the smaller of the two supplementary angles between the force and the motion of the point of application, so that the moment-of-activity exerted by a force would always be a positive quantity.

Clearly then, this notion contained what is later to be found in the employment given by the science of physics to the concept of work.

Having laid down these definitions for the mechanics of machines, Carnot proceeded to state what he called its fundamental theorem:

> Whatever the state of rest or motion of any system of forces applied to a machine, if a geometric motion be imparted to it without altering the forces, the sum of the product of each by the velocity of the point of application, evaluated at the first instant and in the direction of the force, will be zero.

That statement was, of course, simply the adaptation to a system of forces or powers of the principle of conservation of the moment-of-the-quantity-of-motion for a system of bodies in motion. Expressed in the following form,

$$\int Fu \cos z = 0,$$

it was evidently identical with the Equation (AA) into which he had transformed his fundamental indeterminate (F) for the purpose of applying it to the transmission of power.[21] Indeed, it was obvious (Carnot remarked immediately) that strictly speaking this fundamental theorem was nothing more than the principle of Descartes with its defects repaired—i.e., generalized beyond equilibrated pairs to any system of forces with regard both to their direction and intensity.

Following the fundamental theorem, Carnot set out a series of six corollary propositions. The first three concerned the conditions of equilibrium of systems of machines and weights of various sorts on which geometric motions were impressed, and the fourth asserted the balance of torques around each of three mutually perpendicular axes along which forces in equilibrium could be resolved. Except for redeeming his opening promise to give a rigorous derivation of the original lowest center-of-gravity principle of equilibrium, Carnot did not seem much interested in these propositions, for he gave them in summary fashion and mainly for the sake of completeness.

The fifth was different: "Special law concerning machines of which the motion changes by insensible degrees," he called it, and now at last he had come to what he did certainly recognize to be his own, though still without insisting on his originality. The law ran thus:

> *In a machine of which the motion is changing by insensible degrees*, the moment-of-activity consumed *in a given time by the impelling forces* equals the moment-of-activity exerted *in the same time by the resisting forces*.[22]

That would be true if it could be proved that the moment-of-activity consumed by all the forces of the system during the given time was zero. If "F" designated each of the forces, "V" its velocity, and "Z" the angle between them, then the requirement was that $\int FV \cos Z \, dt = 0$; but by the fundamental theorem just stated, it was clear that $\int FV \cos Z = 0$. Hence the corollary followed from the general principle of equilibrium in machines.

It was not, however, in this almost tautological proof that Carnot's interesting contribution consisted, nor yet in the statement of the special law, although he did recognize it to be "the most important in all the theory of the motions of machines properly speaking." For he introduced the principle of continuity in the transmission of power by way of exemplifying an application of this special law, and then developed its consequences in the scholium that concluded the *Essai sur les machines*.[23]

When working out his own thoughts Carnot usually preferred reasoning on weights and asked the reader to suppose that they be the powers applied to a machine and that

m be the mass of each of the bodies,

M   designate the total mass of the system,

[21] *Ibid.*, § XXX, p. 63.
[22] *Ibid.*, § XLI, pp. 75-76.
[23] *Ibid.*, § XLIX-LXIV, pp. 81-99.

g be the force of gravity,

V be the actual velocity of a body m,

K be its initial velocity,

t be the time elapsed since the motion began,

H be the height through which the center of gravity of the system falls in time t, and

W be the velocity acquired in the height H.

Two types of force were involved in the operation of the machine, the pull of gravity and the inertial resistance of the various bodies to change of motion. By definition, the moment-of-activity *consumed* in the whole system by the action of gravity during time t was

$$MgH,$$

which was equivalent to

$$\tfrac{1}{2}\, MW^2.$$

Considering now the forces of inertia, the velocity of m being V and becoming in the next instant $V + dV$, its force of inertia in the direction of V would be

$$m\frac{dV}{dt}.$$

Reverting again to definitions, the moment-of-activity *exerted* by the forces of inertia during time dt was

$$m\frac{dV}{dt}V\, dt, \text{ or, simplifying, } mVdV.$$

The moment-of-activity *consumed*, therefore, during time t by the forces of inertia was

$$\int mV\, dV.$$

Integrating and completing the integral, it became

$$\tfrac{1}{2}\, mV^2 - \tfrac{1}{2}\, mK^2$$

(Carnot sometimes adopted the device of including the constant of integration in the term for initial velocity). Therefore, the moment-of-activity *consumed* in time t by all the bodies of the system would be

$$\tfrac{1}{2} \int mV^2 - \tfrac{1}{2} \int mK^2.$$

Now, by the conditions of the problem, inertia was a resisting and gravity an impelling force. Thus, by the special law just stated in the corollary, it would be true that

$$MW^2 = \int mV^2 - \int mK^2$$

or

$$\int mV^2 = \int mK^2 + MW^2.$$

As usual, Carnot also told verbally what he had just done:

In a weight-driven machine of which the motion is changing by insensible degrees, the sum of the live forces of the system after a given time is equal to the sum of the initial live forces plus the sum of the live forces that would obtain if all the bodies of the system were animated by a common velocity equal to that due to the height through which the center of gravity of the system has fallen.[24]

It followed immediately that (a) in a weight-driven machine in uniform motion, the center of gravity of the system remained at constant height (since then $V = K$, $W^2 = 0$, and $H = 0$); (b) no matter in what manner a weight was raised to a certain height, the forces producing that effect must have been such as to consume a moment-of-activity equal to the product of the weight by the height; and (c) to produce any movement by insensible degrees in a system of bodies, the forces [Carnot said "powers"] to produce that effect must have consumed a moment-of-activity equal to half the quantity by which the sum of the live forces of the system was increased.

From these last two propositions it further followed that in order to raise a weight M from rest to a height H while imparting to it a velocity V, the forces employed to that end must consume a moment-of-activity equal to

$$MgH + \tfrac{1}{2} MV^2.$$

And it seems worth following Carnot's reader through these elementary steps since they end with this expression dimensionally and notationally identical to an energy statement. Conceptually, of course, the range was extremely restricted, but there the expression was, nonetheless, and it would be a generalization and not an alteration of this type of thinking that would extend that range. So far, indeed, all that was new was the type of thinking, for dimensionally and algebraically the equivalence of those two terms was a relation as old and elementary as Leibniz's early discussions on the true measure of force. What Carnot had contributed was couching the relations in terms of the operation of machines, inertial masses transmitting powers over distances, and thinking of live force as the capacity for consuming moments-of-activity of machines in motion, or in later words, of forces doing work over distances and in time.

Now, however, we come to the consequence that was his own both in substance as well as genre. The previous reasoning assumed that motion in the system changed by insensible degrees. Suppose, however, that some discontinuity, some impact or percussion, intervened. Then let

h   be the height from which the center of gravity has fallen at the moment of percussion,

X   be the sum of the live forces immediately before percussion,

Y   be the sum of the live forces immediately after percussion,

[24] *Ibid.*, § XLII, pp. 77-78.

Q   be the moment-of-activity that the live forces must consume throughout the entire movement, and

q   be the moment-of-activity that the live forces must consume up to the moment of percussion.

Finally, suppose for the sake of simplicity that the movement of the system began from and ended in rest. It would be evident from the relation just established that

$$q = Mgh + \tfrac{1}{2} X,$$

and similarly that the moment-of-activity to be consumed following percussion would be

$$Q = Mgh + \tfrac{1}{2} X - \tfrac{1}{2} Y.$$

Now Carnot appealed back to his statement of what he had held in Part I of the *Essai sur les machines* to be the fundamental law of motion and equilibrium in mechanics proper, the conservation of the moment-of-the-quantity-of-motion,[25] and specifically to its first corollary, which found that of all the motions of which a system was capable, that which would actually occur would be the geometric motion such that the sum of the live forces of each of the masses would be a minimum.[26] Therefore, since $X > Y$, the moment-of-activity to be consumed in raising M to the height H was larger than if there had been no percussion, for in the latter case the equation would simply be

$$Q = MgH.$$

It followed that the same impelling forces could raise the same weight to a greater height if percussion were avoided than if it occurred. Carnot deferred an explicit statement of this, his Principle, to the concluding summary. For any machine, the condition of maximum efficiency is that $q = Q$:

"Now, in order to fulfill that condition, I say, first, that any impact or sudden change is to be avoided, for it is easy to apply to all imaginable cases the reasoning developed in that of weight-driven machines; whence it follows that whenever there is impact, there is simultaneously a loss of moment-of-activity on the part of the impelling forces, a loss so real that their effect is necessarily diminished. . . . It is then with good reason that we have proposed that in order to make machines produce the greatest possible effect, they must never change their [state of] motion except by insensible degrees. We must except only those that by their very nature are subject in their operation to various percussions, as are most mills. But even in this case it is clear that all sudden changes should be avoided that are not essential to the constitution of the machine."[27]

[25] *Ibid.*, § XXII, pp. 43-45.
[26] *Ibid.*, § XXIII, pp. 45-48. This is the prin-
ciple that Carnot recognized as analogous to Maupertuis' Law of Least Action.
[27] *Ibid.*, § LIX, pp. 91-92.

Such was Carnot's Principle. It is characteristic that he should have worked it out by example, claimed a generality that he did not actually demonstrate, and then exhibited his awareness that in actual practice the condition of perfect efficiency is an ideal to be approximated so far as the nature of the process allows. To the demonstration he added a final corollary on hydraulic machines, observing merely that since a fluid might be regarded as an infinity of solid but detached corpuscles—and since he had already proved what was usually taken to be merely experimental truth, i.e., the conservation of live force in incompressible fluids wherein motion changes insensibly [that is without splashing or turbulence]—everything that he had demonstrated for systems of hard bodies held equally good for masses composed of incompressible fluids.[28] But that extension was more important than the afterthought it appears to be in the formal structure, and it will be easier to appreciate its significance and indeed the quality of Carnot's performance in the mechanics of machines if we accompany him into the scholium in which he enlarged more informally, and more naturally, on what he hoped his *Essai sur les machines* would accomplish.

E. PRACTICAL CONCLUSIONS

In the concluding discussion containing the statement of Carnot's principle just quoted, he relaxed his formality and spoke his mind. According to Corollary V, the moment-of-activity consumed in a given time by impelling forces equals that exerted in the same time by the resisting forces in a machine that changes gradually in its state of motion. Actually, he observed, that proposition contained nearly all the applicable part of the theory of machines. In practice most working devices were powered by agents—animals, springs, weights—that exerted "dead force" continuously. Most real machines, furthermore, once set in motion soon reached a steady pace of operation such that the forces required to keep the process going balanced the elements of resistance. It now appears, therefore, that this corollary had been the goal of the argument all along, and we would not be forcing his meaning to call it his Work Principle. Indeed, we would not altogether be forcing his language, for in one example of the equivalence between moment-of-activity and effect, he remarked à propos of an arrangement for raising a weight a certain distance, that no machine could be designed by which it would be possible "with the same work [travail] (that is to say the same force and the same velocity)" to lift the object higher in the given time.[29]

In turning now to practice, Carnot maintained the generality of his discussion. Designating by Q $\begin{Bmatrix} \text{the work done} \\ \text{the moment-of-activity consumed} \end{Bmatrix}$ by the impelling forces and by q that $\begin{Bmatrix} \text{done on} \\ \text{exerted by} \end{Bmatrix}$ the resisting forces, then Corollary V, the Work Principle, could be symbolized

$$Q = q.$$

[28] *Ibid.*, § XLVIII, p. 80.   [29] *Ibid.*, § LIV, p. 85.

It followed that there were two related sets of conditions for the most efficient possible operation of machines. The first was that the greatest possible mechanical advantage be secured and the second that no motion be wasted. In order to clarify the former condition, the relation could also be expressed

$$\mathrm{F}\mathrm{V}t = q,$$

where F was the resultant of all impelling forces and V the resultant of their velocities. Achieving the maximum effect would involve varying those three factors. A fourth variable would be the direction in which the impelling forces acted, but it was obvious (though Carnot labored the point in one of his trigonometric excursions) that for best results the force ought always to be applied in the direction of the velocity. As for the intensity of force, the time of application, and speed of operation, no general rules applied and experience would need to govern. For example, a man might turn a crank one foot in diameter for eight hours a day making an effort equivalent to twenty-five pounds at a rate of one revolution every two seconds. If he were to work faster, his output would suffer for he could no longer maintain the twenty-five pound exertion. If he turned the handle more slowly, he increased F proportionately less than V, and the moment-of-activity would diminish. Every source of power would have its own maximum, of which Carnot could say only that it was determined by its physical constitution and that experience alone would find it.

No hidden resources of power lay in the capacities of machines, therefore, and what theory offered was identification of the factors among which economy would dictate the wise proportions to observe. If time were not important, force might be economized at its expense. But (the reader may feel a bit impatiently) so elaborate a discussion can hardly have been required to exemplify the ancient maxim that what is gained in force is lost in time or speed. Carnot clothed the point in new terminology, but gave no new findings, and what must be appreciated is that he evidently thought that rigorous demonstration would help and to that end expressed himself with an engineer's passion and urgency. These matters, so obvious to him, must need the conclusive proof he was giving, else why the persistence of chimerical schemes for perpetual motion under one pretext or another, why the continuing waste of ingenuity and money? It would falsify Carnot's book to underemphasize that, perhaps its most deeply felt point.

The novel considerations that Carnot brought forward, however, pertained mainly to the second set of requirements for maximum efficiency, the avoidance of waste motions. Instead of generalizing the reasoning that had established his Principle in the example of weight-driven machines, so as to cover all possible cases, Carnot turned to developing its relevance for hydraulic machines, and did so in a series of paragraphs that seem particularly striking when compared to the reasoning in the memoir by his son, *Réflexions sur la puissance motrice du feu*, where heat is treated like a fluid in flow.[30]

Looking at the ordinary waterwheel from an ideal point of view, Carnot observed that its design embodied two faults. First, the water was generally allowed to fall onto the blades and thus to transmit motion by percussion. Second, after striking the blade the water ran off with a velocity that was entirely lost to the process. In order that a hydraulic machine be as efficient as possible, its design should be such that the water would lose all its motion to the mechanism, and should do so without turbulence or splashing. It made no difference what form the machine might take. As always, Carnot was dealing in the analysis with idealized machines. He recognized that descending to actuality might make it desirable to depart from these conditions. For simplicity of construction, nothing was likely to prove better than wheels turned by the impact of water.

Practically, moreover, the closer the design approached to satisfying one of the conditions, that is absorbing all the motion of the water in the wheel, the greater would be the impact and the greater the loss of power to percussion. The lesser the impact, conversely, the less would be the proportion of the power of the water transferred to the wheel. The shrewd designer would, therefore, regard these conditions not as goals to be attained but as norms to be approached in the degree that circumstances render the one saving or the other relatively more important. Indeed, it was an ultimate condition of efficiency in machines of all types that no motion be produced extraneous to their purpose, and readers of Sadi Carnot's memoir will also recognize this injunction adapted to the heat engine in his requirement that no differential of temperature be admitted that does not measure itself in a change of volume in the gas confined in a cylinder. As his father originated the example, it showed the most efficient pump to be the one that delivered water into the reservoir at velocity zero.

Expanding in his closing remarks on the concept of moment-of-activity, Carnot introduced a final illustration of its utility that carried his perception of its significance beyond the study of machines into generalized physics. Suppose the problem to be one of a system of bodies mutually attracting one another by forces that vary as any power of the distance. (Only at this late stage did he move into a mechanics of which the model was clearly Newtonian, and then only as an object of an analysis developed out of other considerations and other problems than those of central force systems.) Suppose the system was impelled to move from some given configuration to another. No matter, then, what the sequence in which the individual bodies were displaced; no matter, further, what route they took, provided only that no percussion intervened; no matter, finally, what sort of machines effected the transformations—quite independently of all such incidental means to the end, the moment-of-activity that the

[30] This analogy has often been noticed, e.g., Charles Brunold, *L'entropie* (Paris, 1930), pp. 37-40.

external agents consumed would always be the same, assuming the system to be at rest in its initial and final states.

Now, then, this certainly might be taken for a model of the kind of analysis that the physics of work and energy has found useful ever since those topics became explicit in the 1820's, '30's, and '40's. Its most recognizable offspring again was the heat cycle of Sadi Carnot, which considered a system in view of what had been done to or by it in shifting from an initial to a final state. The family resemblance was most marked in the abstractedness of the system, in the notion that process consisted in the transition between successive "states," in the restriction that this transition be gradual and continuous, in the requirement that all changes be reversible (which for Lazare was still to say that all motions be geometric), in the indifference (given those conditions) to the details (rate, route, or order of displacements), in the relevance of this extreme schematization to the actuality of operating tools, engines, and machinery. Clearly, the relationship between the science of machines and thermodynamics was similar to that between Lazare and Sadi Carnot. It was one of parentage.

CHAPTER III

The Development of Carnot's Mechanics

I T IS fortunate that the two memoirs Carnot submitted to the Academy of Sciences in Paris in 1778 and 1780 have survived in its archives. They give access to the form and content of his thinking about mechanics and the science of machines when he was a young and hopeful officer five or six years out of engineering school. The present chapter traces the development of his thinking from these, its earliest recorded expressions, through to the publication, in 1803, of his *Principes fondamentaux de l'équilibre et du mouvement*. The method is to compare these earlier and later versions to the *Essai sur les machines*, and to continue to an analysis of Sadi's *Puissance motrice du feu* of 1824 in order to bring out its filiation with his father's work. The relevant, theoretical sections of the early two memoirs are reprinted below as Appendices B and C. Readers may wish to refer points in the discussion, or indeed the discussion as a whole, to the original texts.

A. ARGUMENT OF THE 1778 MEMOIR ON THE THEORY OF MACHINES

A notation on the cover page of the manuscript of Carnot's first memoir indicates that he learned of the Academy's announcement of a prize contest from the *Gazette de France* of 18 April 1777. The Academy specified the subject to be:

> The theory of simple machines with regard to friction and the stiffness of cordage, but it [the Academy] requires that the laws of friction and the examination of the effects resulting from stiffness in cordage be determined by new experiments conducted on a large scale. It requires further that these experiments be applicable to machines used in the Navy such as the pulley, the capstan, and the inclined plane.

From a further note it appears that the Academy received the entry Carnot sent on 28 March 1778. Assuming that he had set to work immediately, it took him, therefore, just under a year to design and carry out his experiments and compose the argument. For device he misquoted a line from Lucretius: "Videndum/ Qua ratione fiant et qua vi quaeque gerantur."[1] In setting the competition the responsible members of the

---

[1] Probably relying on memory while working at Cherbourg, he had in mind line 129 of *De rerum natura*, Book I, but spoiled the scansion by inverting the word order; he also displaced the gerundive from the end of line 131. The entire passage reads

   Qua propter bene cum superis de rebus

habenda
nobis est ratio, solis lunaeque meatus
qua fiant ratione, et qua vi quaeque
   gerantur
in terris, tum cum primis ratione sagaci
unde anima atque animi constet natura
   videndum,

Academy would have had no thought that anything of importance to the science of mechanics itself was to be expected in consequence of the entries it might elicit or on the part of the contestants it might attract. The emphasis was on the naval application, and what the Academy clearly wanted was studies of friction. None of the papers having satisfied the judges, it reset the same subject for the contest of 1781, and then awarded the prize to Coulomb, whose investigation gave them exactly that and clearly deserved to win. His memoir remains one of the cardinal contributions to the knowledge of friction.[2]

Carnot's own taste and interest, however, responded primarily to the opening phrase, and he submitted his entry as a "Mémoire sur la théorie des machines." It is divided into two parts, experimental and theoretical. As will appear, the organization of Part II already exhibited the structure within which Carnot's thought about the subject developed throughout his life. Obedient to the Academy's injunction, he did dutifully set to work to determine experimental laws of friction and binding, and Part I of the memoir consists of some twenty folio pages detailing the results. An account of these experiments would be no help in understanding the difference Carnot's work finally made in mechanics, but perhaps the reader will be curious to know what he undertook. Employing the services of an assistant, he spent considerable pains upon two sets of determinations, one concerned with friction and the other with cordage. For the work on friction Carnot constructed the device pictured on p. 64.[3]

The instrument operated in a simple manner for the purpose of determining values for friction between surfaces that (in the so-called first type) slide one across the other and (in the second type) roll one upon the other. The principle of the machine was that the statical moment of a weight Q just sufficient to initiate rotation was the measure of the force of friction (with which it was in equilibrium) between the surfaces of the hemispheres and of the iron shoes *srtg* and *zuxy*. The inner circumference of the rings A and B and the upper blade of the axle HH′ were knife edges. With the retain-

---

et quae res nobis vigilantibus obvia
  mentis
terrificet morbo adfectis somnoque
  sepultis,
cernere uti videamur eos audireque
  coram,
morte obita quorum tellus amplectitur
  ossa.

A translation of the words Carnot chose might read, "Let us look to an account of how all things come to pass and of the forces that govern them."

[2] C. A. Coulomb, "Théorie des machines simples en ayant égard au frottement de leures parties et à la roideur du cordage," *Mémoires de mathematique et de physique présentés a l'Académie Royale des Sciences par divers savans*, X (1785), pp. 161-332. For a discussion of the contest centered upon the winning memoir, and for the importance of Coulomb's work in friction, see Gillmor, *op. cit.*, Chapter IV. The judges were d'Alembert, Bézout, Bossut, Condorcet, and Trudaine de Montigny.

[3] The illustration is in fact from the 1780 memoir, but it is clear from the context that it is the same device of which the drawing has been separated from the 1778 text.

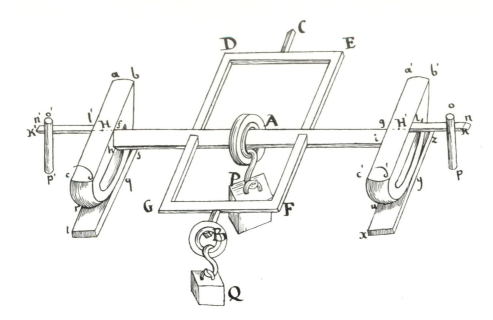

ing pins op and $o'p'$ in place, the axle was free only to turn. A sufficient weight Q would cause the surfaces of the hemispheres to slide across the supporting shoes, and the friction measured was of the first type. With the pins op and $o'p'$ removed, the axle was free to displace. A weight Q would cause the hemispheres to roll on the supports, and the friction measured was that of the second type. Carnot made, too, a determination of the effect of adherence. In measuring sliding friction, the values were always significantly greater if the device was allowed to rest on the supports for an appreciable time than they were when the measurement was made at the instant it was loaded. Adherence appeared to have no appreciable effect, however, upon rolling friction.

Carnot employed increasing weights for P and Q, and tabulated the ratios of the two classes of friction and of adherence to pressure over a range between 100 and 2,000 pounds for the total weight of the movable cradle. In each class of measurement, the ratio of friction to pressure decreased with increasing load. In good engineering fashion, he sought to express the results in formulas that turned out no more beautifully than such empirical expressions usually do. The formula for the ratio x of sliding friction to pressure was

$$x = \frac{A + Bp + Cp^2 + \ldots \text{etc.}}{1 + bp + cp^2 + \ldots \text{etc.}}.$$

Carnot well understood its arbitrariness. He could have given the relation a simpler appearance, he observed, in supposing that

$$x = A + Bp + Cp^2 + \ldots \text{etc.}$$

except that as p increased, so would x, which effect was the contrary of what the experiments revealed. On the other hand, he would fall into the opposite error in supposing that

$$x = A + Bp^{-1} + Cp^{-2} + \ldots \text{etc.}$$

The best he could say for the formula adopted was that it fell as closely as possible between the two extremes.

Further experimentation to determine the influence of velocity on the two classes of friction led Carnot to produce formulas even more empirical in appearance. For sliding friction,

$$X = \frac{\phi + \mu\pi}{1 + \omega\pi} \cdot \frac{1 + bu}{1 + cu},$$

and for rolling friction

$$X = \frac{\phi + \mu\pi}{1 + \omega\pi} + \frac{\phi'}{1 + \omega'\pi},$$

where x was the ratio of friction to pressure; ϕ, μ, ω, ϕ' and ω' were constants depending on the nature and degree of polish of the surfaces; π was the pressure exerted by a single point; u was velocity; and b and c were constants of velocity determined by a method at once cumbersome and inaccurate.

The experiments on stiffness and binding in cordage were of a similar type, and the results no more applicable. Perhaps it will suffice to reproduce the illustration of the pulley-type instrument that Carnot designed and employed in order to determine how the properties of ropes and wires affected the equilibrium conditions of weights of varying magnitude.

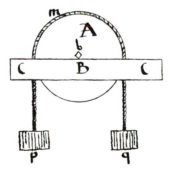

The account of these experiments occupies § 1 through 26, some sixteen folios of a manuscript containing sixty-four folios all told. At the outset Carnot promised that Part II would contain their "application to machines." What Part II actually contains is three sections, the first sixteen folios discussing the principle of "machines en général" in § 27 through 50, and the remaining two, in § 51 through 84, applying the principles respectively to equilibrium and to motion in the cases of the seven types of simple machine. It is the first section of Part II that is most interesting to the historian, for it is one of those relatively rare documents that admits him to the genesis of an approach that was later to become distinctive of a whole new way of seeing and doing a science.

That it was an application of the experiments was a pretense on Carnot's part, however. It is difficult, in fact, to see that they had anything to do with the theoretical discussion of machines in general except to serve as an excuse to qualify the memoir for this particular competition. The excuse is the less convincing in that Carnot had not the time or did not take the trouble to fill up his tables of data completely. There is no mention of these experiments later in his published writings. Nor did there need to be—despite his view of mechanics as an experimental science. The most that can be said, and as a highly likely conjecture this much ought to be said, is that the instrument he constructed for measuring friction may very well have led him to set up problems in terms of torque and moments of rotation around an axis, and therefore towards the emphasis he placed upon the principle of moment-of-momentum in the *Essai sur les machines*.

However that may have been, the theoretical passages that were the heart of the memoir make clear how Carnot thought to approach the subject. He made no preliminary statements of basic principles of mechanics of any sort. Rather, he moved in imagination right into the interstices of hard body and pictured what the inflexible rods and inextensible wires were doing, mini-levers and mini-pulleys, micro-machines that transmit motion there where it starts, corpuscle to corpuscle. Once those interactions were clarified, they would be those ideally operating in the employment of any actual machines. That is what he thought about in the first instance: what one corpuscle was doing to another in a system of hard bodies. A motion was imparted to the system. The mutual interactions of the corpuscles transformed that motion into some other which it was the problem to determine. Designating by

m the mass of the corpuscle,

V its virtual velocity (his expression was "the velocity that it would have taken if it had been free, that is to say without the reaction it undergoes from the rest of the system"),

u the velocity it really did assume, and

y the angle between the directions of V and u,

then the following equation would hold:

$$\int mu \,(V \cos y - u) = o,$$

which asserts that the sum of the products of the quantity of motion (mu) of each of the corpuscles multiplied by the velocity that it lost evaluated in the direction it took ($V \cos y - u$), is equal to zero.[4] For the formation of Carnot's idiom, the significant

[4] App. B, § 29.

term was the one in parenthesis. It makes evident that from the very beginning he thought of interactions within a system in d'Alembert's terms of velocities lost or gained by its constituent parts, and thought of live force (which in later times would be seen as the energy involved in such interactions) as the product of momentum by a velocity or motion thus "lost" or "gained." The actual demonstration involved the physics of hard-body collision and equating the forces of action and reaction in pairs of contiguous molecules: in actual practice the equality of action and reaction was always the basic law of mechanics for Carnot.

In this memoir, Carnot claimed for the above equation the status of a fundamental theorem. It did duty for all the variants that he later elaborated in the *Essai sur les machines*—its determinate Equation (E)[5] for the general case, its transformation by way of geometric motions into the indeterminate Equation (F),[6] its deduction from Equation (F) of the principle of conservation of moment-of-the-quantity-of-motion with respect to any geometric motion, its derivation of the power Equation (AA)[7] concerning the forces applied to a system, its general principle of equilibrium and motion for machines[8] in terms of the balance between moments-of-activity consumed and exerted by impelling and resisting forces. Carnot drew upon the general corpus of mechanics for all these refinements to his problem. In the early memoir only more or less definite foreshadowings of them occur scattered along a series of some seventeen corollaries and remarks. Most were so elementary or trivial that one would never read them twice, and no emphasis was given to the ones that in the retrospect of Carnot's later writings seem significant, the kernels that later opened out into main positions. The most useful procedure will be to point out what elements he did then have of his later findings, and what the differences were.

The greatest difference in emphasis was that when writing of force in his early memoir Carnot began with the effects of accelerative or motive force in producing motions. In the *Essai*, it will be recalled, accelerative force remained largely hidden from attention until near the end, when it appeared as one of the factors in moments-of-activity, i.e., as the force that does act across a distance or in a time. In the 1778 memoir, however, the first corollary concerns motive force. The (virtual) velocity V is now said to be the resultant of the motive force of the corpuscle m at a given instant and of its velocity u' in the previous instant. Evidently, then,

$$V \cos y = u' \cos z + p \, dt \cos x.$$

(Carnot was somewhat careless with his designations in this memoir. He identified z as the angle between u' and u, but not p as "*force motrice*" nor x as the angle between the directions of p and u. That is what he meant, however.) The basic equation would then become

$$\int mu \, (u' \cos z - u) - \int mup \, dt \cos x = 0.$$

[5] Above, p. 42.   [6] Above, p. 44.   [7] Above, p. 51.   [8] Above, p. 54.

Carnot reserved to a later section[9] the proof that the resultant of several forces multiplied by the cosine of the angle that it makes with any given line equals the sum of the component forces each multiplied by the cosine of the angle that it makes with the same line. But it was in this discussion that he first employed this mode of representing and combining the directions of magnitudes of the type called vector quantities in later mechanics.

The second corollary[10] then stated that if there was impact, the term $p\,dt$ became negligible compared to the value of u, so that

$$\int mu\,(u'\cos z - u) = 0.$$

From this equation, Carnot observed, all the laws of impact could easily be deduced, although to do so would have been a digression from his purpose. For our purposes the interest lies mainly in the evidence that it was natural for him to begin with accelerative forces acting over time or distance. There was no hint of anything like a new concept to embrace that point of view. It is only what the problem involved from the outset.

Interestingly enough, a corollary about what were to become geometric motions was also couched rather in terms of force and activity than of imaginary displacements in a system. In language it resembled rather the dynamical treatment that Carnot finally gave the idea in his *Principes fondamentaux* than the kinematical emphasis of the *Essai sur les machines*. It would appear, indeed, that the terms in which he thought about the problems developed from these conventional notions of force, through an intermediary stage where momentum was the subject, into the final analysis of power. The idea appeared in a way subsidiary to equilibrium. From the above fundamental equation he had derived the following consequence[11] for the case in which motion changes by insensible degrees (which he here described as that in which u differs infinitesimally from u'):

$$\int mp\,ds\cos x - \int mu\,du = 0$$

(since $u' - u = -\,du$, and supposing $ds = udt$) or

$$\int mup\,dt\cos x - \int mu\,du = 0.$$

If equilibrium was considered to be the state of the system in which the forces p mutually "destroy" each other, this expression would reduce to the unhelpful identity, $0 = 0$. It was possible, however,[12] to look at the situation in another way, and to conceive that there had been imparted to all parts of a system in equilibrium "motions such that no change resulted in the reciprocal action of the different parts of the system."

[9] App. B, § 53. [10] App. B, § 31. [11] App. B, § 32, Cor. 3. [12] App. B, § 34, Cor. 5.

These would be motions that would be received "without alteration" by the bodies on which they were impressed. The forces p would remain in mutual equilibrium. It would then be perfectly feasible to imagine these motions acquired in time, and the case of equilibrium would be assimilated to one in which the system changes by insensible degrees, where the equation for that case,

$$\int mp \, ds \cos x = 0,$$

was not a mere identity. Though still unnamed, geometric motions were born of this passage and of Carnot's proclivity for thinking even of equilibrium as a state of opposing actions—most characteristically backing into a grasp on the concept of work by defining the situation in which forces did no work, or did merely putative work.

From that equation, Carnot next derived, in the case of systems in internal equilibrium, a series of corollaries that amounted to the conservation of linear momentum, the conservation of moment-of-momentum, the principle of Descartes, and the conservation of live force. To none did he give prior standing over the others. The first two he did not even identify in those terms. Two brief corollaries extended the discussion to elastic bodies, provided that the force of elasticity was included in the value of p.[13] More interesting, they extended also to fluid bodies, which it was legitimate to consider as if composed of hard or elastic corpuscles.[14] In a long digression,[15] Carnot derived the hydrostatical principle of equal pressure in fluids from his equation, and came finally to what his mind always naturally fixed on, the operation of machines handling weight, whether as load or source of power. After all, raising weights was the purpose of the greater number of all machines actually used.[16]

It becomes immediately apparent that these were the considerations from which Carnot's approach took its distinctive features. Designating by

M   the total mass of the system,

H   the height from which its center of gravity falls,

t the time of fall of the center of gravity of the system,

V   the velocity a body acquires in falling the height H,

h the distance which a molecule m falls in time t, and

K   the initial velocity of the molecule m (although Carnot overlooked identifying this),

then since $ds \cos x = dh$, the basic expressions become

$$\int mp \, ds \cos x = \int mp \, dh = Mp \, dH$$

(the first summation is of individual molecules). Thus

$$2 \iint mp \, ds \cos x = 2 \int Mp \, dH = 2 \, MpH$$

[13] App. B, § 40, Cor. 11.

[14] App. B, § 41, Cor. 12.

[15] App. B, § 42.

[16] App. B, § 43, Cor. 13.

(the second summation is of elements of path), but

$$2\,MpH = MV^2,$$

and, therefore,

$$\int mu^2 = \int mK^2 + MV^2.^{[17]}$$

That elementary sequence illustrates very simply the arithmetical reasons that fifty years later led Coriolis to equate the newly named quantity work to one-half the live force, so that still later kinetic energy became defined as half the product of mass times velocity squared.

Thereupon, Carnot drew out as a passing consequence the principle that he was to propose for a fundamental axiom in the *Essai sur les machines*: in order to prove that the weights applied to a system are in equilibrium, it sufficed to show that the center of gravity did not descend. Now followed the remark that this principle could be applied to any machine by the device of exchanging a weight suspended over a pulley for any other force. Suppose that besides weights, other powers should be applied to the machine in order to move them, say manpower or horsepower. The general equation would then become

$$\int mu^2 = \int mK^2 + 2 \int\int mp\, ds \cos x.$$

That last term, in turn, might be decomposed into two parts, of which one was MV^2, and the other was "double the sum of the products of each of the moving forces [i.e., the additional powers] multiplied by the element of path that it describes in the direction of that force." If F represents one such force, and u its velocity, then

$$\int mu^2 = \int mK^2 + MV^2 + 2 \int\int Fu\, dt.$$

Here was the germ of the work concept, therefore, and right alongside it that of a generalized process. Suppose that all bodies in the system were at rest at the beginning and end of the movement. In that case $u = 0$ and $K = 0$, and the expression became

$$2 \int\int Fu\, dt + MV^2 = 0,$$

or if H was the height to which the center of gravity was to be lifted, then the relation might be expressed

$$\int\int Fu\, dt = MpH,$$

which exhibited that in order to raise the center of gravity of a system of bodies to any height it was necessary that the quantity $Fu\, dt$ always be the same, whatever the route chosen, whatever the machine utilized, whether the parts of the system be raised separately or together, and whether a single force or several together be employed.[18]

[17] App. B, § 43, Cor. 13.

[18] App. B, § 47, Cor. 17. It would be perfectly possible to see embedded in this Carnot's principle of continuity, but there is no evidence that he then saw it there.

A further passage[19] identified the factors, F, u, and t to be varied in achieving mechanical advantage and gave examples of processes in which it would be prudent to economize one factor or another. Carnot observed that it would result in a saving of force if each part of the system completed the process with zero velocity, or at least with as small a velocity as possible.

The second and third sections, comprising § 51 through 85, of the theoretical part of the memoir dealt respectively with equilibrium and with motion in the case of the seven classes of simple machines. Actually, Carnot remarked at the outset, "cords" and the lever were the only truly simple machines. It was even possible to reduce the former to the latter, but seven types of machines were customarily considered to be simple: cords, the lever, the pulley, the crank, the inclined plane, the screw, and the wedge. His treatment adapted his equation, $\int m\, p\, ds\, \cos x = 0$, to the characteristics of each in turn in order to state its law of equilibrium while "taking account of friction and the stiffness of cords as far as I could."[20] The expressions were artificial and the discussion forced, and perhaps that was its most important feature since the central purpose of the *Essai sur les machines*, a few years later, precisely was to achieve a general science of machines and to obviate the necessity for deriving laws of equilibrium and motion for each particular type of machine. The third section on motion was more summary, though no clearer, and ended with an appeal to the possibility of combining his principle with that of d'Alembert in order to reduce all questions about the theory of machines to problems of analysis.

Certain features of this discussion, however, did hold importance for Carnot's later work. He devoted the fullest treatment to the law of equilibrium for the "funicular machine," and it was there that he introduced his convention for projecting forces (he did not then speak of other magnitudes) upon other directions in terms of the cosine of the angle, balanced the moments of the resultants of clockwise against counterclockwise forces, and finally considered the projection of the machine onto a plane in which the projection of the applied powers became a two-dimensional force system. (These matters had so central a part in Carnot's later work that it seems wise to extend the reproduction of the theoretical portion of the 1778 memoir in Appendix B in order to include them.[21] It would seem that Carnot's thinking about vector quantity may have originated in his mental picturing of the distribution and transmission of forces by the cords of machines worked by ropes.)

In summary, it may be said of this first memoir that Carnot had in it the germ of his central ideas, but not as yet differentiated from a central principle of equilibrium into concepts themselves. He merely alluded in passing to the well-known principles of mechanics, such as conservation of momentum or conservation of live force, without emphasis on one as more relevant than others to his purposes. If any single concept

[19] App. B, § 49.   [20] App. B, § 51.   [21] App. B, § 52-60.

may be called fundamental for him at this stage, it would be force in the sense of motive force or pressure, but only tacitly so—it was no more than what he meant when he alluded to force. Missing (except as fugitive hints or questionable clues) were two of the central features of the *Essai sur les machines*, the distinction between impelling and resisting forces and the notion of a generalized science of machines to be disengaged from the application of laws of mechanics to the classes of machines taken successively. The notion itself was not there, much less the recognition that the way to bring it to pass was to treat machines as themselves inertial bodies endowed with mass. (Seeing that is what impelled him in the *Essai sur les machines* to go over to terms of displacement and work.) It would be fairly farfetched, though perhaps just barely possible, to say that one discussion that took into account the weight of the lines in a funicular machine represented the first hint of what Carnot later developed into the main emphasis of his theory of machines.[22] It would be more reasonable to imagine that it was writing on theory of machines while repeatedly taking account of friction, instead of neglecting it, that led Carnot into the way of considering friction (like the weight of the machine) in the light of a force against which the motive forces had to work. In the concluding passage to Section I of the theoretical part Carnot said expressly that the force of friction and other "resistances" might be considered as "active" forces.[23]

B. ARGUMENT OF THE 1780 MEMOIR

In reopening the subject for an award in 1781, the Academy reiterated its phrasing of the problem except that it now specified that the friction to be investigated should be that between the moving parts of simple machines: "The theory of simple machines with regard to the friction of their parts and the stiffness of cords. . . ." A note in Carnot's own hand on a cover sheet indicates that he completed it at Béthune on 15 July 1780. In resubmitting his memoir, anonymously as was the Academy's practice, he identified it by the same motto from Lucretius and observed in a note that having been accorded an Honorable Mention in the first competition, he had now "diminished its imperfections as far as I could."[24]

In fact, he had enlarged and transformed it into a memoir that in important respects is the draft of the *Essai sur les machines* of 1783 even to the wording of many central passages. Indeed, comparison of the two prize memoirs makes it apparent that, though the genesis of Carnot's mechanical thought belonged to his education and to the years at Calais and Cherbourg, its formative period came between his first failure to win the prize (the announcement was made at Easter 1779) and the completion of the revised memoir in July 1780. During the interval of approximately a year he must

[22] App. B, § 60.
[23] App. B, § 50.

[24] "Mémoire sur la théorie des machines," 1780, folio 1, App. C, below.

have been reading, thinking, and living little other than the theory of machines.

True, he was still bound by the terms of the Academy's contest, and the organization followed that of the earlier memoir. The 1780 memoir also opened with experiments in Part I and went on to a Part II pretending to be their application to practice. The latter again consisted of two sections, one on "machines in general" and a conclusion on laws of equilibrium and motion in each class of simple machine. In deference (no doubt) to the Academy's predilections, Carnot did elaborate and complete his program of experiments. He introduced it with a broad discussion of friction in general, and this time classified the experiments according to whether the bodies touched point-to-point, line-to-line, or surface-to-surface. He devised new experiments, improved the account of the old ones, and dropped the most cumbersome of the formulas for variation of friction with velocity. Instead, he substituted a tabulation of coefficients of frictional interactions between certain actual materials: iron, copper, red chalk, beechwood, ash, yew, and elmwood. It remains difficult, nevertheless, to see that any of this material, or any of the succeeding section on cordage, had a life longer than that of this memoir, which there is no evidence that anyone ever read again between the time that the Academy awarded the prize instead to Coulomb and the writing of the present study. The attempt left Carnot himself without illusions. The concluding paragraph to the experimental part reads:

> I do not, therefore, conceal from myself that a great deal remains to be done before the knowledge both of friction and stiffness in cordage leaves nothing more to be desired. If someone has given a fully satisfactory account of the topics I have just considered, he will have rendered a great service to practical mechanics, and certainly so difficult a task could be worthily rewarded only by the recognition of the Academy.[25]

As was to be expected, therefore, the interest lies again in the first section of Part II, which consists of the second version of his theory of machines. The discussion was lengthier than in the first memoir, sixty paragraphs occupying some forty folio pages, though still brief compared to the *Essai sur les machines*. In it he achieved notable modifications—in clarification of topics, in the enunciation of principles and definition of quantities, and in the articulation he had given to what the historian can now recognize to be a new mode in the science of mechanics and not merely the groping of a young engineer with a rather special set of interests and a point of view still latent. Indeed, without needing to speak of discovery in the sense of things or relations altogether strange and unknown, we may be justified in taking this text for the most reliable index to the aspects of Carnot's work that were most characteristically his own. The year since the failure of his first memoir was not sufficient for him to have read

---

[25] App. C, § 100.

LAZARE CARNOT, SAVANT

extensively in the literature or to have mounted elaborate experiments. It did suffice, however, for him to see through his problems, to seize their generality, and to elaborate his point of view in coherent propositions. Resubmitting his memoir to a jury of the Academy, he might have been expected to put into it that which would have reflected most credit on his capacity to contribute to the subject. On the other hand, he was not then writing, as he tried to do a few years later in the *Essai sur les machines*, for extrascientific readers for whom his own findings needed to be embedded in elementary exposition of the basic commonplaces of mechanics itself.

On the hypothesis that what was in the 1780 memoir, therefore, was largely Carnot's own, let us see what the arrangement and development of the main motifs were. First, he specified the subject to be the properties common to all machines and began the discussion, not *in medias res* as in the first memoir, but with the operational derivation of the center-of-gravity principle for equilibrium in a generalized machine, the same with which he later, in the *Essai sur les machines*, repaired the deficiency he there exhibited in the traditional statement.[26] As for the so-called Principle of Descartes, Carnot here stated it in passing as a consequence without identifying it by that or any name, and went on to assure the generality of the center-of-gravity principle by remarking that all forces could in principle be reduced to the action of weights.

A brief and clear diagnosis followed of the impediment that mechanics had created for the science of machines in abstracting from their mass. Let machines be considered instead as inertial bodies, and their science like that of all mechanics would be reduced to the problem: given the virtual motion of a system of bodies, what actual motion would ensue in consequence of their mutual interactions? The problem thus posed simply and directly, so were the conditions for a solution. The first assumption, though not so stated, was the comparative anatomy of hard body and elastic body. The second was the sole principle that Carnot always held to have been requisite for a resolution: the equality in opposition of action and reaction, a law simple, "incontestable," and of universal applicability.[27]

On that basis alone Carnot proceeded to derive the two fundamental Equations (E) and (F) of the *Essai sur les machines*.[28] The wording was identical except that the latter he here called an arbitrary rather than an indeterminate. In the *Essai sur les machines* he corrected himself and held that it was the values to be attributed to the geometric motions u of Equation (F) that were arbitrary, not the relation itself. More striking are passages adumbrating the concept of geometric motion, identical in phrasing to those in the *Essai sur les machines*. Their role was that of auxiliaries allowing the derivation of the general indeterminate (F) from the determinate (E) restricted to elastic interactions. Only one feature of the later introduction of geometric motions was missing, the afterthought of the *Essai* wherein Carnot included in the class of

---

[26] Above, pp. 36-38.   [27] App. C, § 111.   [28] Above, pp. 42, 44.

{ 74 }

geometric motions those that, while involving no virtual displacement, did no work.[29] At first reading, it is surprising that Carnot should not have seen the need for that provision. It seems closer to the dynamical emphasis that foreshadowed the concept of geometric motions in the 1778 memoir[30] than does the kinematical way in which he introduced the idea explicitly here and in the *Essai*. Nevertheless, the final development he gave the notion in the *Principes fondamentaux* of 1803 was a proto-energetic one. It may be, therefore, that the successive installments on geometric motions exemplify three phases through which Carnot's thinking about the whole subject passed. In the first phase, that of the 1778 memoir, it was motive force that was primary; in the second, that of the 1780 memoir and much of the first part on mechanics of the *Essai sur les machines*, it was displacement; and in the third, that of the latter part of the *Essai* concerned with machines and of the *Principes fondamentaux*, it was power or its exercise in moment-of-activity or work.

Such must almost surely have been the order in which he evolved the science of machines, for comparison point by point of the 1780 memoir with the *Essai sur les machines* shows that in the latter published version he superimposed the more theoretical work passages and principles upon the arguments of the unpublished memoir without rethinking or reshaping them. The contrast comes out most strikingly in the absence from the memoir of the feature that Carnot added to and emphasized most strongly in the *Essai sur les machines*, the conservation of the moment-of-the-quantity-of-motion, or moment-of-momentum. That principle was there introduced, it will be recalled, in a curiously repetitive manner. First, the conservation of angular momentum appeared unnamed and merely as a sample solution to the problem resolved by employing the indeterminate Equation (F). It would be reasonable to suppose that in that form it represents the early phase of Carnot's thinking. Then there followed in the *Essai sur les machines* (but not in the 1780 memoir) the digression just recalled about the pseudogeometric status of motions that do no work, upon which Carnot came back to conservation of the moment-of-the-quantity-of-motion now magnified and elevated to the status of the fundamental conservation law. Stating it in the form of three propositions,[31] it will be further recalled, he went on to derive from it well-known theorems of mechanics followed by his own equation designated (AA):

$$\int F u \cos Z = 0,$$

in which F, representing force in the sense of motive power, was substituted in effect for the product $m\,U$, the quantity of motion "lost" to internal interactions in the so-called fundamental Equation (F)

$$\int m u\, U \cos z = 0.$$

[29] Above, p. 43.   [30] Above, p. 68.   [31] Above, pp. 46-47.

Not so in the 1780 memoir, which said nothing of moment-of-the-quantity-of-motion. Having defined geometric motions in point of displacement only, Carnot simply obtained Equation (F) for the general solution and expressed it in various ways adapted to different needs. One proposition resolved the quantities into Cartesian coordinates. A second exhibited the analogy to conservation of live force in hard-body interactions and included a term for its loss in impacts. A third concerned motion changing by insensible degrees over time and reverted to the expression Carnot had obtained in the 1777 memoir, i.e.,

$$\int mup \, dt \cos z = \int mu \, d(\text{V} \cos y).$$

A fourth concerned the case in which the value of u is the real velocity V and found that the expression then reduced to conservation of live force. All these consequences came out rather parenthetically, however. Carnot's discussion in the 1780 memoir placed no emphasis on the principles of mechanics and made much less of the later claim that geometric motions permitted transforming live-force conservation in elastic bodies into an expression applicable to generalized hard-body interaction. These matters figured here rather as a sequence of problems than a deductive chain of principles, for it was a feature of the way in which Carnot's ideas developed that he tended in later work to dress up in the guise of principles what he had initiated as the solution to problems.

Also missing in the 1780 memoir was the analysis of power which in the *Essai sur les machines* made the transition between Part I on the principles of mechanics and Part II on Carnot's own subject, "machines strictly speaking." The omission is consistent, for it was in order to prepare the ground for those considerations that Carnot would then introduce the moment-of-the-quantity-of-motion into the *Essai sur les machines* in the interval between 1780 and its publication in 1783. All these differences, therefore, reflect the limitation on his thinking at the time of the 1780 memoir: there at the point where he was later to move on to the concept of power and work, he fell back on the traditional notion of dead force.[32] When the question was one of live forces applied to machines, he observed, the theory was related to that of impact, which sufficed to solve such problems. But the theory of machines conventionally understood envisaged only the forces employed to move them, i.e., dead weights, animal power, and wind or water. Accordingly he would limit himself to these forces for the remainder of the theoretical part of his memoir.

Nevertheless, limited though Carnot's vision may thus have been at the stage of the 1780 memoir, the evidence is interesting in showing that it was in connection both with such conventional problems and also with the later, more elaborate notions that he developed the concept of work. For if he did not have it fully disengaged in 1780,

[32] App. C, § 128.

neither was it entirely missing. The quantity itself, moment-of-activity in 1783 in the *Essai sur les machines*, was called "quantity of action" in the 1780 memoir. There it was always coupled with impelling or resisting forces (for those terms also made their appearance) as the quantity that they produced or consumed. Carnot never referred to it without the qualifier and expressly warned the reader against confusing his quantity of action consumed or produced with that which mathematicians following Maupertuis meant by quantity of action of a system. Of action in that sense he, Carnot, would make no use in the present memoir,[33] and it must have been the obvious inconvenience that led Carnot to modify the terminology in the *Essai sur les machines*. It may, further, have been the emphasis there placed on moments in general, particularly that of the quantity of motion, that led him to redesignate his basic quantity "moment-of-activity" and to consider it in itself and not merely as a measure of the impelling or resisting forces it was consuming or producing.

Lacking a fully disengaged concept of power or work, together with the principle of conservation from which to deduce it, Carnot gave the sequence of propositions about machines themselves more episodically in the 1780 memoir than in the *Essai sur les machines* and failed to draw out of them his own principle of continuity, which appeared only in the form of a practical injunction in passages constituting the draft of the later scholium. In the *Essai sur les machines*, it will be recalled, he stated a single theorem for machines:

> Whatever the state of rest or motion of any system of forces applied to a machine, if a geometric motion be imparted to it without altering the forces, the sum of the product of each by the velocity of the point of application, evaluated at the first instant and in the direction of the force, will be zero.[34]

Or

$$\int Fu \cos z = 0,$$

from which he drew all the corollaries comprising the science of machines "strictly speaking," the fifth being what we have called the Work Corollary:

> *In a machine of which the motion is changing by insensible degrees,* the moment-of-activity consumed *in a given time by the impelling forces* equals the moment-of-activity exerted *in the same time by the resisting forces*.[35]

The 1780 memoir had yet to achieve that unification of his point of view. In the absence of conservation of the moment of the quantity-of-motion, and therefore of work, he stated two parallel theorems for the science of machines, the first for equilibrium and the second for motion. Thus the former:

[33] App. C, § 132.   [34] Above, p. 53.   [35] Above, p. 54.

"When a machine is in equilibrium, if an arbitrary geometric motion be imparted to it without altering the applied forces in any way, the quantity of action produced in the first instant by the impelling forces will be equal to the quantity of action produced in the same infinitely short time by the resisting forces."[36]

That proposition he then stated in various ways (one of which is virtually identical to the phrasing just cited from the *Essai sur les machines* except that it did not embrace a machine in motion) and drew from it several corollaries. In those concerned with weight-driven machines, the equivalence of $M g H$ to $\frac{1}{2} M V^2$ came out repeatedly.

The companion theorem concerning machines in motion was stated more awkwardly:

If the actual motion of a machine is suddenly converted into any other geometric motion and the machine is left to itself, the conservation of live forces will obtain throughout the ensuing motion no matter what changes there may be in the motive forces.[37]

The corollaries with which Carnot followed this proposition were all correct, and the second of them essentially amounted to the Work Principle of the *Essai sur les machines* in the following form:

When a machine is in uniform motion (i.e., when each point of the system has a constant velocity), the quantity of action produced in a given time by the impelling forces is equal to the quantity of action produced in the same time by the resisting forces.[38]

The statement was weaker than in the *Essai sur les machines*, but already Carnot identified it as the most useful of all the propositions in the theory of machines. The historical question that poses itself, therefore, is what difficulties still remained in his mind that prevented his having combined motion and equilibrium into a single principle as he seemed quite naturally able to do in the *Essai sur les machines*?

Several possible explanations suggest themselves. One is that lacking moment-of-momentum and thinking in terms of displacements, he had not seized on all the advantage that his geometric motions conferred in authorizing him to neglect internal forces of the system. That might be why he followed the theorem with this otherwise unnecessary justification—since live forces were conserved in a system of which the motion was changing by insensible degrees, and since when a geometric motion was substituted for a real one, nothing was changed in the mutual interactions of the system, it followed that motion must still be changing by insensible degrees, and hence

[36] App. C, § 133.   [37] App. C, § 140.   [38] App. C, § 143.

live forces must still be conserved "at least for some time" after such a transformation, no matter what motive forces should be substituted for those that were really influencing the parts of the system.

Another possibility is that Carnot had not quite succeeded in generalizing in his own mind relations that he had obtained from considering the behavior of weights under gravity. That might be why the corollaries are clearer than the statement or discussion of the theorem itself, notably the first which concerns weight-driven machines:

> If the actual motion of a weight-driven machine be suddenly changed into any other geometric motion whatever, and the system be left to its own forces, the ensuing sum of the live forces at any instant is equal to the sum of the initial live forces (i.e., immediately after the change of motion) plus what the sum of the live forces would be if each point of the system had a velocity equal to that due to the height from which the center of gravity has fallen since the change of motion.[39]

Perhaps, finally, the obligation to conduct the argument toward a frictional application concealed the desirability of establishing the work principle in a unified treatment of motion and equilibrium. The reason he advanced in the 1780 memoir for considering it the most useful for the theory of machines was that in practice such ordinary devices as mill-wheels, pulleys, and capstans, once set in motion, soon reached a steady rate of operation. A final corollary brought all real machines within the scope of the discussion by specifying that friction was always to be classified as a resisting force, since its direction inevitably countered the real motion of the points whereon it acted.

It would be misleading, however, to emphasize shortcomings on the formal level of argumentation, and it is more illuminating historically to appreciate that what Carnot lacked as principle in the 1780 memoir, he already had as maxim. The final paragraphs of the theoretical section of the 1780 memoir were largely similar and in certain places identical with the scholium that concluded the *Essai sur les machines*. An interesting passage reveals the latency of his point of view. He had reverted to machines of which the state of motion changes by insensible degrees and pointed out that they had in common with periodic machines (i.e., pendulums) that live force was conserved and that in a given time, therefore, the quantity of action produced by the impelling forces equals that produced by the resisting forces. Even worthier of remark, a quantity one-half the increase in the sum of the live forces is identical with the quantity of action produced by the force of inertia of the bodies comprising the system. It followed that:

[39] App. C, § 142.

"If we consider all forces, whether active or passive, applied to a machine in motion, including even the inertia of its matter taken itself as a real force that resists change of state in the bodies, the quantity of action produced by the impelling forces in a given time will always equal the quantity of action produced in the same time by the resisting forces whatever be the motion of the machine."[40]

Here then, noted in passing in the 1780 memoir, was precisely the remark that he stated rather as a theorem in the *Essai sur les machines*, and that there appeared to be the principle in virtue of which he unified the discussion. Indeed, he even noted in the memoir that the entire theory of machines whether at rest or in motion could be reduced to that single statement. He simply did not yet do it, which shows that what he really needed to develop was less the content of theorem or propositions than the line of reasoning that connected them.

As for his own principle of continuity, it followed upon a discussion virtually identical with that in the *Essai sur les machines* about varying the factors of time, force, and velocity in achieving mechanical advantage. A footnote distinguishes between the "absolute work or labor of the agent" ("le travail absolu ou la peine") and "what I call here the work of the moving or impelling force,"[41] the difference residing in the extra force consumed by the inertia of the machine itself. The things that he mainly had in view in all these remarks were the classical simple machines and their variants used on shipboard. It was for this reason, Carnot said further, that he had written only about those of which the motion changes by insensible degrees. "For when the purpose is to produce the greatest effect possible from a machine, there is a decisive advantage to be gained in excluding percussion or sudden changes in the state of motion." Carnot went on to show how the intervention of impacts necessarily involved loss of a corresponding quantity of action. The argument was in much the same terms as that reproduced three years later in the *Essai sur les machines*, but the mode in the 1780 memoir was one of explanation rather than demonstration or proof.[42] It is, indeed, the most revealing feature of the document that this point, which later appeared to be the chief finding of Carnot's mechanics, so much so that it was called his law or principle, he first wrote down not as discovery of an unknown truth, but as a justification for limiting his subject to machines that in their operation embodied it.

For the rest, the 1780 memoir presents little interest. Like that of 1778 it concluded with a section purporting to apply the experimental results to determining expressions for each class of simple machine. Carnot had simplified the treatment over that of the earlier memoir by combining consideration of motion and equilibrium in each class of machine—all the more curious does it seem, therefore, that he did not do this

[40] App. C, § 148.   [41] App. C, § 153, notes.   [42] App. C, § 157.

in the preceding theoretical discussion—rather than by going through all seven classes once for the case of equilibrium and again for that of motion. He recognized that his organization had the effect of separating the theoretical from the practical study of machines in general. No doubt the terms of the Academy's competition imposed on him the necessity thus to descend to particular cases for the practical part. But it does not appear that he had formed the project of such a science of machines as would dispense him from specifying particular conditions of motion and equilibrium characteristic of each class in practice. The separation of theory and practice was necessary, he observed, "in order to avoid the confusion that would arise from mingling them."[43]

We can see that he had in mind with greater or less cogency all the relations or theorems that would comprise such a science. What he had still to accomplish was to graduate in his thinking from forces to powers and to seize on the conservation of the moment-of-momentum in order to make full use of his own idea of geometric motions in abstracting from the effects of internal forces in a system and hence overcoming the disjunctions between hard and elastic body, pressure and impact, dead force and live force. In a way that was conceptually subsidiary but probably more important in actuality, he needed to generalize from the manipulation of weights, whether as load or power, in order to embrace within his reasoning all processes involving work.

Such, at least, was the development that he gave his subject between the failure of his second memoir to win the academic prize in 1781 and the publication in 1783 of his *Essai sur les machines*, in which all the experimental and particular passages were eliminated, and the theory of machines alone was presented rather as an adaptation of the science of mechanics itself. It must, therefore, have been during this interval that he reviewed the whole science as it was known to an engineer in the late eighteenth-century, though how and through the agency of precisely which writings in that science he then seized on the unifying possibilities inherent in the principle of moment-of-momentum, the sources do not permit us to say: for the rest of it, all the development he gave was latent in his own approach in the two memoirs successively submitted to the Academy.

C. ARGUMENT OF THE *PRINCIPES FONDAMENTAUX DE L'ÉQUILIBRE ET DU MOUVEMENT* (1803)

The remaining discussion of Carnot's science of machines will concern its immediate development and influence. Let us take first the revision that he himself gave the subject in his *Principes fondamentaux de l'équilibre et du mouvement*, published just twenty years after the *Essai sur les machines*, and in the same year with the *Géométrie de position*. Most of the later writers who allude to Carnot's mechanics make reference to the *Principes fondamentaux* rather than to the *Essai sur les*

---

[43] App. C, § 191.

machines, which, if mentioned at all, is not always distinguished from the second, more renowned version.[44] Meanwhile, the development of the eighteenth-century French school of rational mechanics had culminated in the publication in 1788 of Lagrange's *Mécanique analytique*, a work that imposed itself immediately as the mathematical chef d'oeuvre of its subject. Lagrange's book was no isolated monument. On the contrary, the last two decades of the eighteenth century were a period of intensive activity in the formalization of mechanics, so much so that Carnot himself remarked in the preface to his *Principes fondamentaux* that in the interval since his earlier writing, there had appeared so many others "on all aspects of mechanics, so beautiful and so extensive that my own can scarcely be remembered."[45] (So far as can be determined, his modesty was accurate. The present writer has never seen a single mention of the *Essai sur les machines* in the contemporary literature of mechanics, nor any evidence that it was ever read.)

Why, then, was a second edition wanted? Carnot justified it on the grounds that the *Essai sur les machines* had contained some ideas that were new when they first appeared, and that it was in any case always useful to envisage the fundamental truths of science from various points of view. Of the new version he said further, "Several scientists have strongly urged me to furnish it."[46] Unfortunately, we do not know who they were. Did they do so because Carnot was now a prominent statesman recently emerged from political eclipse and influential in the Institute? It would be unrealistic to suppose that such facts carried no weight, although neither is it necessary to impute base and much less baseless flattery to Carnot's advisers. Statesman though he now was, he was a statesman with a certain scientific reputation in virtue of his mathematical writings. He had published his book on the calculus in 1797 and his first book on geometry in 1801.

Are we to suppose, then, that the new engineering emphasis in technical education might have been thought to create a public for a book on mechanics from the point of view of a former engineer, since become famous? So it almost certainly was in Carnot's own mind, for the *Principes fondamentaux*, unlike the *Essai sur les machines*, is at once textbook and treatise in the fashion created by the *Ecole polytechnique*. The former quality flattens Carnot's already unassuming style and puts the book in the shadow when contrasted to the brilliant light in which the works of Lagrange

[44] See, for example, Charles Brunold, *L'entropie, son role dans le développement historique de la thermodynamique* (Paris, 1930), pp. 37-38; D.S.L. Cardwell, "Some factors in the early development of the concepts of power, work, and energy," *The British Journal for the History of Science*, 3 (June, 1967), 209-224; René Dugas, *History of Mechanics*, trans. J. R. Maddox (Neuchâtel, 1955), pp. 324-331; Jouguet, *Lectures de la mécanique*, II, 72.

[45] *Principes fondamentaux*, p. v.
[46] *Ibid.*, p. v.

and his school are illuminated by their formal elegance. At all events it must have been some such combination of personal and institutional motivations that encouraged Carnot to revise his science of machines. The explanation cannot be that his approach was an answer to the problem structure in the science of mechanics itself, for (as will appear) if his work was no longer unknown, it was another twenty years before these concerns began to make any difference.

To turn to the content of the *Principes fondamentaux*, it will be convenient to take it in its textbook aspect first, for Carnot had completely reorganized his presentation. The division between fundamental principles of mechanics and the theory of machines "properly speaking" had disappeared. The subject was still presented in two parts, but the distinction was now between its experiential and its rational aspects. The former reads like a glossary of mechanics. It comprised mainly definitions of the quantities and expositions of elementary laws of statics and dynamics for which Carnot claimed the sanction of experienced fact. Part II by contrast contained propositions that could be derived from those first principles merely by ratiocination —of them more in a moment, for they were those that were peculiarly Carnot's own either in conception or in function.

In observing that Part I now reads like a primer, the historian should intend nothing denigratory. It was a good primer. No franker source exists if he wishes to know what persons literate in mechanics at the end of the eighteenth century understood physically by its basic conceptions, such as force, motion, action, velocity, acceleration, etc.[47] Like many of the simplest matters generally taken for granted, these are not always easy to come by historically. In this respect, the *Principes fondamentaux* exhibits a notable improvement in clarity over the *Essai sur les machines*, in which the modern reader often has to judge of Carnot's comprehension of terms and dimensions by the use he made of them rather than by an explanation. In the later work, for example, the discussion of projection and combination of directed quantities was explicit. It is extremely curious, however, that Carnot seems to have gone all the way back to his earliest, 1778, memoir for these proto-vectorial considerations.[48] Was it his intervening work in geometry that encouraged him to see the merit of this representation, which he had merely taken for granted in the intervening work in mechanics? It may be so. At all events Carnot suggested the following notation for representing the projection Aa' of one velocity Aa upon an intersecting straight line AB:

$$\overline{Aa'} = \overline{Aa} \cos \overset{\frown}{\overline{Aa} \quad \overline{AB}},$$

wherein the last term denoted the angle, a convention that would have been obvious

---

[47] See above, pp. 35-36, n. 7.

[48] See below, App. B, § 52-54. The geometric figures missing from the 1778 MS may be reconstituted from figures 2, 3, and 4 of Pl. I of the *Principes fondamentaux*, which fit the argument of the 1778 memoir.

although cumbersome if generally adopted. Here, for example, is his expression for the relation between initial velocity $(\overline{MW})$, altered velocity $(\overline{MV})$, and the velocity $(\overline{MU})$ "lost" when a body was constrained to change its motion:

$$\overline{MW} = \overline{MV} \cos \widehat{\overline{MW} \; \overline{MV}} + \overline{MU} \cos \widehat{\overline{MW} \; \overline{MU}}.[49]$$

Writing of moments, Carnot now pointed out that the moment of a motive force, being the product of a force by a line, could always be reduced to a live force, and that the moment of a quantity-of-motion could be reduced to the product of a live force and a time. He now had clear that the latter was properly designated quantity-of-action of a mass. The concept of work he still called moment-of-activity, but his way of seeing these quantities seems to represent a certain development in the differentiation of the notion of energy (live force) from that of work (moment-of-activity). He distinguished between moments-of-activity "consumed" or "absorbed" (instead of consumed or produced). By the former he still meant the product of the force by the path described by its point of application, evaluated in the direction of the force; whereas the latter was now the product of the force by the velocity of the point of application, evaluated in the direction opposite to that of the force. It still comes out that the two quantities are identical numerically and of opposite sign since the angles are supplementary, but the distinction seems closer than it had done in the *Essai sur les machines* to that in later usage between the work done on or by a system.[50] For there is the beginning of a distinction here between notions that in the next fifty years would evolve into the concepts of kinetic and potential energy. He laid it down that live force "properly speaking" had the dimensions MV^2 (as we would write it) and that when it took the form of the product of a motive force by a line (PH as he wrote it, or *mgh* as we would write it), it could be given the special name of "latent live force." In the latter case it was dimensionally identical with moment-of-activity (or work), and the significant thing is that Carnot saw it as conceptually distinct. For a system of bodies in which forces were operating over time, the following expression

$$mp \cdot u \, dt \cos k$$

would give the moment-of-activity consumed during time dt by the force mp with respect to the velocity u (k being the angle between u and p). The moment-of-activity consumed by the whole system throughout the entire motion would then be given by

$$\Sigma \int mp \cdot u \, dt \cos k$$

(of which expression one might note that Carnot had begun to distinguish between summation and integration, and also that he had lapsed into his old notation). That expression being a latent live force, it could be reduced to a quantity of the form MV^2:

---

[49] *Principes fondamentaux*, § 43, p. 26.   [50] *Ibid.*, § 60-61, pp. 38-39.

"whence it is easy to appreciate that the notion we have just given of moments-of-activity is encountered frequently in the theory of equilibrium and motion, either in the form of live force properly speaking, or in that of latent live force."[51]

From the definitions Carnot turned to a summary of what he now called "hypotheses that can be admitted as general laws of equilibrium and motion," which he pretended to found on experience and reasoning, and which were in fact simply the classical laws of statics and dynamics. It would elucidate nothing not already evident about Carnot's reasoning to enumerate these propositions. Two remarks only may be curious to note. The first reflects on Carnot's own sense of historicity. He here attributed the principle of virtual velocities to Galileo, even calling it the "principe de Galilée," and went on to credit the lowest center of gravity condition for equilibrium to Torricelli as its consequence before rectifying it in his own operational statement. The attribution to Torricelli was in fact correct, though not that to Galileo, and one has the impression that even in the rather trivial respect of acknowledging the history of the concepts, Carnot's reworking of the book was shaped by the example of Lagrange. The second remark about Carnot's presentation of the elements is that he concluded it by deriving from the conservation of live force the relations that he needed in the various cases of change of motion. As usual, W was initial, V final velocity, and U the velocity lost in impacts. Then for hard bodies in the general case,

$$\int MW^2 = \int MV^2 + \int MU^2,$$

for hard bodies in which the motion changes by insensible degrees,

$$\int MW^2 = \int MV^2 \text{ (since U is infinitesimal)}.$$

For elastic bodies, he comes at it in reverse order. Since

$$\int MV^2 = \int MW^2 - \int MU^2,$$

and since the velocity U lost at impact is restored on rebound,

$$\int MV^2 = \int MW^2.$$

These are statements remarkable only in that he could now thus simply set them down one after the other.[52]

In Part II of the *Principes fondamentaux* Carnot turned to a formal development of the so-called hypotheses considered as laws of nature. It is difficult, on the whole, to think the presentation felicitous. Ostensibly the goal was to achieve analytical expression for the laws, and the result was that the distinctiveness that Carnot brought to mechanics by directing attention to the study of machines was obscured by his attempt (or so it would seem) to imitate in some degree the style of Lagrange. The sequence of theorems swept so wide of the subject of machines that its dénouement there would

[51] *Ibid.*, § 64, p. 41.   [52] *Ibid.*, § 131-132, pp. 104-105.

{ 85 }

come as something of an anticlimax except that the climax itself, a generalization or vindication of the Principle of Least Action, was somewhat beside the point.

At the same time it would be wrong to underestimate all the theoretical aspect of the *Principes fondamentaux* just because this part of the book may not have been an entire success or because he had already made all the significant points in the *Essai sur les machines*. He did achieve a greater clarity, most notably in the passages defining geometric motion:

> any motion that, when imparted to a system of bodies, has no effect on the intensity of the actions that they exert or can exert on each other in the course of any other motions imparted to them, will be named geometric.[53]

Clearly, he had modified the definition, for he states it entirely in terms of the function of geometric motion, which it is now easier to recognize as what in later parlance would be called virtual displacements. Neither in the 1780 memoir nor in the *Essai sur les machines* had Carnot adapted his concept of geometric motions from the principle of virtual velocities. In the *Principes fondamentaux*, however, he went on to recognize the analogy between such motions and that principle in the use Lagrange made of the latter. The difference was that virtual velocities being infinitesimal were inapplicable in problems involving the sudden changes of motion produced by impact, whereas geometric motions, being finite, generalized the principle so that it might apply to all cases, whether the change of motion be continuous or discontinuous, whether the bodies be hard, elastic, or fluid.[54]

It is because of these passages that the historian will detect in the concept of geometric motion the forerunner if not the common ancestor of reversible processes and vector analysis, the former in that reversibility was the criterion by which the independence of such motions from the rules of dynamics might be recognized, and the latter in that these motions were, therefore, determined by the geometry of the system alone. Carnot nourished a prophetic if not always lucid vision of the prospect for a science that should be "intermediary between ordinary geometry and mechanics." Both in the *Principes fondamentaux* and in the *Géométrie de position* he alluded to hopes for thus uniting mechanics and geometry in a science yet to be created. It was mainly for lack of a theory of geometric motions that analytical difficulties frequently impeded the solution of mechanical problems. Carnot thought this to be specially true of hydrodynamics,[55] and the remark goes to strengthen the case for seeing in geometric motions the origin of reversible processes, since it was in connection with the fluid model of heat transfer that Sadi introduced the concept of reversibility.

As an earnest of such a theory, Carnot opened the "rational" part of his *Principes fondamentaux* with a series of nine theorems about the properties of geometric

[53] *Ibid.*, § 136, p. 108.   [54] *Ibid.*, pp. ix-x, 115 (§ 144).   [55] *Ibid.*, § 145, p. 116.

motions in hard-body interactions. The sequence will prove somewhat puzzling to the reader not yet oriented to his purposes in mechanics itself. When he did finally come to that, he unfolded the argument in a manner different from that of the *Essai sur les machines*. Both editions prepared the ground for the application of mechanics to machines with the proposition that in the motion of a system of hard bodies, the net effect of mutual interactions of the parts is zero. In the *Essai sur les machines* (it will be recalled) Carnot stated that result in the guise of the general solution to the problem, "Given the virtual motion of any system of hard bodies, . . . find the real motion it will assume in the next instant,"[56] and he symbolized the solution in the Equation (F)

$$\int mu \ U \cos z = 0.$$

which he had obtained through transforming an expression equivalent to conservation of live forces by means of the auxiliary idea of geometric motions. But in the *Principes fondamentaux*, having elaborated the concept of geometric motions in a series of propositions, he stated the same thing in the form of a theorem:

> In the impact of hard bodies, . . . the sum of the product of the quantities of motion lost by each of the bodies multiplied by its velocity after impact evaluated in the direction of the quantity of motion lost, is equal to zero.[57]

And he symbolized the relation in a mere corollary

$$\Sigma \ MUV \cos \widehat{U \ V} = 0.[58]$$

The difference is indicative of that between the two books: in the *Principes fondamentaux* Carnot set out in a didactic order what in the *Essai sur les machines* he had established in a problematic one. A consequence was that he himself began the process of obscuring the distinctiveness of his own approach while improving its clarity. There was about the *Principes fondamentaux*, moreover, not only an imitative quality in its relation to Lagrange but a regressive quality in its relation to theory. Carnot now tended to look back to the Principle of Least Action for the significance of his work concepts rather than forward to energy considerations. That he should have done so is natural enough for he never claimed to be an innovator, but to bring it out will require exemplification in a certain amount of detail.

Following the theorem just cited, a further proposition stated that in hard-body collision the sum of live forces prior to impact equals that following impact plus what the sum of the live forces would be if each of the bodies were moving freely with the velocity lost in impact:

$$\Sigma \ MW^2 = \Sigma \ MV^2 + \Sigma \ MU^2.[59]$$

---

[56] *Essai sur les machines*, § XX, p. 40, and above, p. 32.
[57] *Principes fondamentaux*, § 168, p. 139.

[58] *Ibid.*, § 169, p. 143.
[59] *Ibid.*, § 175, pp. 145-146.

The point to notice here is that Carnot did not yet hold steadily in mind a concept of live force equivalent to kinetic energy. Sometimes what bodies were said to lose in collision was velocity or motion; sometimes it was live force itself, as in a corollary bringing out that the sum of live forces after collision was bound to be less than beforehand—but the case was restricted to hard body. He did have clearly in mind the convertibility of the quantities expressed in the dimensions of live force [kinetic energy] into those of moment-of-activity or latent live force [work or potential energy], but he was thinking of these matters dimensionally rather than conceptually. Only the concept of work did he have fully developed in all but name, and sometimes even in name. The role of energy he had to compound out of forces acting over distance (usually though not exclusively when it was a question of the measure of the force) or in time (usually though again not exclusively when it was a question of the operation of a machine process), and that necessity may explain why, the more he reflected, the more prominent the Principle of Least Action became in his thinking.

As he had done in the *Essai sur les machines*, Carnot turned to Least Action for the purpose of identifying the real motion among the infinity of geometric motions of a system of hard bodies. The motion that actually occurred would be that for which the sum of the products of the masses by the square of the velocity lost by each, was a minimum, or to put it more generally, such that

$$\delta \int MU^2 = 0.$$

Carnot considered that to be very beautiful and the fundamental law of hard-body collision. It might, he went on to show, be extended to cover elastic and even partially elastic bodies. When thus generalized it took the form,

$$\delta \int MUX = 0$$

(where X is the distance traversed by mass M with velocity U in time t), and this quantity, he pointed out, was that which Maupertuis had first denominated Action. Indeed, Carnot's view of his own accomplishment was that in the *Essai sur les machines* he had been the first to establish its applicability to the case of sudden or discontinuous change of motion in hard-body collision. Maupertuis had envisaged the matter so vaguely that he made no distinction between continuous and discontinuous change of motion. Euler had then distinguished the former case and made of it a rigorous, if still metaphysically justified, principle applicable to motion under central forces of attraction.[60] Now in the *Principes fondamentaux* Carnot thought himself to be exhibiting how superfluous metaphysical considerations were, and how generally the principle applied—only he did think preferable the form

$$\delta \int MU^2 = 0,$$

and that is the tantalizing aspect in later eyes since it does look very like the energy statement it could not yet be. He had not abstracted a notion of live force equivalent

[60] *Ibid.*, pp. vi-ix, 163-164 (§ 188).

to energy from the properties of bodies: live force was lost completely in hard-body collision, partially lost in partially elastic collision; and Carnot was relying more fundamentally than ever on his workless geometric motions to bridge these differences. At the same time, he gave greater emphasis to the work principle at the expense of moment-of-momentum. In the *Essai sur les machines*, it will be recalled, conservation of the moment of the quantity-of-motion was the fundamental principle of mechanics sufficient for derivation of all laws of motion and equilibrium in all cases whatever.[61] Not so in the *Principes fondamentaux*, where he invoked in its place a series of propositions about moments of rotation, moments of percussion, and d'Arcy's Principle of Areas,[62] and singled out instead as the fundamental conservative quantity the moment-of-activity of the general system with respect to any of its possible geometric motions, i.e., conservation of work.[63] But that appears to be something of an afterthought following the enthusiastic vindication of Least Action.

A final section headed "Considerations on the Application of Moving Forces to Machines" gathered together the remarks about machines that Carnot had interspersed in the earlier edition to explain his reasonings, and combined them with the analysis of mechanical advantage and exhortations about the employment of power from its concluding scholium. The effect for the reader who knows only the *Principes fondamentaux* is to reduce to the level of an appendix what the reader of the earlier drafts can see to have been the motivation of the work as a whole. He had added nothing new to the earlier discussion about the factors of mechanical advantage or the way in which they might profitably be varied according to the economics of the process. He did here make one of his few references to the work of others, dissenting from Daniel Bernoulli's argument that a power source yielded much the same result whether the operator chose to augment the force or the speed. In the case of manpower and animal power, experiments by Coulomb had invalidated that hypothesis for the reason that it took no account of fatigue or boredom.[64]

A little further on Carnot also attributed to Daniel Bernoulli the maxim that in any proposed operation, the first thing to do is examine it in order to specify what

---

[61] Above, pp. 46-47.

[62] One wonders whether the papers of the chevalier Patrick d'Arcy had been an important source of inspiration to Carnot, but this is the only place he mentions them (§ 195, p. 174). The relevant memoirs are: "Problème de dynamique," *Mémoires de l'Académie Royale des Sciences, Année 1747* (Paris, 1752), 344-356, with an addendum, 356-361; "Suite d'un mémoire de dynamique," *ibid., Année 1750* (Paris, 1754), 107-108; "Théorèmes de dynamique," *ibid., Année 1758* (Paris, 1763), 1-8.

[63] *Principes fondamentaux*, § 197, p. 176.

[64] Carnot cited Coulomb's memoir on the employment of manpower, "Résultat de plusieurs expériences destinées à déterminer la quantité d'action que les hommes peuvent fournir par leur travail journalier, suivant les différentes manières dont ils emploient leurs forces," *Mémoires de l'Institut*, II (Paris, an 7—1799), pp. 340-428. Coulomb first read this memoir to the Academy in 1778. See Gillmor, *op. cit.*, Chapter II. Carnot cited further a memoir of Euler, "De machinis in genere," *(Novi) Commentarii Academiae Imperialis Scientarum Petro-Politanae*, III (1751), p. 254.

effect intrinsically pertains to that operation, and avoid so far as possible producing any side effects. No work of Bernoulli was cited, but Carnot now said that it was pursuant to this principle that all shocks or sudden changes that are not essential to the construction of the machine are to be avoided, for shock always involves loss of live force and consumption to no purpose of a portion of the moment-of-activity developed. The only thing surprising here is that Carnot should thus have casually read back to Daniel Bernoulli the finding generally credited to himself, which remark illustrates further that what was involved in the development of engineering mechanics was less the knowledge of how things work than the articulation of the principles that all informed people knew more or less explicitly to be involved in processes.

In general, the argument about machine processes is briefer and easier to follow in the *Principes fondamentaux* than in the *Essai sur les machines* or in the draft memoirs, and the concluding passage deserves quotation in order to exhibit further the gradual clarification in ideas that this edition reflects in what can be seen as the unwitting approach to energy considerations by way of the concept of work. After observing that the effect produced is always a live force, real or latent, comparable to the product PH of a weight P by a height H, or a force by a line, and that the moment-of-activity is, therefore, always the quantity to be economized for maximum effect whatever the process, whether it be a weight to be raised; a mill-wheel to be turned; a void to be created in the atmosphere, the sea, or some confined fluid; a machine to be started; or a system of bodies attracting each other in any proportion to the distances; and also whatever the motive agent, whether weights, wind, water, men, or animals in any combination at all, then:

> Whatever change may be occasioned in the system, the moment-of-activity consumed in a given time by the external powers always equals one-half the amount by which the sum of the live forces increases in the system to which they are applied during that same time, minus one-half the amount by which that same sum of live forces would increase if each of the bodies moved freely on the curve it would describe, supposing it to be subjected at each point on the curve to the same force that actually does affect it—

provided always that the motion changes gradually and that if spring-driven machines are involved, the springs at the end of the process are left in the initial state of tension.[65]

D. COMPARING THE WORK OF SADI CARNOT

Let us turn now to the memoir by Sadi Carnot, *Réflexions sur la puissance motrice du feu et sur les machines propres à développer cette puissance*, published in 1824,

---

[65] *Principes fondamentaux*, § 293, pp. 261-262.

the year after his father's death and twenty-one years after the *Principes fondamentaux de l'équilibre et du mouvement*. The similarity has often been remarked between Lazare Carnot's observations there on hydraulic machines and his son's model attributing motive force or power to the passage of heat considered as a real fluid falling from a higher to a lower level of temperature.[66] In fact, however, when Sadi Carnot's memoir is read in direct succession to Lazare's writings, the son's inheritance appears fuller than the mere adoption of an hydraulic model for the flow of heat. Indeed, the *Puissance motrice du feu* may be taken both for the foundation stone it certainly became in the science of thermodynamics and for the final item in a series of Carnot memoirs on the science of machines beginning with the prize essay of 1778. It is the latter relation that concerns us here.

There is no reason to think that Sadi Carnot would have objected to such an attribution. It is one that may be supported on biographical as well as substantive grounds. Graduated from *Polytechnique* in October 1814 (after having fought with many of his classmates in the brief and vain defense of Vincennes in March), Sadi completed his training with two further years of study at the school of military engineering in Metz. Until 1819 he led the garrison life of a second lieutenant. He then arranged to go on inactive duty in Paris in order to devote himself to study and technical research. All the while his father and younger brother were in exile in Magdeburg, where Sadi was able to visit them for a few weeks in 1821.

It was evidently after this visit that Sadi began concentrating his attention on the principles of heat engines, and first the steam engine. In addition to the published *Puissance motrice du feu*, there is extant the manuscript of a brief "recherche" entitled "Investigation of a formula for the Motive Power of Steam." Scholars hold differing opinions about the date of this memoir. To the present writer it seems probable that it was composed somewhat prior to the *Puissance motrice*.[67] It was in a genre of which there were many examples in the years around 1820, and which will be discussed briefly in the next chapter. Sadi obtained a value for the motive power of one kilogram of steam expressed as a function of temperature by employing Clément's law for the pressure of saturated vapors and Dalton's table relating vapor pressure to temperature. The discussion was clear and the derivation clever. In the interests of generality, he identified three stages in the operation of a steam engine: as the steam passes into the cylinder, it may be considered as expanding isothermally in the first phase and adiabatically in the second while the third stage was isothermal compres-

[66] Brunold, *op. cit.*, pp. 37-40.

[67] Recherche d'une formule propre à représenter la puissance motrice de la vapeur d'eau —the manuscript has been published by W. A. Gabbey and J. W. Herivel, "Un manuscrit inédit de Sadi Carnot," *Revue d'histoire des sciences*, 19 (1966), 151-166. For a discussion of the question surrounding the date, see note 73 below.

sion in the condenser. (These terms were not yet coined.) But though recognizing elements of Sadi's analysis, the reader of the *Puissance motrice du feu* will see this schematization as a partial one. The cycle was complete only with respect to the return of the piston to the starting point and not with respect to the temperature of the steam.

What appears to be a residue of that incompleteness marred the reasoning in the opening part of the argument of the *Puissance motrice du feu* itself. It seems probable that the closer attention Sadi might reasonably be expected to have paid to his father's work on receiving in that same year the news of Lazare's death was what showed him the way to overcome that incompleteness and to put his argument on a fully general basis. We know from Hippolyte's memoir of his brother that when he, Hippolyte, returned to Paris after Lazare died in 1823, he found Sadi at work on the manuscript and was made to read and criticize important passages in point of their comprehensibility to general readers.[68] The brothers could scarcely have failed to talk then of their father's science.

The very word "Reflections" in Sadi's title recalls the ruminative vein of Lazare's book on the calculus, while in analysis and subject matter the genre was that of his science of machines. Like the *Essai sur les machines*, the *Puissance motrice du feu* was a treatise nonetheless rigorous for being verbally expressed and nonetheless general for being an adaptation of science to the principles underlying the employment of machinery. Despite the gratifying state (Sadi began) to which the development of steam engines had attained in practice, there existed no theory in the light of which their further improvement might be guided. Was there any limit to the power that ingenuity might draw from heat? The question was rhetorical and implied that there must be. What, then, were the optimum conditions for the design and operation of heat engines? Might not other materials prove preferable to steam for developing the expansive force of vapor? Atmospheric air, for example—and this possibility, to which Sadi recurred often enough to confirm its appeal to his imagination, indicates that he had at some point found suggestive the model air engines of Niepce and Cagniard, of which Lazare had written enthusiastic accounts many years earlier.[69]

Those questions could not previously have been answered (to paraphrase further the *Puissance motrice du feu*) for the reason that the availability of motive power in heat had never been considered in generality but only with respect to particular types of machines. In order that a fundamental theory might be found, it would be necessary to abstract from all particular mechanisms and from the properties of all par-

[68] Hippolyte's biography is contained in a reprinted edition of the *Réflexions sur la puissance motrice du feu* published in Paris in 1878. References in the notes that follow are to the pagination of the facsimile of the first, 1824, edition published by Blanchard in 1953.

[69] Above, pp. 27-29.

ticular materials and to consider how heat might produce motion as a problem independent of all contingencies. The theory must confine itself neither to steam engines nor even to vapor engines: it must embrace every conceivable heat engine and must model itself upon the theory of the classic machines. For there was a science that did indeed have the character of a complete theory (so held his father's son), resting as it now did upon the principles of mechanics itself. And only when the laws of physics should be sufficiently extended to embrace all the mechanical effects of heat acting upon bodies of any sort would the theory of heat engines be in a comparably satisfactory state.[70]

Sadi Carnot's memoir has been so fully commented, studied, and paraphrased in recent years that no need exists to summarize the entire contents or to enlarge on its later importance in the history of thermodynamics. For the present purpose it will suffice to identify those elements of the argument that derived from the work of Lazare. Sadi began by calling attention to a circumstance that always accompanied the production of motion by a steam engine. He chose to see it as a restoration of equilibrium in the caloric, by which he meant that heat is always transferred from a hotter to a colder body, from boiler to condenser. The process, he emphasized, involved the movement and not the consumption of caloric, and Sadi's analysis depended upon adopting the point of view that availability of motive power from heat presupposed some prior disruption of equilibrium in the distribution of caloric, and that reciprocally wherever a difference of temperature occurred, there existed the potentiality of drawing motive power from the transportation of caloric that would restore the state of equilibrium. In principle the agent might be anything, a metallic bar, for example, the reason being that in any object a change in temperature always involved a change in volume that might be harnessed. The choice depended on efficiency.

The question of how the motive power of heat might vary with the nature of the agent chosen to realize it, whether steam, air, metallic bars, or whatever sort of body, could be discussed decisively only for a given amount of heat and a given drop in temperature. Suppose that a body A was maintained at a temperature of 100° and a body B at a temperature of 0°. What would be the motive power (or work—the term itself was only five years from adoption) that could be delivered by transferring a given quantity of heat from A to B? Was it more efficient to employ one substance than another? The obvious advantage of vapors was that the temperature of gases would rise on compression and fall on expansion so that it was possible to disturb the equilibrium of the caloric in the same substance as often as desired. Moreover—and here Sadi gave the first hint of the analytical use he was about to make of the idea of a reversible process—it would always be possible to consider that steam might theo-

---

[70] *Puissance motrice du feu*, pp. 6-9.

retically be employed in a manner the inverse of that of a steam engine for the purpose of disturbing the equilibrium in the caloric, i.e., for transporting heat from a colder to a hotter body by the expenditure of motive power.

In the initial outline of such a process, Sadi made it occur in the three stages of his unpublished "Recherche": (1) In the first, the body A discharged the function of a boiler generating steam at its own temperature, i.e., isothermally. (2) In the second, the steam expanded (adiabatically) in the cylinder until its temperature fell to that of body B. (3) In the third, the steam was condensed at constant pressure by contact with body B, which, therefore, was filling the office of the cold water injected into a condenser, except that it remained at constant temperature and did not mix.

Having thus to a degree idealized and schematized the functioning of a steam engine, even as Lazare had done for ordinary machines, Sadi observed that these operations "could have been done in one direction and also in the inverse order." Steam could, first, be formed by employing the caloric of body B at its temperature; second, compressed until it rose in temperature to that of body A; and third, condensed by further compression at the temperature of body A. In both directions, therefore, the first and third stages were isothermal and the intervening step adiabatic—though it is to be emphasized that these terms had not then been coined. The former sequence of operations produced motive power [or work] and transferred the caloric [or heat] from the higher temperature of body A to the lower of body B. The inverse operation expended motive power [or work] and returned the caloric [or heat] from B to A. But if the same quantity of vapor were involved and if no motive power nor caloric had been lost, then:

> the quantity of motive power produced in the first case will be equal to that expended in the second, and the quantity of caloric transported in the first case from body A to body B will be equal to the quantity restored in the second from body B to body A, so that one could perform an indefinite number of similar operations without there finally being either motive power produced or caloric transported from one body to another.[71]

Evidently, then, Sadi was introducing the idea of a reversible process in the same place in his argument that Lazare had done in his, at the outset, and for the same reason: as an auxiliary in the reasoning to permit comparison of the initial and final states of a system by eliminating from consideration internal changes of work or energy. In the *Essai sur les machines*, it will be recalled, Lazare's initial definition of geometric motion had been in terms of reversibility.[72] If (as seems probable) Sadi had the idea of the three-stage cycle set out in his "Recherche" manuscript before his father's death,[73]

[71] *Ibid.*, p. 20.   [72] Above, pp. 42-44.

[73] It is agreed that Sadi Carnot must have composed the manuscript of the "Recherche"

at some time between November 1819 and March 1827. In publishing the text (above, n. 67) Drs. Gabbey and Herivel concluded that it is specially significant that he incorporated the idea of a reversible process only in the text of the *Puissance motrice du feu* discussed with Hippolyte upon the latter's return from Magdeburg.

The next comparison is equally telling. Discussing ideal hydraulic machines in the *Principes fondamentaux*, Lazare had brought out that it was a condition for maximum efficiency that there be no motion in the millstream that was not transmitted to the wheel, since any residual velocity in the water could in principle be harnessed on egress to produce an additional effect.[74] Sadi for his part went on from the formulation of reversibility in ideal steam machines (for perhaps the literal translation of "machine à vapeur" helps bring out the carry-over of ideas) to argue that if there existed any method for employing heat more advantageous than the pair of processes just described, it would follow that some greater quantity of motive power could be drawn from the flow of caloric in the first or forward process. It would then be possible to divert the excess or a portion of it to the job of driving the reverse process, that is to restoring the caloric from body B back up the temperature scale to body A. If

a date prior to 1824, when the *Puissance motrice du feu* appeared, was more probable than a later one. Mr. James Challey suggests that 1823 is the most probable date of composition because of Sadi Carnot's use of the "dyname" as the unit of motive power, Dupin having coined the word in a report to the Academy in April of that year ("Sadi Carnot," *Dictionary of Scientific Biography*, III (1971), 83, n. 3). For reasons indicated in the argument of this section, I agree. Dr. Robert Fox disagrees, however. He kindly provided for me the proofs of his article "Watt's expansive principle in the work of Sadi Carnot and Nicolas Clément," *Notes and Records of the Royal Society*, 24 (1970), 233-253. In that paper he explores the relations between Clément and Sadi Carnot very carefully. The main point is to establish the importance to Sadi's work of contemporary power technology. This it does admirably. Dr. Fox argues further that the manuscript "Recherche" represents an attempt by Sadi Carnot to compute a formula for the motive power of steam that would be applicable in actual engines as the highly abstract and unrealistic reasoning of the *Puissance motrice du feu* could not be, specifically because in the manuscript memoir Carnot computed the motive power that would be developed in the phase of adiabatic expansion between any two temperatures in-

stead of restricting it to the unreal case of a span of 1°. Dr. Fox may well be right. I doubt it, or doubt at least that this is the whole explanation because it reverses the configuration both of his work and his father's, which moved from the analysis of machine processes to theory rather than from theory to specific engineering application. Further, it seems more likely that the inclusion of the three-stage cycle applicable only to steam in the early passages of the *Puissance motrice du feu* reflects rather the carry-over of an unperfected stage of his analysis into the final product, than it does that he should afterwards have reverted to this imperfect analysis in order to base an applicable calculation on it. There is a possibility that we are both right, and that the manuscript memoir, which does indeed have all the appearance of having been finished, represents the final development that Sadi Carnot, after publishing the *Puissance motrice du feu*, then gave to the idea of a three-stage cycle with which he had begun his thinking. If so, it remains a problem that he did not publish it. Since almost all his manuscripts were burned after his death from cholera, it is unlikely that the question will ever be decisively resolved.

[74] *Principes fondamentaux*, § 273, pp. 248-249.

that could be done, it would amount to an indefinite creation of motive force without consumption of caloric or indeed of any agent at all. It would amount, in short, to perpetual motion. Such a conclusion being contrary to "the laws of mechanics and sound physics," it was inadmissible. Excluding that impossibility afforded Sadi the basis for a preliminary statement of what he called the fundamental theorem: "The maximum motive power resulting from the use of vapor is also the maximum motive power that can be realized [from heat] by any means at all." A moment's reflection, consequently, would exhibit the condition for realizing the maximum motive power in general: it was "that there should not occur in the bodies employed to realize the motive power of heat any change of temperature that is not due to a change of volume."[75]

The reader familiar with later thermodynamic reasoning will immediately notice that thus far the argument was not complete: it omitted the step of an adiabatic compression on the return process. The vapor in this initial illustration being steam, Sadi could hardly have incorporated such a stage since to vaporize water by compression without input of heat would have been physically unimaginable. He himself recognized the difficulty and, in moving on to a full and rigorous demonstration employing an air engine, admitted that in this preliminary sketch the vapor had not been supposed to be entirely restored to its initial state.[76] Commenting on the discrepancy, T. S. Kuhn has conjectured that Sadi began with a steam engine because of its greater familiarity to his prospective readers.[77]

It may well have been so, and there may be further reason for supposing also that the partial cycle was the residue of an earlier stage in Sadi's own thinking in which he had not advanced that far beyond his father. His own justification was not in point of the familiarity of the steam engine but rather of a principle of continuity reminiscent of Lazare's restriction of live-force [or work] conservation to motions changing by insensible degrees. Having drawn his consequences about maximum motive power, Sadi defended the reasoning by claiming the right to suppose the temperature differential in the flow of caloric to be infinitesimal. In that way he could posit the reverse process and restore the caloric from body B to body A without violating his stipulation that two bodies at different temperatures were not to be in contact. (The skeptic could rightly have pointed out that neither would work have been done, and that, of course, was the flaw in this preliminary statement of the argument.) The argument might then be extended to the finite case by supposing the change of temperature to consist in a sequence of infinitesimal steps.

---

[75] *Puissance motrice du feu*, pp. 22-23.

[76] *Ibid.*, pp. 36-37.

[77] "Sadi Carnot and the Cagnard Engine," *loc. cit.*, p. 571 n. It might be remarked in passing that Sadi himself did not, as Professor Kuhn might be read as implying, speak of a reversible process in terms of two ideal engines coupled together and working in opposite directions.

It is plausible to think that these reasonings, and the conclusions so far drawn, reflect the direct carry-over from Lazare's science of machines into Sadi's early thinking on heat. His having resorted to the argument from gradual change is persuasive, since reversibility plays the part of geometric motion. So also is the ensuing paragraph in which Sadi explicitly invoked the analogy between the flow of heat and the flow of water. That passage is usually cited and often quoted to exhibit the relation of his work to Lazare's. Nothing said here is in any way meant to deny the relation, but only to show that it was one wider than this model and that it extended to the very mode and terms of the analysis. If so, it was probably at this juncture that Sadi started to go beyond his father. Two aspects of the internal evidence make that plausible. In the first place, he here began citing the extensive experimental determinations of his own generation on the thermal aspects of the physics and chemistry of gases, matters on which he enlarged in the main body of his memoir. In the second place, it was also precisely here that he put forward a second demonstration of the "fundamental proposition"—i.e., that the utilization of vapor was the means of realizing the maximum motive power from heat—and did so in a fully general form. It was on the basis of that demonstration, the full Carnot cycle, that he stated the theorem later regarded as his fundamental contribution, and did so in these words: "The motive power of heat is independent of the agents put to work to realize it; its quantity is determined uniquely by the temperatures of the bodies between which the caloric in the final result passes."[78]

The present purpose being to exhibit the relation of Sadi's work to his father's, it will be appropriate to stop short of discussing the importance of the Carnot cycle in the later development of thermodynamics. That story has often been told and more often commented. Perhaps it will serve the interests of completeness, however, to outline the way in which Sadi proposed it himself, for although the four successive stages of a gas undergoing an isothermal and an adiabatic expansion restored to its initial state by an adiabatic and an isothermal compression were indeed contained in it, he imagined it occurring in six steps. His diagram of an idealized system exhibited the elements of any heat engine employing a generalized vapor—though Sadi now suggested air as what he had in mind. The body A represented the boiler, an inexhaustible reservoir of caloric. The body B represented the condenser, though for air it would have been better to say a sink for caloric. Neither A nor B changed temperature. B was at a lower temperature than A. Caloric flowed without loss from A to the cylinder to B, and the gas was contained in the cylinder under a weightless, frictionless piston. The operations ensued as follows:

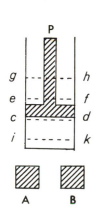

[78] *Ibid.*, p. 38.

1) Caloric flowed from A to the cylinder.

2) The piston rose from *cd* to *ef* while A supplied caloric and the temperature remained constant [i.e., an isothermal expansion].

3) Caloric from A was cut off, but the gas continued to expand while cooling until the piston had reached *gh*, whereupon the temperature had fallen to that of B [i.e., adiabatic expansion].

4) The gas was compressed until the piston returned to *cd* and the displaced caloric flowed into B so that the temperature remained constant at the lower level [i.e., isothermal compression].

5) The flow of caloric to B was cut off, and compression continued until the piston was at *ik*, at which point the temperature had risen to that of A [i.e., adiabatic compression].

6) The body A was again connected to the cylinder and the gas expanded at constant temperature to return the piston to *ef* [i.e., isothermal expansion].

Sadi pointed out that these steps could equally well have been imagined in the reverse order:

"The result of these first operations has been the production of a certain quantity of motive power and the removal of caloric from the body A to the body B. The result of the inverse operation is the consumption of the motive power produced and the return of caloric from the body B to the body A; so that these two series of operations annul each other, after a fashion, one neutralizing the other."[79]

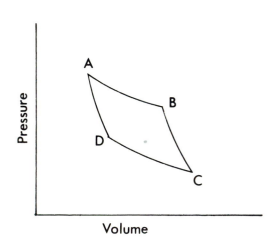

Pressure / Volume

Nevertheless, although the concept of reversibility was now associated with a complete cycle both of gas and piston, Sadi was not thinking of his cycle quite in the way that is schematized in modern physics tests. There the graph represents an isothermal followed by an adiabatic expansion, and the system is restored to the initial conditions by an isothermal followed by an adiabatic compression. The work delivered may then be denoted by the area enclosed by the graph.

Not quite so for Sadi, who in step 6 returned his piston past the initial point *cd* to *ef* and then had the steps of the cycle repeating in the sequence 4, 5, 6, 3, 4, 5, 6, 3, 4,

[79] *Ibid.*, p. 36; the diagram in the text is a somewhat consolidated schematization of the figures Sadi gives facing page 118 of his memoir.

5, etc.—i.e., beginning each time with the adiabatic expansion. For what he wished to argue was not the equivalent of the modern cycle, as it would have been had he said that the motive power [work delivered] was proportional to the volume *cdki* of the cylinder (equivalent to the area enclosed in the graph), but rather that for any given position of the piston, the temperature was higher during expansion than during compression and hence that power could be drawn from the cycle. It comes to the same thing, but was not entirely the same way of seeing what he had proved. Thus Sadi on expansion and compression:

> During the former, the elastic force of the air is found to be greater, and consequently the quantity of motive power produced by the movements of dilation is more considerable than that consumed to produce the movements of compression. Thus we should obtain an excess of motive power—an excess which we could employ for any purpose whatever. The air, then, has served as a heat engine; we have, in fact, employed it in the most advantageous manner possible, for no useless re-establishment of equilibrium has been effected in the caloric.[80]

This result is as far as we need to follow Sadi in order to exhibit the full inheritance of the son from the father. No doubt the most significant items were the development of the idea of geometric motion into that of a reversible process and the application to the analysis of fully cyclic processes. Before that there was the exclusion of perpetual motion, axiomatic in Sadi, demonstrative and tutelary in Lazare; the generalization from principles of operation in particular types of machine to the principles of machines in general, which amounted to making physics out of the industrial reality of the age; the restrictive mode of reasoning in which the maximum possibilities inherent in operations were determined and then the conditions for realizing them defined; the curious combination of quantitative mode and verbal expression, such that the reasoning moves from the ideal case of changes occurring continuously and infinitesimally to the physical reality of discontinuous and irreversible changes of state in a system; the discussion of force in terms of what it can do, taken usually over distance, when it was a question of its measure, and over time, when it was a question of its realization in mechanical processes. It cannot be said that either Carnot brought this tacit distinction to a decisive differentiation between proto-concepts of energy and work. Yet both of them assumed the conservation of the quantity measured by the dimensions of work and energy, Lazare balancing his accounts between moment-of-activity produced and moment-of-activity consumed or live force expended, Sadi between motive power produced and caloric transported.

There was even the long wait for recognition, and the question whether what developed was less the theories of father and son than the state of knowledge and

[80] *Ibid.*, pp. 34-35.

interest in the subject that permitted their cogency and applicability to be recognized, by which time it was no longer clear that it was from the authors that the subject was really learned by those who used it. Sadi's treatise was largely ignored until Clapeyron put it into mathematical form in a memoir of 1834,[81] and it did not occupy the attention of physicists until the 1840's, when the problem of a mechanical equivalent of heat was taken up by Joule, Mayer, and Thomson.

[81] "Mémoire sur la puissance motrice du feu," *Journal de l'école polytechnique*, 14 (1834), 153.

CHAPTER IV

The Carnot Approach and the Mechanics of Work and Power, 1803-1829

T HE WAY in which the work of Lazare and Sadi Carnot made itself felt reveals a similarity between their writings as interesting as their substantive filiation and the analytic and stylistic parallels. The failure of Lazare's *Essai sur les machines* to attract contemporary attention has already been discussed. Although more often mentioned since, the *Principes fondamentaux* fared little better when it appeared in 1803. It fell into the same obscurity, lasting another fifteen years. The book was occasionally mentioned prior to 1818, but rather by way of noticing its existence than because its point of view affected the treatment of problems. When at about that time Lazare Carnot's mechanics of machines began to be comprehended in print, the recognition was of his distant paternity of principles or applications that might be taken for granted. Indeed, it does not appear that, apart from Sadi himself, others of the next generation of engineering mechanics actually read Lazare's writings, for what they credited to him was their own much simplified employment of the concepts and not his enunciation of them.

The explanation cannot well be that either father or son was an obscure or neglected personality (except in Lazare's early years). What seems likely, therefore, is that attention first to machines and then to heat and power developed in a largely verbal, pedagogical, and practical way. The subjects constituted a kind of engineering mechanics *avant la lettre* in which problems were posed, principles tacitly selected, and quantities employed because that was the way to get results. What further seems likely is that both Carnots participated in that development personally rather than through their books. Lazare in his latter years was a kind of Nestor of engineering busying himself judging inventions for the Institute. Sadi kept up with his fellow polytechnicians and was a contemporary and in several cases classmate of those in whose writings recognition that his father's mechanics was of elementary importance now began almost spontaneously to appear: Petit and Navier, Coriolis and Poncelet, Clapeyron and Dupin, Chasles and Barré de Saint-Venant.

It supports this view to notice how largely the literature of early engineering mechanics was couched in the form of commentaries on earlier writings and new editions rather than in memoirs of research or systematic treatises. In this mode it was typical of engineering in general, wherein the sequence of specific researches, whether experimental investigations of novel phenomena or theoretical resolutions of particular problems, carried much less of the traffic than it did in science proper. There was such a literature, of course, and it grew in importance in the contribution of the first generation of polytechnicians. But even then engineering knowledge was growing, and perhaps mainly so, through the gradual codification of practices brought up to the surface for commentary and formalization. For the present purpose there is no need to go back beyond the works from which Lazare Carnot himself doubtless learned the elements of practical constructions, Bélidor's *Science des Ingénieurs* and *Architecture hydraulique*.[1]

The career of Bernard Forest de Bélidor was a colorful example of what, after all, had been open to a man of technical capacity in the Old Regime. Left a military orphan in childhood, he developed from an artillery instructor into an engineering expert and died a member of the Academy of Sciences. The former of his books had a narrower and the latter a wider application than might be supposed nowadays from the titles. "Engineering" connoted mainly fortification and siegecraft, military works in short, while waterworks in the eighteenth century were important in transportation, the provision of power, and the embellishment and supply of cities, castles, and gardens—all problems that later fell into the domain of mechanical and civil engineering.

Bélidor began with an initial parade of founding practice on the principles of the science of mechanics. Most probably it was there that Carnot read of the principle of equilibrium that in much the same connection Bélidor and he both attributed to Descartes.[2] The demand for Bélidor's works never slackened throughout the eighteenth century. When the copper plates wore out through constant reprinting, the publisher, Firmin Didot, decided to have them re-engraved and to reissue both sets, and he engaged the services of C.-L.-M.-H. Navier, a rising light among the new breed of theoretical engineers, to bring the work up to date. The history of that edition is very revealing for the relation between practical construction and theory in mechanics. For despite the immense development of the science of rational mechanics in the eighty years intervening between Bélidor's first edition and Navier's reissue, it seemed wise not to make the slightest change in Bélidor's text lest the work lose its public.

That public consisted of contractors and builders who knew Bélidor for an engineering handbook in which they could look up specifications for such or such a foundation, revetment, doorway, cornice, or other standard construction, instructions

[1] Bernard Forest de Bélidor, *Architecture hydraulique, ou l'Art de conduire, d'élever et de ménager les eaux pour les différens besoins de la vie*, 2 vols. (Paris, 1737-1739). There were further editions in different formats, and finally a new edition "avec des notes et des additions," ed. by Navier (Paris, 1819); *La science des ingénieurs dans la conduite des travaux de fortification et d'architecture civile* (Paris, 1729; Navier edition, 1813).

[2] Above, p. 36, and Bélidor, *Architecture hydraulique*, I, pp. 27-31.

covering not only architectural details but even the most efficient mode of organizing the work gang.

As for the theoretical part of the work, Bélidor's mathematical equipment had been faulty even for his own day, and the formal aspect of his presentation required complete revision. Navier undertook the task by composing for *Architecture hydraulique* a series of footnotes so extensive as to amount to a companion treatise that runs along the bottom of Bélidor's republished pages correcting, supplementing, and clarifying his discussion of principles and laws, yet somehow not supplanting Bélidor himself. Indeed, one is tempted to paraphrase the famous remark that L. J. Henderson lifted from J. T. Merz about the debt that science owed the steam engine. One feels that the science of mechanics owed more to Bélidor than Bélidor to the science of mechanics—it took from him the body that brought engineering mechanics out of it, and what it contributed in the way of refinement seemed insufficiently usable in actual constructions to find a public for a literature all its own.

Navier, a graduate of *Polytechnique* in 1804 and then of *Ponts et Chaussées*, came like Carnot from a Burgundian background. His father, a lawyer and member of the Assembly of Notables in 1787, having died prematurely of "chagrin" (according to Prony) in the Revolution,[3] Navier was brought up by an engineering uncle, one Gauthey, whose *Traité des ponts* he then edited and published. His properly theoretical work belongs to the history of elasticity and the early adoption of Fourier analysis in mechanics. But he also engineered actual constructions: the Pont de Choisy, the Pont d'Argentueil, the Pont d'Asnières, and the Pont des Invalides, a suspension bridge built in the style of the Tweed bridge and torn down in 1826 when a reservoir on the Butte de Chaillot burst, and the resulting flood undermined one of the piers. His writings, apart from particular memoirs on theoretical problems, practical constructions, and railway projects, developed out of courses he gave at *Polytechnique* and *Ponts et chaussées*. Just as he edited others, so in mechanics proper, others edited him. Thus, a *Résumé des leçons d'analyse*, consisting of lectures at *Polytechnique* for three successive semesters (1832, 1832–33, and 1835–36) was formally published under the editorship of J. Liouville, who rearranged it and appended his notes.[4] Similarly the course on mechanics evolved from a résumé for 1831, 1832, and 1833, and reached treatise form in 1841.[5] Finally, a famous course on the science of machines given at *Ponts et chaussées* and first distributed in 1826 evolved in 1864 to a third edition of its first volume under the title *De la résistance des corps solides*,[6]

[3] Gaspard Prony, "Notice biographique sur M. Navier," extrait des *Annales des ponts et chaussées* (Paris, 1837).

[4] (Paris, s.d.). A second edition "Revue par M. Ernest Liouville" (apparently the famous mathematician) was published in 2 vols. in 1856.

[5] C.-L.-M.-H. Navier, *Résumé des leçons de mécanique* (Paris, 1841).

[6] *Résumé des leçons données a l'école des*

and was furnished with extensive notes and appendices by Barré de Saint-Venant, which themselves form an indispensable source for the history of mechanics, particularly the theory of elasticity and related topics.[7]

Together with Navier, Coriolis and Poncelet constituted the ruling triumvirate in the engineering mechanics of their generation, those who graduated from a *Polytechnique* still under the direct influence of its founders. Their writings fall into the same pattern. The most important single treatise was Coriolis' *Du calcul de l'effet des machines* in 1829, which gave Carnot's moment-of-activity its modern name of work. Of that, more in a moment. Coriolis then circulated in multigraphed form notes of his course on mechanics at the *Ecole centrale* for 1830–31 and again for 1836–37.[8] Evidently he intended a complete treatise of applied mechanics combining the science of machines with an up-to-date exposition of the fundamental principles of rational mechanics. He died before completing that synthesis, and the publishers had instead to reprint most of the 1829 memoir together with the later material on the mechanics of solid bodies.[9]

Poncelet for his part circulated his course at the *Ecole d'application* at Metz in multigraphed form beginning in 1826. He observed à propos of Coriolis' 1829 treatise that his own course contained similar material, but that ill health had prevented him from publishing it.[10] The course itself, or perhaps extensions of it given by Poncelet between 1827 and 1829, was then edited for publication by a captain of engineers, one Gosselin, and published in 1828 under the title *Cours de mécanique industrielle*.[11] A second edition appeared in 1841, and a third (this was edited by X. Kretz) in 1870. Poncelet himself thought it useful to publish a review of his own scattered writings in a pamphlet undated but apparently of 1835 or thereabouts.[12] Finally, in 1874–76 appeared a two-volume edition of *Cours de mécanique appliquée aux machines*,[13] this too edited by Kretz, containing a useful introductory résumé of

---

ponts et chaussées, sur l'application de la mécanique à l'établissement des constructions et des machines (Paris, 1826).

[7] *De la résistance des corps solides*, 3rd ed. (Paris, 1864), avec des notes et des appendices par M. Barré de Saint-Venant.

[8] Gaspard-Gustave de Coriolis, *Leçons sur la mécanique, 1830-1831, 1836-1837* (s.l.n.d.; 2 vols.; *Bibliothèque nationale* [V. 14276-14277]).

[9] Coriolis, *Traité de la mécanique des corps solides et du calcul de l'effet des machines* (Paris, 1844).

[10] J.-V. Poncelet, "Note sur quelques principes de mécanique relatifs à la science des machines." Extrait du *Bulletin universel des sciences* (October, 1829), Sect. 1: (*Bibliothèque nationale* [Vp. 5866]).

[11] Paris, s.d. (*Bibliothèque nationale* [V. 5562]); see also *Bulletin des sciences mathématiques* 12 (1829), 323, n. 1.

[12] *Notice analytique sur les travaux de M Poncelet* (*Bibliothèque nationale* [4° Ln²⁷ 16475]).

[13] An earlier edition of this work was published in Liége in two volumes in 1845. It is not clear whether it had Poncelet's approbation. "Cet ouvrage n'a jamais été publié en France ni en Belgique," says a legend on the title page, and indeed the *Bibliothèque nationale* has no copy.

the evolution of the entire school. The original patron of the school, Monge himself, had difficulty in actually writing and finishing treatises of his own, and in fact never accomplished it. His immediate disciples put together his *Géometrie descriptive*.[14] A work that he purportedly intended to publish on the mechanics of machines had instead to be undertaken by Hachette, a wheelhorse who replaced him at *Polytechnique* when he went off to swell Bonaparte's scientific train in Egypt.[15] It was written, says the dedication to Monge, "assisted by your advice."

In these early decades of the nineteenth century, two different modes may to a degree be distinguished in the conception of French treatises of practical mechanics. On the one hand, the literature that we have just been considering, despite its tendency for books to remain unfinished by the author, was generally theoretical in spirit and continued in increasingly mathematical form the intention of the two Carnots, i.e., to draw from the science of mechanics a theory of machines and of power at once general and applicable. That was the novel emphasis in content, and it was inspired by the actual development of machine-driven technology. Alongside this literature, there also continued to appear treatises and memoirs of a type that may be called manuals. To accounts of old machines they added those of new ones, and to the expression of classic principles they brought the newer formulations of rational and analytical mechanics. The distinction between the two traditions does not turn on the presence or absence of mathematics. Where a difference in the mathematics was relevant (it was not always so) it was on the kind of mathematics, the theoretical tradition being associated with the new geometry rather than analysis.

For the background of this second, more conventional approach, there is for present purposes no need to look further in the history of mechanics than to *Traité élémentaire de méchanique* of the abbé Bossut.[16] Bossut had divided his subject into statics and dynamics. Statics was the science of equilibrium, which results from the mutual destruction of forces. The topics were general equilibrium principles; determination of centers of gravity; application of the foregoing to stating equilibrium conditions in each type of machine taken successively; and finally friction. The topics of Part II on motion were inertia, uniform motion, acceleration, fall, the motion of centers of gravity, and collision theory. The book was an excellent primer, and exhibits the subject matter at an elementary level as it was conceived when Lazare Carnot was an engineering student at Mézières.

In the field of practical engineering and applied mechanics, the most pretentious

---

[14] Gaspard Monge, *Géométrie descriptive* (Paris, 1799); for Monge's interest in the science of machines, which derived from his early days at Mézières, see René Taton, *L'oeuvre*

scientifique de Monge (Paris, 1951), 311-314.

[15] J.-N.-P. Hachette, *Traité élémentaire des machines* (Paris, 1811).

[16] Paris, 1775, and numerous later editions.

work was Prony's *Nouvelle architecture hydraulique*[17] intended evidently to supplant Bélidor, which purpose it failed to accomplish. Prony was then (1790) inspector at the *Ecole des ponts et chaussées* and one of the foremost engineers of the revolutionary period. It is not this work, however, that will make apparent to later readers the justice of the high reputation he undoubtedly enjoyed. It purported to be a general treatise of mechanics in a form applicable to the arts. In fact, the exposition of the standard topics of mechanics was intrinsically unclear and was continually interrupted by tables containing empirical formulas and descriptions of favorite designs for pumps, prime movers, roof vaultings, and incidental devices of all sorts. The second volume concerned steam engines—and here a further characteristic of this school of practical mechanics becomes evident. In it motors were thought of as what one employs "to overcome resistance," and successive writers normally harked back to the Coulomb memoir on friction, that which won the prize of 1781 over Carnot,[18] in estimating the magnitude of resistance. Prony never mentioned the work of Carnot in these volumes on practical mechanical engineering. The course he gave on mechanics at *Polytechnique* drew on Euler, Laplace, and Lagrange, and for hydrodynamics on Bossut.[19]

As time went on, and Carnot became famous, other writers did allude to his *Principes fondamentaux*. Those of the Coulomb-Prony succession did so, however, in the lip-service of prefaces or appendices. In 1810 a prominent engineer of the Lyonnais, André Guenyveau, a graduate of *Polytechnique* in 1800, published an *Essai sur la science des machines*.[20] It is apparent that the subject was not yet synonymous with Carnot's lead. The body of the treatise passed from calculations about the state of equilibrium to practical application by drawing in Coulomb's results for friction.[21] An interesting afterthought was printed in smaller type than the body of the text. The author there observed that motion might be communicated in two ways, by impact or by pressure, and from the difference he deduced Carnot's principle of the superior efficiency of continuous transmission. He did that without crediting it to Carnot, however, who was merely mentioned in the preface as one who had written a general treatise on machines. Hachette accorded him similar recognition in the preface to the 1811 *Traité élémentaire des machines* already mentioned.[22] There Carnot's *Principes fondamentaux* came in for a brief encomium. Hachette observed that its closing pages contained the entire theory of machines and of motive forces applied to them, and called it the work of the "deepest of thinkers and the most

[17] Two vols.; Paris, 1790 and 1796.
[18] Above, p. 63, n. 2.
[19] Published as "Mécanique philosophique" in tome III of *Journal de l'Ecole polytech-nique*, cahiers, 7 & 8 (Paris, an VIII, 1800), pp. 1-477.
[20] Lyons, 1810. [21] *Op. cit.*, pp. 2-3.
[22] Above, p. 105, n. 15.

experienced of engineers. It is to the science of machines what the recent work of M. Poisson is to rational mechanics properly speaking."[23] But nowhere did the findings or analysis of Carnot, or indeed of Poisson, figure in the actual text or treatment of Hachette's course, which was no more than a compilation of designs for particular machines, a book of recipes for engines, pumps, gears, pulleys, and linkages of many sorts.

A more sophisticated contribution to this literature of applied mechanics was a memoir of 1818 by Petit "Sur l'emploi du principe des forces-vives dans le calcul de l'effet des machines."[24] No mere engineer, Petit was a rising young physicist—part author of the famous law of Dulong and Petit—who died young before giving his full measure. At first reading one is tempted to classify this memoir on the theoretical side of the division between the handbook approach and the tradition inaugurated by Lazare Carnot's mechanics of machines. It was exact, it was analytical, and it took the principle of live force to be that one among the general properties of motion that was most readily adaptable to the "calcul des machines." Live force furnished in any particular case the most natural evaluation both of the motor and of the effect produced, so that the equation determining the relation between these two quantities would furnish the direct solution of the problem.

That application of the principle of live force had for a long time been well known to geometers, so observed Petit at the very outset, but then surprisingly went on to say: "The theory of machines, envisaged from this point of view, remains to be created almost in its entirety."[25] Could he have said that because, unlike Hachette, he did not even know of Carnot's work? Or was it that, knowing it, he did not regard it as an adequate theory of machines from the point of view of live force? Or was it that by machines, he, like Sadi but unlike Lazare, was beginning to mean engines or motors? The first of these alternatives seems the most likely, although it is not impossible that the explanation lies in a combination with the latter two, for, as will appear, Petit was not in fact yet looking at the problem from the point of view of Lazare Carnot's theory; and if he did know it, he did not yet see, as Sadi was to do, its applicability to the study of heat engines.

Like Prony and Hachette, Petit considered that the fundamental function of a motor was to overcome resistance. He intended his essay to be an examination of particular cases of that general problem. For a machine in equilibrium all that need be taken into account was the intensity of the forces. For a machine in motion, however, regard had also to be paid to the distance traversed by the points of application. Petit illustrated the special applicability of the principle of live force to calculating the

[23] Hachette, *op. cit.*, p. xix. [24] *Annales de chimie et de physique*, 8 (1818), 287-305.
[25] *Ibid.*, p. 287.

efficiency of machines through the equivalence dimensionally of MgH to ½ Mv^2. First, he adduced the classic example from the law of falling bodies. If the resistance was a weight of mass M raised to a height H, then the effect produced by the machine in raising it would be MgH. But since the velocity acquired by a body falling from height H was given by the relation $V^2 = 2\,gH$, the effect MgH equalled ½ MV^2. No matter what the type of resistance, it would always be possible thus to express the effect produced by a machine in dimensions that reduce to a live force, the product of a mass by the square of a velocity. Hence a motor could always be regarded as containing a given quantity of live force. What was true of the resistance was also true of the motor: its expression also could always be reduced to a live force. Hence, the ratio of the live force employed by the motor to that communicated to the resistance [the terminology here is faithful to the text] would give an expression for the efficiency of the motor and permit determining the conditions under which it would be a maximum. The rule was: "The live force communicated to the resistance is equal to that which the motor contained, less the live force lost in sudden change of motion, plus that retained by the motor after acting."[26]

Now then, although this reasoning and the consequent statement read very like Lazare Carnot's principle of continuity, which indeed would follow from them, Petit did not actually state such a principle. Probably he would not have thought to do so, and in order to see what differentiated their several sets of conceptions, we will need to follow the examples to which Petit applied his rule. For though the "moteurs" to which he alludes at the outset were the traditional waterfalls, springs, and animals, those in which he exemplified the rule were engines or prime movers, and the central thrust of the memoir was directed to devices powered by the expansion of vapor.

In such engines the live force developed by the fluid might not be computed in the direct manner of that developed by the flow of incompressible fluids such as water. Instead, we are to imagine that the expansible fluid was contained in a horizontal cylinder closed at one end and fitted at the other with a frictionless piston. Then let a be the length, b the cross-section of the cylinder, and h the height of a column of water of which the weight would be in equilibrium with the "elasticity" [he did not say pressure] of the gas. Since the thrust of the confined gas would give the piston an accelerated motion, it was easy to obtain equations for the live force generated by the expansion of the gas. The first case Petit considered was that of a movement in the absence of any external pressure on the piston.

1)　Letting v be the velocity acquired by the piston after time t, and x the length of the gas-filled volume of the cylinder at that moment, the "elasticity" of the gas would then reduce to $\frac{ha}{x}$, and its motive force would be equal to $g\frac{\delta\,h\,a\,b}{x}$, g representing the

[26] *Ibid.*, p. 290.

gravitational constant and δ the density of the water. But motive force was customarily represented by $m\frac{dv}{dt}$ or by $m\frac{v\,dv}{dx}$. Since m was the mass of the piston, it followed that

$$mv\,dv = g\frac{\delta\,a\,b\,h}{x}\,dx.$$

Integrating and choosing the constant so that v was zero when $x = a$, he found

$$mv^2 = 2\,g\,\delta\,hab\,\log\frac{x}{a}.$$

That is to say, when an elastic fluid occupying a volume a, and exerting a pressure equal to that of a column of water of which the height was h, expanded in the absence of external resistance, the live force that developed at the volume x would be capable of raising to the height h a mass of water of volume $a\,\log\frac{x}{a}$.

For the two other cases, perhaps it will suffice to state the laws without the algebra. 2) If instead of no external pressure, the piston had to overcome a constant external pressure equivalent to the weight of a column of water of height h', then the live force developed in expanding from volume a to volume x would be capable of raising to the height h a mass of water of volume equal to $a\,\log\frac{x}{a} - \frac{h'}{h}\,(x - a)$. 3) If, finally, a piston was subjected to two uniform pressures exerted in opposite directions by gases of which the elasticities [following Petit's terminology] sustained in equilibrium columns of water of height h and h', the live force communicated to the piston would be capable of lifting to the height $(h - h')$ a mass of water equal in volume to that which the piston had traversed.

In an example of the application of these laws, Petit imagined the problem of comparing the live force that a given quantity of heat could produce, supposing it to be employed first to vaporize water and second to heat a mass of air. In the first case, a quantity of water weighing one gram at $0°$ was vaporized at the temperature of $100°$. It would occupy approximately 1,700 cubic centimeters as steam and exert a pressure equivalent to that of a column of water ten meters high. If it was then completely condensed, the live force developed would be capable of raising a weight of 17 kilograms one meter. Turning now to the problem of heating a volume of air, the same heat required to vaporize one gram of water would heat 666 grams of water through $1°$; or, taking the Bérard-Laroche value for specific heat of air of 0.267, it would heat 2,500 grams of air at a pressure equal to that of a column of water of 10 meters. The elasticity of the air would then increase by 0.0375 meters, and the live force produced would lift to a height of 0.0375 meters the weight of a volume of water equal to that occupied by the 2,500 grams of air. That volume was 1,925 cubic decimeters, so that reducing everything to the same units, the live force was sufficient to raise a weight of 72.2 kilograms to a height of one meter—more than quadruple the live force in the first case.

It is very interesting historically that a paper containing these conclusions and written by one of the most intelligent physicists of his time should have appeared at just this date. What does it reveal?

First of all Petit's paper was evidently an important element in the immediate background of Sadi Carnot's thought—indeed, it was the application of the Carnot approach that enabled him to correct Petit's error by thinking of heat engines in terms of reversible process instead of overcoming resistance. One of the first points Sadi raised in the *Puissance motrice du feu* was the possible superiority of air over steam engines, and further on he cited the Petit memoir after demonstrating that it makes no difference whether steam, air, or alcohol vapor be employed: what determines the motive power of a heat engine is temperature differential. Sadi their attributed Petit's erroneous conclusion to "an altogether incomplete method of considering the action of heat."[27] So, indeed, it was, but the passage in which Petit summarized his conclusion is nonetheless striking in its modest disclaimer of exactly the central deduction that it was the purpose of Sadi's memoir to draw. Thus Petit:

> Although I do not pretend to deduce from the comparison I have just made any consequences relative to the best method of employing the action of heat as motive force, it is nevertheless permissible to think that some advantage could well be drawn from perfecting those machines that, like the pyréolophore of MM. Niepce, employ air suddenly expanded by heat as a motor.[28]

That these confusions should have appeared in a paper of just this date, composed by one of the leading physical minds of his generation, is most instructive to the historian. For not only did Petit compare the energy consumed in vaporizing water to the work done by an expanding gas, but the paper exhibits to particular advantage the very necessity for a concept of work and its differentiation from energy. True, Petit had grasped the significance of live force for evaluating the operation of power machinery, but so had everyone interested—"la force vive est celle qui se paie," in Montgolfier's words.[29] Nevertheless, he was still very far from a concept of energy, and was indeed confusing work with energy as did everyone else in what it is convenient to call the aggregative tradition in mechanics.

Since early in the eighteenth century it had been a standard practice to evaluate the capacity of a machine or an engine in terms of the height to which it could lift a

[27] Sadi Carnot, *Puissance motrice du feu,* p. 86 n.

[28] Petit, *op. cit.,* p. 294. For the Niepce engine, see above, pp. 27-28.

[29] The allusion is in Hachette, *op. cit.,* pp. xiv-xv. For the widespread recourse to the theoretical model of a water-pressure or "col-umn of water" engine in the pertinent literature, and for its importance in the background of Sadi Carnot's analysis, see D.S.L. Cardwell, "Power technologies and the advance of science," *Technology and Culture,* 6 (1965), 188-207.

given weight. The quantities of MgH and $\frac{1}{2}Mv^2$ being dimensionally reducible to each other in all such practical situations, Petit like most of his contemporaries tacitly came to regard them as alternative expressions for live force Their conception of what that term conveyed did not yet distinguish between (kinetic) energy and work, the one being the capacity to perform the other and the equivalence being a convertibility rather than a fundamental identity.

That distinction originated rather in Lazare Carnot's theory of machines in the difference between live force and moment-of-activity—not that he had it cleanly or consistently, but what distinguished the two modes in mechanics was that the one failed to perceive it and the other began to develop it. But Sadi need not have been so condescending about Petit, for in one respect he missed a point implicit in his father's convertibility of live force into moment-of-activity: the confusion between heat and energy in Sadi's theory was a further stage of that between work and energy in Petit and the others.

Navier was anything but condescending about Petit's paper. A few issues later he published in the same *Annales de chimie et de physique* a historical note reviewing the history of the live-force principle in the theory of machines.[30] It had, he said at the outset, the same purpose as Petit's: to publicize a set of notions that could be very useful in mechanical (i.e., engineering) practice by recalling what had already been done on that subject and attempting to revive certain procedures and results that had been almost forgotten. Navier had in mind here precisely Lazare Carnot's theory of machines, and its seems probable that it was this paper, followed more massively by his edition of Bélidor's *Architecture hydraulique*, that brought it centrally into view in a generation finally prepared by experience and environment, not just to allude to it as the writing of a famous man, but to develop it into real engineering physics and mechanics.

First, however, Navier gave a résumé of the history of the principle from the beginning. He credited the earliest precise conceptions on the effects of machines to Galileo, who had established that a given force in a given time could achieve only a determinate effect and that its measure was the product of a weight and the height to which it was lifted.

That product of force into distance sufficed for estimating the effect of machines so long as they were supposed to have reached a state of uniform motion. When such was not the case, however, an auxiliary principle was needed in order to evaluate from the variation in motion the forces that had produced them and vice versa. The earliest principle to fulfill that requirement was the conservation of live force. It had

[30] "Détails historiques sur l'emploi du principe des forces vives dans la théorie des machines, et sur diverses roues hydrauliques," *Annales de chimie et de physique* 9 (1818), pp. 146-159.

been enunciated by Huygens, although it amounted merely to an extension to a system of bodies of Galileo's proposition that a heavy body sliding freely down any curve always acquires the same velocity in descending from any given height.

The earliest paper that Navier cited as an attempt to evaluate the effect of machines in actual engineering was a contribution on water-wheels fitted with blades that Parent had published in 1704.[31] Parent had got his results, however, from the simple observation that effort expended on the blades varied with the relative speed of the wheel compared to that of the current. There ought to be some definite proportion between these velocities that would yield maximum power, and the problem was to determine it. Parent assumed quite arbitrarily that the effort or force exerted on the wheel was proportional to the square of the relative velocity, and found that the speed of the wheel ought to be one-third that of the current. As it chances, the assumption is not far from correct, and the result a good approximation. Nevertheless, Parent's absolute value for a maximum effect was quite inaccurate, nor could such a value be determined precisely except by calculating the respective quantities of live force that the water communicated to the wheel and that it conserved in the runoff.

According to Navier, the first work actually to apply the principle of live force to the theory of machines was Daniel Bernoulli's *Hydrodynamica* of 1738. The association between the subject matter of hydrodynamics and the principle of conservation of live force began there. It took the form of the calculation that, abstracting from friction, the weight of water raised multiplied by the sum of the height to which it was lifted plus the height due to its final velocity was equal to the power employed to run the machine. Bernoulli's point of view was entirely neglected, however, in the famous writings on practical mechanics (or engineering) that appeared contemporaneously—Desaguliers' *Physique*[32] and Bélidor's *Architecture hydraulique*. Scientists themselves ignored it: Euler made no use of it in memoirs on the reaction wheel, the centrifugal force wheel, or the Archimedean screw.

Not until a brief memoir by Borda some thirty years later did anyone carry on the application of the principle of live force to machines.[33] Adopting Bernoulli's approach, Borda improved on it in one respect. Bernoulli had given as the expression for the live force lost in percussion the quantity, $m \ (v^2 - v'^2)$ whereas Borda recognized that in accordance with the laws of collision it ought to be $m \ (v - v')^2$.[34]

---

[31] Antoine Parent, "Sur la plus grande perfection possible des machines," *Histoire et mémoires de l'Académie royale des sciences . . . année 1704* (1706), 323-338.

[32] J. T. Desaguliers, *Cours de physique expérimentale*, 2 vols. (Paris, 1751), trans. from *A course of experimental philosophy* (London, 1734).

[33] J.-C. de Borda, "Mémoire sur les roues hydrauliques," *Mémoires de l'Académie royale des sciences . . . 1767* (1770), 270-287.

[34] Borda followed this memoir with a second in the following year, "Mémoire sur les pompes," *Mémoires de l'Académie . . . 1768* (1770), 418-431.

Some years later, in 1781, Coulomb published a memoir on windmills employing the same method for evaluating the output of machines and calculating the effect of percussions.[35]

According to Navier's résumé, at the period of Daniel Bernoulli's *Hydrodynamica* the principle of live force was not considered to have been rigorously demonstrated, but was so considered by the time that Borda and Coulomb published their memoirs. What or who brought about that change he does not say, observing merely that Borda and Coulomb gave only particular applications in their memoirs and that no general theory then existed that could integrate the evaluation of machine processes with the principles of the science of mechanics.

In Navier's view it was Lazare Carnot's *Essai sur les machines* of 1783 followed by the *Principes fondamentaux* of 1803 that made the junction. It will be germane to set down exactly what it was that Navier attributed to Carnot. (He made no distinction between the contributions of the two books, but saw them as successive versions of a single work.) First of all, Carnot demonstrated in a general manner the theorem on the evaluation of the loss of live force in the collision of inelastic bodies [in Navier that term had replaced hard body], a theorem that Borda and Coulomb had noticed only in the particular case that concerned them. Secondly, Carnot laid down in consequence the principles that must be observed in the design of hydraulic machines in order to achieve maximum efficiency—i.e., that the water should communicate all its motion to the machine, or at least retain no more than the minimum required to flow off; that there should be no turbulence or splashing in the fluid or percussion between the parts of the machine; and finally that the "impelling forces" should create no motion in the system that did not serve the purpose, the example being that of pumping water to a raised reservoir where it should arrive ideally at zero velocity.

No witness could now be better placed than Navier then was to judge of Carnot's work in mechanics. The items just cited were the specific results he singled out. Even more important in his view, however, was Carnot's way of looking at machines, and it is extremely interesting that Navier goes on to say that Lagrange himself adopted it in the *Théorie des fonctions analytiques*. There Lagrange explained in very precise and elegant fashion what may seem naïve to the modern reader when he encounters it in Carnot's *Essai sur les machines*—that the action of almost any force could be assimilated to that of gravity or of springs.[36] The evaluation of the live force produced by the former was easy—the product of a weight and a height. As to forces that act in

[35] Coulomb, "Observations théoriques et expérimentales sur l'effet des moulins à vent," *Mémoires de l'Académie . . . 1781* (1784), 65-81.

[36] Lagrange, *Théorie de fonctions analytiques, Oeuvres de Lagrange*, ed. J. A. Serret, vol. IX (Paris, 1881), pp. 409-410. For the appreciation of Carnot's theorem that Lagrange included in the second edition, see p. 407.

the manner of springs, and here what was mainly in view were those produced by the alternate expansion and condensation of vapors, the rules were those established by Petit in the article just discussed.

For himself, Navier informed the reader, he would in a few months be submitting to the Academy the notes and additions he had prepared for the republication of Bélidor's *Architecture hydraulique*. Describing his purpose, he employed virtually the same phrase as that which he used in characterizing the importance of Carnot's *Essai sur les machines*: "I have tried to assimilate these considerations in the most elementary way to the first principles of mechanics."[37] It is significant that he had made no such comprehensive attempt in the notes and corrections he supplied for the republication in 1813 of Bélidor's *Science des ingénieurs*. Actually, such an enterprise would have been just as appropriate to that work. It must, therefore, have been developments in the interim that led him to undertake the ambitious and comprehensive commentary amounting to an analytical theory of machines accompanying the later edition.

Since his historical remarks were concerned mainly with reminding readers of what he said mechanics had largely forgotten, the work of Carnot on the theory of machines, it seems reasonable to suppose that the renewed appreciation in his own mind was what led Navier to edit *Architecture hydraulique* in the light of the theory of machines. In his notes to that work, he demonstrated the principle of conservation of live force for a single mass point and extended the result to a system of points by means of a conventional argument from d'Alembert's Principle. The discussion was in terms of virtual velocities. From the conservation of live force, he then derived Carnot's theorem, which he now described as completing the principle of conservation of live force in the case of discontinuous change of motion, and of which he gave the statement

> that the sum of the live forces in the system after a sudden change is less than what it was beforehand, and that *the system has lost a quantity of live force equal to what it would have been if all the bodies were moving with the velocities that they lost in this change of motion.*[38]

Earlier on Navier credited another device to Carnot, one which except for this remark the modern reader might not think to notice as distinctively his, the trigonometric practice of evaluating the projection of any directed quantity on another as the product of its multiplication by the cosine of the angle between them,[39] which may be taken to have been a small step in the direction of vector analysis. It is a possibility

[37] Navier, *op. cit.*, p. 152.
[38] Navier's notes in Bélidor, *Architecture*

hydraulique (1819 edition), pp. 111-113.
[39] *Ibid.*, pp. 17-18.

that gains plausibility from the probable importance of Carnot's geometry in the background of that subject.[40]

Since our purpose is not to recount the history of mechanics, but only to identify the points of entry of Lazare Carnot's engineering science into mechanics, the remaining episodes may be briefly summarized. They culminated in 1829 in the composition of a book generally recognized to have been the foundation of nineteenth-century work in engineering mechanics, Coriolis' *Du calcul de l'effet des machines*. In that work Coriolis undertook to gather all the general considerations that might illuminate problems of economizing force or mechanical power. To the historian of science the most interesting passage is that in which Coriolis gave the quantity "work" the name by which it has since been known in physics and distinguished it from live force. His discussion is worth paraphrasing for it exhibits how usage had come into existence and changed.

By the word "work" (travail) Coriolis meant the quantity called variously mechanical power (puissance mécanique), quantity-of-action, or dynamical effect, all of them variations on Carnot's original moment-of-activity. (For its unit Coriolis further suggested the designation "dynamode," which, however, did not take.) A further innovation he described as "slight" although it has proved equally significant. He would call by the name live force the "product of a weight by the height due to its velocity." Under that definition, live force would be only half of the product that had previously been meant by that term, the mass by the square of the velocity, and both Coriolis' reason for making the change and his discussion of how the confusion had arisen are revealing.

Charged with a course at *Polytechnique*, Coriolis proposed the change out of his experience of the burden that a badly defined dimensionality imposed on students. In their problems in mechanical engineering they were forever coming upon the quantity $\frac{1}{2} Mv^2$ in its equivalence to MgH, and they inevitably found it confusing that the name live force should refer to double that most commonly recurrent of quantities. That it had come to do so was owing to the difference between the traditions of rational mechanics and mechanical practice. In times past, rational mechanics never envisaged work and did not mean the product of weight by velocity in employing the term live force. More recently, however, all practical engineers had come to mean by the term live force precisely the work that could be produced by the velocity a body had acquired. Coriolis admitted that rational mechanics could equally well have got on with its accustomed usage if mathematicians preferred, and insisted only that his redesignation would simplify discussion of the theory of machines. Neverthe-

[40] See below, Chapter 5.

{ 115 }

less, the outcome makes evident what it was that by then had become the dominant set of concerns, for surely it was there in the redesignation and not thirty-odd years later in its renaming as kinetic energy that live force became differentiated from work.

As for the term "work," Coriolis observed that it came so naturally to mind that it had already been employed for the same purpose in passing by Navier in his notes on Bélidor and also by Prony in a memoir on a new machine. He might have added Lazare Carnot himself.[41] A little while previously the question of designating units for the study of power and machines had been raised in the Academy. A commission had been appointed. Laplace had been a member, and his opinion was that the Academy would be unwise to take the initiative in proposing names. He thought—was it with distant memories of the commissions on the metric system?—that it could properly do no more than sanction usage once it was established, and Coriolis invoked this weighty support.

The treatise itself was impressive testimony to the development its subject had undergone. It was a comprehensive manual of engineering mechanics and treated the problems arising in the design and employment of power machinery in theoretical terms. Coriolis wrote at a level that presupposed in the reader familiarity with the elements of mechanics and of the differential and integral calculus. He invited readers who did not have these qualifications to persevere, and he had taken pains to insure that the verbal passages alone should suffice to convey the principal notions of the theory of machines. It is difficult to imagine readers who could in fact have profited from that invitation, however, although it is true that the mathematical treatment served to express the propositions and not to generate theorems or even insights. The work was not intrinsically mathematical like eighteenth-century rational mechanics or the generalized mechanics of the later nineteenth century.

Nevertheless, comparison of Coriolis to Carnot's *Principes fondamentaux*, not to mention the *Essai sur les machines*, exhibits an enormous development in comprehensiveness, sophistication, and idiom; its canon had become established in the intervening twenty-five years; and since, except for Navier's notes to Bélidor, there were no treatises that had contributed to that development, it is obvious that it must have come about mainly in an educational milieu, and in the form of courses of instruction in engineering schools: Coriolis taught at *Polytechnique*, Navier at *Ponts et chaussées*, and Poncelet at the *Ecole d'artillerie et du génie* in Metz; and multigraphed lectures formed the literature of the subject.

Inevitably when a number of alert people were working along similar lines under such pressures, questions of priority arose. They were handled with restraint in this instance, and interest us less for themselves than for the evidence the record affords

[41] See above, p. 58, and App. B, § 27, note c, below.

that Navier, Coriolis, and the other protagonists now recognized the pioneering role of Carnot. Coriolis testified in the preface that when he first became interested, he knew only the works of Carnot and Guenyveau, and conceived what were then several original ideas. Simultaneously, however, Petit published his note on live forces followed hard after by Navier's edition of Bélidor, which between them contained everything he had thought to be novel in his own studies, the only differences being in mode of treatment. He already had in draft certain parts of what he was publishing in 1829 and had communicated them in 1819 to Mallet, Bélanger, and Drappier, all graduate engineers, and also to several students. In 1820 he was in communication with Ampère on the same subjects (as, indeed, Petit had already been), and in 1824 with Poncelet.[42]

The latter exchange created the conditions for a priority dispute that was skirted if not quite averted by the good sense and self-restraint of all concerned. In 1825 Poncelet began offering a course in applied mechanics at the *Ecole d'artillerie et du génie* in Metz, the text of which he circulated in multigraphed form in 1826. Besides supplying it to his students, he sent copies to a few interested persons in Paris, and in 1827 an account of its contents was given to the Academy of Sciences by Arago and Dupin.[43] The topics and treatment were in significant instances similar to the contents of Coriolis' treatise.

Coriolis learned of Poncelet's chagrin, and on 1 October 1829 wrote a note to the editor of the *Bulletin des sciences mathématiques* apologizing that in consequence of a misunderstanding between Poncelet and himself over dates of earlier correspondence, certain passages in the original version of the preface to his treatise might be construed as implying that Poncelet had taken advantage of private communications, and advising that persons who already had bought his treatise should substitute for the first two pages of the preface as printed a revised version that he was making available at the publisher's.[44] In Poncelet's view, the shoe was on the other foot. Not that he accused Coriolis directly of having appropriated his, Poncelet's, findings, but in a lengthy note of 8 October he did take issue with the report on Coriolis' treatise composed by Navier, who was secretary of the commission to which the Academy referred it.[45] Navier gave a full and enthusiastic summary. Poncelet neither disputed the merit

[42] A note in Petit's paper (*loc. cit.*, 297-298 n.) remarks that Ampère had been studying the application of the principle of live force to the theory of machines and had told Petit that he, Ampère, had already worked out certain of the findings of Petit's paper.

[43] Arago and Dupin, "Rapport sur un mémoire de M Poncelet intitulé 'Cours de mécanique appliquée aux machines,' " *Procès-verbaux de l'Académie des sciences de l'Insti-*tut de France, 8, 7 May 1827, 527-531.

[44] *Loc. cit.*, 12 (1829), pp. 322-323.

[45] Navier, "Rapport sur le mémoire de M. Coriolis, intitulé 'Du calcul de l'effet des machines,' " *Procès-verbaux de l'Académie*, 9 22 June 1829, 266-272. The commission consisted of Navier, Prony, and Girard. Their report was also published in *Bulletin des sciences mathématiques*, 12 (1829), 103-116.

of the work itself nor enlarged on all the similarities it exhibited with his own. He took exception mainly to Navier's ascription of originality to Coriolis on two specific topics, both of which had already been worked out in his own course, and one of which had then formed the subject of a further note by Cauchy.[46]

The point concerned a general expression for what Poncelet called the "virtual moment of friction in gears" and Coriolis called the work consumed by friction in gears.[47] Poncelet's discussion is interesting in that it constituted a criticism both of d'Alembert's principle and of Carnot's theorem on the grounds that neither took account of the tangential resistances in a system. He described d'Alembert and Carnot, together with Poisson, Navier, and Petit, as the best authors on the subject of impact. He had criticized the theorems, not from the point of view of the inner structure of bodies with its distinction between elastic and hard or inelastic, but from that of the interactions of the moving parts of machinery, from which he then deduced the interactions of the molecules of bodies.

Poncelet's own analysis had shown how to evaluate in general the alterations of velocity and losses of live force caused by impact among the bodies constituting a machine; these occurred in the first instance in consequence of the direct and reciprocal reactions of the bodies themselves, and in the second in consequence of the passive resistance deriving from reaction wherever friction arises between moving parts or bearings. For effects of the first class, Poncelet set up differential equations of motion for each part of a machine, and his criticism consisted in showing that, neglecting tangential resistances and frictions, these equations expressing the normal forces of compression and their reactions reduced to a single linear equation from which, given simple assumptions, first the principle of d'Alembert, and then the theorem of Carnot fell out. The latter he held to be of no practical use, since to apply it presupposed the prior evaluation of the loss or gain of velocity in each part of the system according to d'Alembert's principle. In other words, given Poncelet's own more operational and more general point of view, the two principles were tautological—nor did he appear to see them as historical stages in the approach to his own point of view.

The second criticism was rather of their adequacy. In the collisions wherein external resistances were in some degree a function of the forces of compression within the system, a further term had to be introduced into the expressions. If, as often happens, that function was not rational and the resulting equations were non-linear, then the results were different from those obtained in applying d'Alembert's principle on the equilibrium between the total variation of the quantities of motion taken to be simple forces of pressure or percussion. In practice, to be sure, d'Alembert's principle

[46] Poncelet, "Note sur quelques principes de mécanique relatifs à la science des machines," *Bulletin des sciences mathématiques*, 12 (1829), 323-325.

[47] *Ibid.*, pp. 324-325.

was usually a good approximation, and the remainder of Poncelet's note consisted of an analytic method for handling the radicals that then appeared in the equations to be integrated, and a criticism of Coriolis' looseness in failing to provide for the effects of friction and tangential or passive resistances.

Cauchy for his part approached the same difficulty from the point of view of theory.[48] He attacked Carnot directly:

> In the various treatises of mechanics it is taught that live forces are lost every time bodies undergo a sudden change in velocity, and that this loss of live force is the sum of the live forces due to the velocities that are lost. But this proposition, which has been named Carnot's theorem, is evidently inexact as is the demonstration on which it purportedly rests.[49]

Essentially, Cauchy's criticism came down to a mathematician's refutation of the physicist's or engineer's distinctions between sudden and gradual change of motion and between elastic and hard body. All that sudden change of relative velocity really meant was that it occurred in a time too short to measure; similarly, the only difference between elastic and inelastic bodies was that in the former the interactions between the molecules depended only on the distances by which they were separated, whereas in the latter the interactions depended both on time and distance. Actually, one might equally well describe as a sudden or instantaneous change of motion any change that occurred in a very short time. Were that to be allowed, however, there would be instantaneous change of velocity in elastic collision—to which Carnot's theorem did not apply. In order to evade that difficulty, writers on mechanics had been forced to say that velocity varied continuously in elastic and discontinuously in inelastic collision. But this was nonsense, and so consequently was the statement that the loss of live force was the sum of the live forces due to losses of velocity in sudden changes. Pursuing the argument towards the goal of a properly general statement, Cauchy arrived at what he considered a preferable proposition:

> When in a system of material points, the velocities change suddenly in consequence of molecular actions produced by impacts between certain parts of the system, the sum of the virtual moments of the quantities of motion gained or lost during impact is zero whenever a virtual motion is being considered in which the velocities of the molecules that act on each other are equal among themselves.[50]

—which principle Cauchy then went on to formulate in analytic terms. But although Cauchy regarded this as refuting, and Poncelet referred to it as superseding, Carnot's

[48] A. Cauchy, "Mémoire sur un nouveau principe de mécanique," *Bulletin des sciences mathématiques*, 12 (1829), 116-122. This memoir antedates the Coriolis-Navier-Poncelet interchange. Cauchy read it to the Academy on 21 July 1828 (*Procès-verbaux*, 9, p. 94).

[49] *Ibid.*, p. 116.

[50] *Ibid.*, p. 119.

theorem, the historian of Carnot's scientific work might prefer to point out that both of them in imagining the interactions of the molecules in a system employed the concept—and Poncelet even used the phrase—of geometric motions—in Poncelet's words "a geometric or virtual common motion such that they cease from then on to react on each other and their system assumes a form that is stable for each body, one that at that moment allows the ordinary conditions of liaison."[51]

Cauchy observed of his principle that if the impact terminated at the exact instant when any material point that had acted upon another coincided with that other, then the principle yielded all the equations needed to determine the motion of the molecules of the body at that moment, and that one of those equations, specifically the one obtained in supposing the virtual velocities to coincide with the actual velocities after impact, would contain the consequence that the loss of live force was the sum of the live forces due to the velocities lost.[52] But this consequence, Poncelet was pointing out, justified the principles of d'Alembert and Carnot for what was improperly called the impact of hard bodies, since what was envisaged there was that neighboring molecules should either be joined or else retake invariable distances and that the virtual velocities should be equal and parallel. This came to the same thing as supposing that at the moment of impact the "different molecules of the bodies acquired a geometric or virtual common motion such that they cease from then on to react on each other." In a word, it supposed that after impact, the bodies did not separate or else that they travelled together without any velocity relative to each other along the normals common to their points of contact—and this assimilated Cauchy's analysis to the same terms that he had written of in his course of 1826.

It would appear probable, therefore, that neither Cauchy nor Poncelet had actually read Carnot, for these statements, far from refuting him, in effect repeated in terms of virtual displacements the analysis he had started in terms of geometric motions.

[51] Poncelet, *op. cit.*, p. 323.
[52] Cauchy, *op. cit.*, p. 119.

CHAPTER V

An Engineering Justification of Algebra and the Calculus

THE geometer Michel Chasles, one of the nineteenth-century personages who was both a master and a masterly historian of his favorite subject, discussed the contributions of the "illustrious Carnot" after the descriptive geometry of Monge and before the projective geometry of Poncelet in the sequence to which he attributed revival of the methods of Desargues and Pascal and a consequent renaissance of geometry following its eclipse by eighteenth-century analysis.[1] It was a central feature of the careers of those Chasles singled out that each of them had his place in the tradition of engineering mechanics and occupied himself in the science of machines concurrently with the new geometry.

In theme and discourse Carnot's mathematical writings are as characteristic as his mechanics of the scientific personality of one interested in things for the purpose of operating with them correctly and most effectively. Carnot came at the mathematical problems that concerned him from even farther behind the front lines than he did in mechanics. The questions he posed had more to do with the foundations and procedures than the content of mathematics, and were largely about the meaning of infinitesimal and negative quantities and their significance in the distinction between analysis and synthesis. He put them with the bold simplicity of an engineer for whom any mathematical expression had merely auxiliary standing unless it represented an operation that could be carried out metrically or physically. In the *Géométrie de position* of 1803 he distinguished the notion of a true quantity from that of a "value," which pertained merely to algebraic functions: "Every quantity," he asserted, "is a real object such that the mind can be seized of it, or at least its representation in calculation."[2]

It is evident from the scientific manuscripts surviving at Nolay that the problems occupying Carnot's mind in the years following his return from exile in 1800 were mainly those of geometry and technology. The family archives there contain notes on the inventions referred to his judgment by the Institute, including the Niepce and the Cagniard engines, together with the manuscripts of reports submitted for the *Procès-Verbaux*. They also contain numerous drafts on what actually motivated most of his writings in geometry, the anomaly of negative quantity, together with fragmentary correspondence with his German translators that gives the historian welcome insight both into his purposes and the state of his mathematical reputation. As

---

[1] Michel Chasles, *Aperçu historique sur l'origine et le développement des méthodes en géométrie*, 2nd ed. (Paris, 1875), p. 210.

[2] *Géométrie de position*, p. 7.

remarked in the preface, it was through his exchanges with J.-K.-F. Hauff, who translated the first, 1797, edition of *Réflexions sur la métaphysique du calcul infinitésimal* into German in 1800 that we learned of the origin of that work in the prize essay submitted to the Berlin Academy in 1785. In a reminiscence Hauff acknowledged that when he had initially seen the title listed by the bookseller, he had paid small attention. The author, though widely known to be one of the most distinguished of practitioners—"eines der vorzuglichsten Pratiker"—had no reputation in theoretical mathematics. Hauff would, indeed, have expected little from a French metaphysicist on the subject of geometry, as would the Hofräth Kästner, to whom he happened to mention it. Of this initial prejudice he had been disabused by Lacroix, who informed him that, on the contrary, Carnot's was a mathematical intelligence to be taken seriously. His misapprehension thus corrected by the most widely read of contemporary mathematicians, Hauff set to work to master the argument of the *Réflexions sur la métaphysique du calcul infinitésimal* and, having done so, to translate it, since the work was one that provided all the foundation the subject might require.[3]

There is more distinguished testimony to his contemporary reputation in geometry. In 1810 Professor H. G. Schumacher of the University of Copenhagen published a German translation of the *Géométrie de position*. He had consulted Gauss on the importance of the work and in the preface mentions with gratitude Gauss's good opinion of it—a good opinion not (it may be remarked) easily given.[4] Among admirers qualified to judge later in the century, Chasles was not alone in the appreciation mentioned at the beginning of this chapter. In 1854 Giusto Bellavitis referred to the same work as the origin of his own method of equipollencies, and Felix Klein in the well-known lectures published in 1926 referred to it as a "noteworthy book."[5] It is odd, therefore, that Carnot's mathematical works should have fallen into such obscurity since the earlier twentieth century that few mathematicians can say what they contained.

His writings fall into two groups, one consisting in the geometric works and the other in the successive versions of *Réflexions sur la métaphysique du calcul infinitésimal*. It will be convenient to consider the geometric writings first since their

[3] Hauff's translation appeared under the title, *Betrachtungen über die Theorie des Infinitesimalrechnung von dem Burger Carnot* (Frankfort-am-Main, 1800). The relevant papers are in carton 28 in the Carnot family archives at Nolay.

[4] *Géométrie der Stellung oder über die Anwendung der Analysis auf Géométrie*, 2 vols. (Altona, 1810), II, Vorrede.

[5] Ettore Carruccio, "Bellavitis," *Dictionary of Scientific Biography* (New York, 1970), I, 590-592. Klein, *Vorlesungen über die Entwicklung der Mathematik im 19. Jahrhundert* (Berlin, 1926), I, 79. Carnot introduced the concept and word "equipollence" in § 82, pp. 83-84, to signify the equivalence, not just of values, but of any mathematical objects such as points or curves that could be substituted for one another.

relation to his work in mechanics is more explicit, though not necessarily deeper, and then to go on to an account of his justification of the calculus. That was the subject of his earliest work in mathematics, the Berlin dissertation of 1785, and also of his last: he published the final revision in 1813.

A. GEOMETRIC ANALYSIS AND THE PROBLEM OF NEGATIVE QUANTITY

Carnot published *De la corrélation des figures de géométrie* in 1801, a year after leaving ministerial office under Napoleon. It contains the initial working out of the views with which he reentered scientific work after nearly a decade in public life. He opened the subject with the observation that between the parts of a geometric figure two sorts of relations obtain, relations of magnitude and of position. His topic was limited to the latter. Its problems were to determine whether a point was situated above or below a given line, to the right or left of a given plane, inside or outside a given circumference or curved surface, etc. The method would be to compare every figure he was investigating to another of known properties, to exhibit the modifications point to point by means of a systematic notation, and thus to "establish the correlation of the figures."[6]

The approach will be familiar to the student of Carnot's mechanics. The reader was to imagine any system whatever of variable quantities, whether or not geometric, and to consider it in two different states. The one taken for the basis of the comparison he called the primitive and the other the transformed system. In the analysis of the latter the difference between any two quantities was "direct" when the greater and the lesser corresponded to their analogues in the primitive system. In the contrary case, the difference was "inverse." (In these comparisons, Carnot always meant real quantities and not fictitious values of merely algebraic significance.)

As was his wont, Carnot began developing his point of view with an elementary

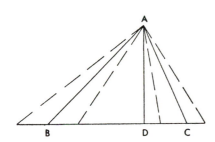 example. In the triangle ABC, a perpendicular $\overline{AD}$ is lowered from the apex to the base $\overline{BC}$. In the initial construction, the point D falls between B and C. This is the primitive system. It may be transformed by moving the point C to the left. In effect the figure constitutes a system in which the segment $\overline{CD}$ is a variable and $\overline{BD}$ is a constant. So long as the point C approaches but does not reach the point D, it remains true of both primitive and transformed systems that $\overline{BC} > \overline{BD}$. Their difference $\overline{CD}$ is a direct quantity, and $\overline{BC}$ and $\overline{BD}$ are in direct order in the two systems. By definition, therefore, the correlation is direct. Let the point C pass to the left of D, however, and then in the transformed system

[6] *Op. cit.*, § 2, p. 1.

$\overline{BD} > \overline{BC}$. Their difference $\overline{CD}$ is inverse, and the quantities $\overline{BC}$ and $\overline{BD}$ are in inverse order compared to the primitive system. The correlation, therefore, becomes indirect.

By correlative systems Carnot meant those that could be considered as different states of a single variable system undergoing transformation by insensible degrees. But it is important to notice that the question was not simply one of stretching or rearranging the parts. It was not required that all correlative systems should actually have been evolved out of the primitive system. It sufficed that they might be assimilated to the primitive system through mutations that could be imagined to occur by insensible degrees. We are back with Carnot's favorite reasoning device: a comparison of systems between which the nexus of change is a continuum. Such were correlative systems, and the quantities that corresponded were correlative quantities.

When the correlation was direct, any train of reasoning that was valid for the primitive system would hold for the correlative. For example, in the above system ABCD, we know that in the right triangle ABD,

$$\overline{AB}^2 = \overline{BD}^2 + \overline{AD}^2$$

and that in the right triangle ACD,

$$\overline{AC}^2 = \overline{CD}^2 + \overline{AD}^2.$$

Subtracting,

$$\overline{AB}^2 - \overline{AC}^2 = \overline{BD}^2 - \overline{CD}^2.$$

Furthermore,

$$\overline{BD} = \overline{BC} - \overline{CD}.$$

Substituting,

$$\overline{AB}^2 - \overline{AC}^2 = \overline{BC}^2 - 2\,\overline{BC}\cdot\overline{CD}.$$

This same reasoning would be true of all the transformed systems as C moved towards D so long as CD was a direct quantity.

Once C had passed the point D, however, CD would become an inverse quantity and the correlation indirect. The final equation would then be

$$\overline{AB}^2 - \overline{AC}^2 = \overline{BC}^2 + 2\,\overline{BC}\cdot\overline{CD},$$

in which the sign of the last term was inverted in the correlative system. It was—to generalize the example—the characteristic of indirectly correlative systems that the formulas of the primitive system might be applied to the transformed system by virtue of changing the signs of all the variables that had become inverted.

Reciprocally, these procedures of correlation might be used for solving problems, and that is why they held the appeal for Carnot that led him to develop their application to analysis in general. Reasoning on the same example, suppose it was given that the three sides were in the proportion,

$$\overline{AC} = \overline{BC} = 2/3\,\overline{AB},$$

and the problem was to find the segment $\overline{CD}$. It is unknown whether C falls to the right or to the left of D. Trying the former hypothesis first, would give

$$\overline{AB}^2 - \overline{AC}^2 = \overline{BD}^2 - \overline{CD}^2,$$

which works out

$$\overline{CD} = -\,1/12\;\overline{AB},$$

a negative value, and hence anomalous as a solution. What the minus sign actually signified was that the initial formulation had been wrong, and that in fact C fell to the left of D. When the right assumption is made, the problem yields a positive value in the result, and if the calculation has been valid, this confirms the correctness of the choice among the possible conditions of the problem.[7]

So much by way of sample from Carnot's *Corrélation des figures* will show how the anomaly of negative quantity led him into the problem-solving reaches of geometry, the branch of mathematics to which he was in any case best suited by temperament and training. Apparently Carnot's contemporaries did not find it easy to form a clear idea of precisely what his theory of negative quantity was amid all the geometry with which he argued it, for to the last of his original publications, the essay of 1806 developing the theory of transversals, he appended a brief "Digression sur la nature des quantités dites négatives."[8] He there explained that he had written it at the behest of several "savans du premier ordre"—one may guess that Lacroix may well have been among them—who had urged him to disengage his theory of negative quantity from its extensive detail "eliminating from the discussion everything that was not strictly elementary."[9] In 1810 Schumacher attached a German version to the first volume of his translation of the *Géométrie de position*, and quoted a letter from Carnot to the effect that many duties and weakening health no longer permitted him to return to these subjects, and that in any case this epitome was the best he could do.[10] We may conveniently follow Schumacher's example and guide ourselves on this abstract of the argument before passing on to an account of other aspects of the larger of the works that it informed, the *Géométrie de position*.

Doing algebra involves manipulating negative values at every turn. What might they signify? Clearly the signs $+$ and $-$ merely indicate operations to perform, and addition presents no difficulty since $+\,a$ can always be regarded as $0 + a$. As for negative quantity, however, two schools had developed in the course of the many discussions the problem had evoked in the eighteenth century. One interpretation accepted the notion of quantities less than zero. The other considered that a minus sign meant that a quantity was to be evaluated in the direction opposite to that of a positive

[7] *Ibid.*, § 33, pp. 18-19.

[8] *Mémoire sur la relation qui existe entre les distances respectives de cinq points quelconques pris dans l'espace, suivi d'un essai sur la théorie des transversales* (Paris, 1806).

[9] *Ibid.*, p. 96.

[10] *Geometrie der Stellung*, I, Vorrede.

quantity. Carnot found both explanations unconvincing. As for the former, he asserted roundly that the notion of something being less than nothing was absurd. It was here that he introduced his distinction between a quantity properly speaking and an algebraic value, the latter being a merely fictitious entity introduced for purposes of calculating. If it was permissible in a calculation to neglect quantities of no magnitude (as it was), then it ought surely to be justifiable to ignore those less than zero if such there were. But everyone knows that negative terms cannot be neglected. Thus, whatever they signify, it cannot be quantity.[11]

Advocates of the second view, on the other hand, never defined what they meant by a direction opposite to the positive. Sometimes they adduced the amount of a debt opposed to that of a credit. Sometimes they referred to a left ordinate opposed to a right one. But these were examples, not definitions, and they left difficulties. Why, for example, is it impossible to take the square root of a negative quantity? There is no difficulty in extracting the square root of a debt or a left ordinate. Why, to consider a related point, did negative quantities dominate in the multiplication of unlikes, giving their sign to the product? Moreover, there were exceptions. The secant to an arc of a circle in the third quadrant cannot be distinguished from the secant to the opposite arc in the first quadrant in either magnitude or direction, and it should accordingly be positive in sign. In fact it is negative.

In the place of such interpretations, none of them satisfactory, Carnot proposed two senses in which negatives could be properly understood. In the more obvious, a negative quantity was the magnitude of a value governed by a minus sign. Strictly speaking this usage was correct only when the value was preceded by a positive one of greater magnitude, and was, therefore, of relatively trivial importance. The deeper and more revealing sense was that a negative quantity is a magnitude governed by a sign that was wrong. What Carnot meant by wrong may be illustrated in the problem just cited in which a negative result indicates that the initial assumption about the correlation of the figures was mistaken and that in the conditions of the problem BC was smaller than BD, whereas it had been initially taken to be larger.[12]

In the "Digression" Carnot distinguished between the two senses of negative quantity by means of a trigonometric example. Consider the formula

$$\cos (a + b) = \cos a \cdot \cos b - \sin a \cdot \sin b \ldots \tag{A}$$

Here the last term was negative in the first or literal sense. In the second sense, however, the same term was positive so long as a, b, and $(a + b)$ were each less than 90° because the sign had to be correct if the equation were to be exact in the first quadrant. But if $(a + b)$ were greater than 90°, the equation would be incorrect. For if

---

[11] "Digression," *loc. cit.*, n. 8, p. 98.   [12] Above, p. 123.

the problem were investigated directly and synthetically, it would turn out that

$$\cos\ (a + b) = \sin a \cdot \sin b - \cos a \cdot \cos b \ldots \qquad\text{(B)}$$

If Equation (A) were to apply to this case, it would be necessary to regard the first term, cos $(a + b)$ as governed by a sign contrary to what it ought to have been, for changing it would have given

$$-\cos\ (a + b) = \cos a \cdot \cos b - \sin a \cdot \sin b \ldots$$

which would have reduced to Equation (B). In effect, therefore, extending Equation (A) to the case in which $(a + b)$ was greater than 90° subjected the term, cos $(a + b)$, to the wrong sign and made of it a negative quantity in the second sense. Obviously these two meanings of negative were quite different, and for the second sort of quantity, those bearing a wrong sign, Carnot preferred the term "inverse quantity" in contradistinction to "direct quantities" governed by their right and proper sign. The usage is that of the correlation of geometric figures, which would frequently be the means of establishing whether a quantity was plus or minus, direct or inverted.

How, then, did these inversions get into the process of calculation? Carnot said by error, the kind of mistake or misapprehension about the basic conditions of the problem illustrated in the example of correlation cited above. If in setting up a calculation we mistake a credit for a debt, then the algebraic expression of what is to be paid or received is opposite to what it should be, and we recognize the error practically through knowledge of whether money is owed or owing. Similarly, if we mistake a right for a left ordinate in formulating a problem, we get the wrong sign in the result and know it to be so since the solution makes no sense physically. In general, "Any inverse quantity can be considered as the difference of two direct quantities of which the greater has been taken for the smaller, and reciprocally."[13]

A skeptic might be forgiven for asking the use of introducing into calculation quantities governed by a false sign. The answer should be that it is often necessary to risk doing so in order to formulate a problem at all. The ordinate of a curve might be required without the geometer's knowing in which quadrant it was located. He would then simply make an assumption. That would permit him to give an absolute value. If then he had been wrong about the sign, the error would show up as an absurdity in the result, and he could change signs without redoing the calculation. The procedure turns out to involve no real risk of final inaccuracy, therefore. Since only a positive value ever represents an intelligible solution to a genuine problem, a negative solution merely reveals an inconsistency between the actual initial conditions and the hypothesis that had been adopted in framing the equations.

[13] "Digression," *loc. cit.*, n. 8, p. 102.

Trigonometry was to Carnot in mathematics what hydrodynamics was in mechanics—the part of the science in which his mind ran on the problems exemplary of the general issues.[14] Discussing the transition from an inverse to a direct quantity and vice versa, he invoked the algebraic principle that a variable in changing sign passes through zero or infinity. The point of view was one that naturally embraced, and in his own mind probably grew out of, the change of sign in trigonometric functions. Carnot thought that phenomenon to be the decisive one in destroying the position of those who held negative quantity to signify nothing more than ordinary magnitudes taken in reverse direction. Why, for example, should the cosine of an obtuse angle necessarily be an inverse quantity? It was so only because the primitive system, i.e., that on which the reasoning in trigonometry had originally been founded, was one in which all angles were acute. The instrument best adapted to analyzing relations of this sort was the geometry of position, and that was why it was the approach to geometry most suited to clarifying its application to analysis and algebra in general.

The *Géométrie de position* was published in 1803, the same year as *Principes fondamentaux de l'équilibre et du mouvement*. It is an imposing book, a book more important for its approach to geometry than for the theoretical point it purported to be establishing. It cannot be said that Carnot's theory of negative quantity entered into the texture of later mathematical thinking, but writing the *Géométrie de position* carried him far beyond his original intention of giving a second, somewhat fuller edition of *Corrélation des figures*. He enlarged the examples and treated the rules of correlation in an altogether more comprehensive manner. Two years seem a short time for the development he gave the subject, and it would appear both from internal evidence and from the number of drafts of particular problems in the family archives that he must have had much of this material long in mind and that he worked very hard to set it out in order. No longer limiting the scope to correlations of particular geometric systems, he now proposed to compare and unify the two main types of geometric relations and thus to associate in a single treatment relations of magnitude with relations of position.[15]

[14] It is interesting that the first published hint of the correlation of figures, and also the first original scientific work to which Carnot seems to have turned after the Revolution, appeared in the form of a "Lettre du cit. Carnot . . . au cit. Bossut, contenant quelques vues nouvelles sur la trigonométrie." It is appended to Bossut's *Cours de mathématiques* II: "*Géométrie et application de l'algèbre à la géométrie*" (Paris, 1800), pp. 401-421. Bossut recounts that just as his book was about to go to press, Carnot, then serving as Minister of War, mentioned to him several ingenious and stimulating ideas on trigonometry. At Bossut's persuasion, he agreed to write them up for publication in a way suited to an elementary text. He had leisure to give only the results, in any case, leaving the proofs to the students as an exercise. The subject is that which he developed fully in section 4 of *Géométrie de position*, the relation between the linear quantities of a triangle eliminating all angular quantities from the expressions.

[15] *Géométrie de position*, p. 1. In a letter to Lacroix of 4 Nivôse an 10 (25 December 1801) Carnot observes that he is spending his

Once past the first section on negative and imaginary quantities, the work contains much substantial geometry and trigonometry, which, like any actual mathematics, does not lend itself to summary as do questions of principle, method, and lineage. The second section proposes a notation and mechanism for exhibiting the modification that figures undergo through changing the relative positions of their significant elements—points, lines, angles, arcs, areas, etc. Consider, for example, the correlative triangles ABCD and A′B′C′D′ (in principle the same figures as in the earlier work):[16]

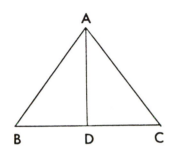 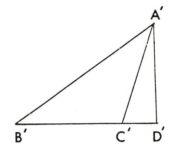

For the points, two categories of correlation are relevant, correlations of construction and of position. Points that correspond under the correlation of construction are

first system   ABCD

second system   A′B′C′D′.

The correlations of position, however, are

first system $\overline{BDC}$

second system $\overline{B′C′D′}$.

(The bars signify that the points are on a straight line.)

As for lines, angles, arcs, and areas in the two systems, further correlations are needed to complete the comparison, specifically correlations of absolute value and of signs. Given the correlation of construction just cited, correlations of absolute value may be written down directly:

$$\text{first system:} \left\{ \begin{array}{l} \overline{AB},\ \overline{AC},\ \overline{BC},\ \overline{BD},\ \overline{CD},\ \overline{AD},\ B\hat{A}C, \\ \overline{A′B′},\ \overline{A′C′},\ \overline{B′C′},\ \overline{B′D′},\ \overline{C′D′},\ \overline{A′D′},\ B′\hat{A}′C′, \end{array} \right.$$

$$\left\{ \begin{array}{l} A\hat{B}C,\ A\hat{C}B,\ B\hat{A}D,\ C\hat{A}D,\ \overline{\overline{ABC}},\ \overline{\overline{ABD}},\ \overline{\overline{ACD}}. \\ A′\hat{B}′C′,\ A′\hat{C}′B′,\ B′\hat{A}′D′,\ C′\hat{A}′D′,\ \overline{\overline{A′B′C′}},\ \overline{\overline{A′B′D′}},\ \overline{\overline{A′C′D′}}. \end{array} \right.$$

(The double bars signify an area or surface, and such expressions as B A C the angle BAC.)

From the correlation of positions, it is evident that there is an inversion in the order

leisure developing the ideas in the *Corréla-tion des figures*, which had suffered from the haste with which it was composed, for a new edition containing a chapter on the applications of algebra to geometry and another on the theory of curves, neither of them a treatise in itself, but applications of his theory of positive and negative quantity and of transversals. *Bibliothèque de l'Institut de France,* MSS. 2397.

[16] This example is worked in great detail on pp. 81-87.

of the points D, C and D′, C′. This inversion establishes the correlation of signs, and the general tabulation of correlations of the two systems is thus:

$$\text{first system:} \quad \left\{ + \ \overline{AB}, \ + \ \overline{AC}, \ + \ \overline{BC}, \ + \ \overline{BD}, \ + \ \overline{CD}, \ + \ \overline{AD}, \ + \ B\hat{A}C, \right.$$
$$\text{second system:} \quad \left\{ + \ \overline{A'B'}, \ + \ \overline{A'C'}, \ + \ \overline{B'C'}, \ + \ \overline{B'D'}, \ - \ \overline{C'D'}, \ + \ \overline{A'D'}, \ + \ B'\hat{A}'C', \right.$$
$$\left\{ + \ A\hat{B}C, \ + \ A\hat{C}B, \ + \ B\hat{A}D, \ + \ C\hat{A}D, \ + \ \overline{\overline{ABC}}, \ + \ \overline{\overline{ABD}}, \ + \ \overline{\overline{ACD}}. \right.$$
$$\left\{ + \ A'\hat{B}'C', \ + \ A'\hat{C}'B', \ + \ B'\hat{A}'D', \ - \ C'\hat{A}'D', \ + \ \overline{\overline{A'B'C'}}, \ + \ \overline{\overline{A'B'D'}}, \ - \ \overline{\overline{A'C'D'}}. \right.$$

From inspection, we can tell that the primitive system ABCD has been transformed into the correlative A′B′C′D′ in consequence of a movement of the point C on $\overline{CB}$ so that it is now situated between B and D. The table also tells us that none of the quantities $\overline{AB}$, $\overline{AC}$, $\overline{BC}$, $\overline{BD}$, or $\overline{AD}$ passed through zero or infinity, and consequently that these are direct quantities in the correlative system and there carry a positive sign. Not so the quantity $\overline{CD}$, however, which did pass through zero in becoming $\overline{C'D'}$, and therefore becomes inverse and bears (as the table shows) a minus sign. So it is, not to insist further on the obvious, for all the other quantities in the two systems—their relations and signs may be found by inspection of the table, which gives at a glance the transformation undergone by any quantity of the primitive system in becoming its correlative.

The tables of correlation that Carnot established for certain more complex systems cover several pages of text. Among the more obviously useful was a complete tabulation of the correlations in all four quadrants of what he called lineo-angular quantities, i.e., sines, cosines, tangents, etc., which exhausts the entire theory of the signs of trigonometric functions. He appears to have thought of these complications as the nucleus of a sort of engineering handbook of geometric systems that would permit resolving problems by considering unknown systems as correlatives of the set of primitive systems of which the properties were known. The formulas expressing the relations were to contain only real and intelligible expressions—no imaginary and no inverse quantities. Then the procedure was to ascertain for a given correlative system what mutations the primitive system underwent in the transformations that produced it, a change that he always considered to occur continuously and one that never modified the basis of the original construction but only the relative positions of the elements.

The body of the *Géométrie de position* consists of various sorts of problems resolved by the technique of forming tables of positional properties. We shall have to limit ourselves to a few among a great many examples. Thus, "Three right lines being drawn in the same plane, and given the angles formed by the directions of any two of them with the direction of the third, find the angle formed between the first two directions."[17] That constitutes the first problem in a lengthy section devoted to the techniques of determining relations among systems of straight lines without having recourse to trigonometric functions. A further section applies the method to various

[17] *Ibid.*, § 189, p. 255, Problem XXIV.

problems of elementary geometry—e.g., given a circumference and two tangents, construct between them a third tangent equal to a given line. A final section adapts the correlation of figures to the solution of problems by the transformation of coordinates.

The contents consist in good, hard geometry approached algebraically. As time went on, it was the train of concrete problems and theorems, rather than the theory of negative quantity inspiring it, that won esteem for Carnot as one of the founders of nineteenth-century geometry. The most original propositions concern the theory of transversals, on which Carnot published a further small treatise, his last, four years later.[18] Felix Klein singled out for appreciation the theorem sometimes called Carnot's in the nineteenth century. It asserts the equality of the products of the segments created by a transversal intersecting the three sides (prolonged as need be) of a triangle: the product of any three segments always equals the product of the remaining three provided no two segments with the same angle of the triangle or point of the transversal for extremity enter as factors into either product.[19] Otherwise, Klein complained of the elementary nature amounting almost to triviality of many of the propositions, while acknowledging that the book was full of good ideas. It was, he said acutely enough, a reflection of Carnot's personality: honesty without genius. Chasles' final judgment was that Carnot's work formed only a series of powerful inductions and that he never did achieve a rigorous, absolute and fundamental demonstration of the principle of correlation of figures.[20]

Carnot himself recognized the affinity of his method with that of the porisms of the ancients (for he was well informed about the history of geometry).[21] He acknowledged that the greater part of the problems in this, his fullest work, belonged to elementary geometry but considered the characterization no matter for apology. That having been the only geometry known to a Napier, a Viète, a Fermat, a Galileo, a Huygens, that having been the mathematics preferred by a Halley, a Maclaurin, and a Newton, its fundament might even be allowed to offer certain advantages. Elementary though it was in subject matter, it was certainly not so in point of difficulty or intellectual value. It was important, moreover, to take a wider and more generous view of its matter. It was absurd, for example, that the science of projections, the descriptive geometry renewed by Monge, should form a science separate from the elements of

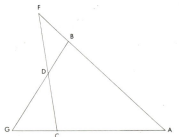

[18] Above, n. 8.

[19] Felix Klein, *Vorlesungen über die Entwicklung der Mathematik im 19. Jahrhundert* (Berlin, 1926), I, 79-80. This proposition appears as Theorem V, § 219 in *Géométrie de position*, pp. 276-278, the triangle GBA being intersected by the transversal CF,

$$\overline{AF} \cdot \overline{BD} \cdot \overline{GC} = \overline{AC} \cdot \overline{BF} \cdot \overline{GD}$$

[20] *Traité de géométrie supérieure* (Paris, 1852), p. v.

[21] *Géométrie de position*, p. 483.

geometry. To plead that its problems were applied geometry was a quite inadequate excuse, for in actual fact to separate principles categorically from applications was an impossibility. Euclid's *Elements* nowhere contain instructions for the techniques of projection that, applications though they might well be, embodied principles as theoretical as any in the elements themselves. A similar judgment could be made of the study of polygons and polyhedrons, which might be thought comprised in trigonometry, but so implicitly that to disengage it amounted to starting a new science.

Indeed, Carnot's handling of geometry exhibits throughout what we have come to recognize as his hallmark, the gathering of elementary subject matter under a point of view at once general and operational and in this combination of qualities novel. The statement of the problem of mechanics that opened the *Essai sur les machines*[22] is echoed in the definition of the problem of which the study constituted geometry:

> *In any system of right lines whether lying in the same plane or not, and some of them, or else of the angles formed by the assembly either of the lines themselves or of the planes containing them, being given in sufficient number so that the entire figure is determined; then find all the rest of it.*

Putting the problem of his book thus, Carnot permitted himself one of the few self-assertive asides that his scientific writings contain, remarking, "It is a long way from what is ordinarily called the elements of geometry to this general problem."[23]

More than the style carries over from mechanics into Carnot's vision of a new, a more vigorous geometry. At the outset he distinguished the "géométrie de position" from "géométrie de situation," the subject (since called topology) started by Leibniz although no comprehensive treatise yet existed. Carnot thought that the term "géométrie de transposition" would have been more appropriate since movement among the parts of a system was an essential aspect of all its problems. Although it was distinct from his own subject, he imagined a potential relation between them. A geometry of transposition would be to the geometry of position what motion was to rest. At that it would be but the least part of a very extensive, highly important subject, one that never had been treated for itself, the general theory of motion abstracted from consideration of the forces that produce or transmit it. The new science, in a word, would be that of geometric motion.[24]

To the reader of the *Essai sur les machines*, it will be obvious that Carnot had more in mind than mere kinematics here, and indeed that he was coming at his favorite notion from a point of view complementary to that of his mechanics. He gave a further hint of what he had in mind at the conclusion of a series of theorems about a point called the center of mean distances, better known to mechanics as the center of gravity. It seemed to Carnot that it would contribute to the vitality of geometry to

[22] Above, Chapter II, p. 32.
[23] *Géométrie de position*, p. xxxiii.
[24] *Ibid.*, p. xxxvii.

recapture that topic from mechanics. For geometry was impoverished in being limited to its conventional elements and to determinations of equilibrium, that is to say, to statics. Strictly speaking, it ought to be geometry that considered motions per se, mechanics being rather the science of the communication of motion. The idea of motion was, after all, as simple as that of dimension. It might even be genetically inseparable. Pedagogically, a line is said to be the path of a point and is in fact traced out by a pencil in motion. In turn the motion of a line is said to produce a plane and that of a plane, a solid.

Why should not the method be extended in order that geometry consider abstractly what is produced by the motion of a solid in space? For mechanics cares nothing about that as motion. It limits itself to what happens in encounters with other bodies along the way. More comprehensively, three categories of spatial relations between bodies might be imagined. In the first, bodies are entirely separate: conventional geometry pertains to these relations. Second, they may interact by collision, pressure, or traction: this is the proper domain of mechanics. Third, they may be in contact at several points without in any way interacting: and it was determination of these relations that would form the subject of a new science of geometric motions that would be transitional between the other two. To determine those conditions presented a problem independent of mechanics since, by definition, no motion was communicated or destroyed within the system, and Carnot here restated the criterion in terms of reversibility that he had given in the *Essai sur les machines* in 1783: "When such a motion exists in a system of bodies, the contrary motion is always possible, which is not the case when the motion is not geometric."[25]

Once the theory of such motions was fully penetrated, it could be counted on to rid mechanics and hydrodynamics of their analytical difficulties, because it would then be possible to base both sciences entirely on the most basic general principle of communication of motion, the equality in opposition of action and reaction, always the starting point of Carnot's analysis. From these aspirations it becomes clear that the appearance in the same year of both *Géométrie de position* and *Principes fondamentaux de l'équilibre et du mouvement* was the result of no mere coincidence or fit of industry. They were parts of the same program. The introduction to *Géométrie de position* promises the reader another work on the idea of geometric motion, "the theory of which is the transition from geometry to mechanics,"[26] and the *Principes fondamentaux* redeems the promise, recasting the entire theoretical part of the science of machines in terms of geometric motions. Not that he flattered himself actually to have created the science to which all this was a summons—his research extended only to the principal properties of these hypothetical motions.[27]

[25] *Ibid.*, § 298, p. 338.
[26] *Ibid.*, pp. xxxv-xxxvi.
[27] *Principes fondamentaux*, § 145, p. 116.

B. THE COMPENSATION OF ERROR IN INFINITESIMAL ANALYSIS

From the Berlin dissertation of 1785, it is obvious that Carnot's thinking about infinitesimal analysis long antedated any recorded interest in the problem raised by the role of negative and imaginary quantity in finite analysis. In the essay on that document that Professor Youschkevitch has contributed to the present work, which completes its discussion of Carnot's mathematical work, he compares this earliest version to the first edition of the *Réflexions sur la métaphysique du calcul infinitésimal* of 1797 and to the further revision that Carnot gave the book in its second edition of 1813.[28] These views, too, he was developing concurrently with his preoccupation with the mechanics of machine processes, and although his treatment has never commanded great interest on the part of professional mathematicians, its immediate translation into many languages and republication from time to time for over a century, are evidence of its appeal to readers professionally concerned with mathematics. The text of Carnot's final version, that of the 1813 edition, was unchanged in later printings and forms the basis of the account that follows.[29]

Such a reader of mathematics, one whose own education in the calculus recapitulated like that of many an engineering student its progress from l'Hôpital to Cauchy or thereabouts, may peruse the opening passages of Carnot's *Réflexions* with the gratifying feeling that here is an author declaring what one thought one had noticed oneself but never dared to come out and say. To Carnot, it was obvious that the infinitesimal calculus had produced the speediest and most fortunate of revolutions in the mathematical sciences. Decomposing bodies right down to their elements, it seemed to give entrée into the internal structure and very organization of things. (Notice that Carnot began his analysis in mathematics as he did in his earliest mechanics memoir in the instinct that the texture of reality is corpuscular.)[30] Yet those who handled this keen and powerful tool left unclear what those elements actually consisted in, those "singular beings that at one time play the role of true quantities and at another are to be treated as absolutely null and that seem by their equivocal properties to occupy some middle ground between magnitude and zero, between existence and nullity."[31] Not that the difficulty had impeded discovery, for mathematicians had robustly gone on to employ infinitesimals without penetrating the question of their nature or dissipating its obscurity. Philosophers for their part remained divided not only in their views, but even in their way of seeing the problem. In this situation, what he was proposing was to bring the existing points of view

---

[28] Below, pp. 149-168.

[29] Page references are to the most recent edition, that published in Paris in 1921 in the series "Les maîtres de la pensée scientifique" (Gauthier-Villars; 2 fascicules, ed. Maurice Solovine). Paragraph numbers apply equally to other, earlier French editions.

[30] Above, p. 66.

[31] *Op. cit.*, § 1, I, p. 2.

together, show the relation between them, and justify the procedures of the calculus. In good eighteenth-century style, he began genetically.

The very notion of the calculus, in Carnot's view, had arisen pragmatically out of the necessity to make approximations in calculating. Given the normal difficulty and frequent impossibility of formulating the conditions of a problem precisely in solvable equations, it would have been natural to neglect embarrassing quantities when they were so insignificant that eliminating them might be supposed to introduce only trivial error into the result. Such was the ancient geometric practice of treating curves inaccessible to theory as if they were polygons composed of a very large number of very short sides. To take the most familiar example, that of a regular polygon inscribed in a circle, when the number of sides is increased sufficiently, it becomes possible without appreciable error to attribute to the circle the properties of the polygon. A particular calculation might then be simplified by neglecting the length of one of these now very small sides by comparison with that of some given line, say the radius. It might be a great convenience to be able to substitute the length of the latter for the sum of the radius and the side. If the latter was small enough, the error would not be worth calculating.

But how legitimate was such a shortcut?

In arguing that the metaphysics of the calculus rested upon its rigorous legitimacy, Carnot continued the illustration. Suppose the problem is to construct a tangent at the point M to the circumference MBD of a circle of center C and radius r [Carnot denotes the radius by a but it will denature nothing to alter that]. The axis DCB serves as abscissa and D as origin, so that DP = x, the corresponding ordinate MP = y, and the point M has the coordinates (x, y). TP then is the required subtangent.

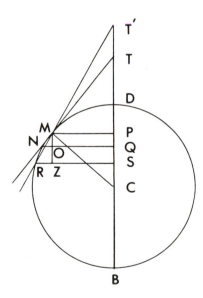

To find it, consider the circle to be a polygon with a very large number of sides. Let MN be one of the sides prolonged to intersect the axis at T′. Evidently MN will be the tangent in question since by definition it does not penetrate the interior of the polygon. If, then, the perpendicular MO be erected intersecting at point O the line NQ parallel to MP, we should have

$$MO:NO::TP:MP$$

or

$$\frac{MO}{NO} = \frac{TP}{y}$$

Further, since the equation of the curve for the point M was

$$y^2 = 2\,x\,(r - x),$$

for the point N, the equation would be

$$(y + NO)^2 = 2\,r\,(x + MO) - (x + MO)^2.$$

Subtracting the latter equation from the former and reducing,

$$\frac{MO}{NO} = \frac{2\,y + NO}{2r - 2x - MO},$$

so that, setting the two values for $\frac{MO}{NO}$ equal to each other and multiplying through by y,

$$TP = \frac{y\,(2y + NO)}{2r - 2x - MO}.$$

If MO and NO were known, we should have the value of TP and the construction of the tangent. In actuality, however, these quantities MO and NO were by hypothesis smaller than the side MN, itself very small. They were, therefore, negligible by comparison to the lengths, $2y$ and $2\,(x - r)$, to which they were added. Hence, the equation would be reduced to

$$TP = \frac{y^2}{r - x}.$$

The result would thus be acceptable in practice even though not absolutely exact. But—and this is the central tactic of Carnot's argument—the relation was not only close to being true: it was entirely true, and could be demonstrated geometrically. There was no need to appeal to the doctrine of infinitesimals. The point could be proved even to someone who had never heard of that kind of analysis. The radius of a circle being perpendicular to the tangent at its extremity, the triangles CPM and MPT were similar, whence

$$CP:MP::MP:TP$$

so that again

$$TP = \frac{MP^2}{CP} = \frac{y^2}{r - x}.$$

In short, the synthetic geometric demonstration confirmed the validity of the analytic method of infinitesimals. The advantage of the latter was that once allowed, any curve could be considered as divided into a very large number of very small sides, any surface into strips, any body into corpuscles, and any quantity into parts of itself. But it was not to be thought that the immense facility thus won for calculation rested on any inexact procedure. Infinitesimal analysis was to be viewed not as a method for merely reducing error to tolerable insignificance, but as a rigorous process that when rightly applied compensated perfectly for errors it admitted into the reasoning in order to simplify the solution of problems.

Returning to the example and to the equation

$$TP = \frac{y\,(2y + NO)}{2r - 2x - MO}$$

it was necessary to neglect NO and MO in order to get the solution. To do so was deliberate. The way in which the problem was formulated must necessarily have introduced some error, since strictly speaking a circle could never equal a real polygon. And since we know from the geometric demonstration that actually

$$\text{TP} = \frac{y^2}{r-x}$$

what we have done in neglecting NO and MO in the infinitesimal analysis has exactly compensated the error introduced into the statement of the problem in order to facilitate the reasoning.

A complementary way of looking at the problem confirmed what Carnot meant by the compensation of errors. Consider the circle now as a true curve rather than a polygon, let the line RS be drawn parallel to MP, and construct the secant RT' through R and M. Obviously, then

$$\text{T'P:MP::MZ:RZ,}$$

and

$$\text{T'P} = \text{TP} + \text{T'T} = \text{MP}\,\frac{\text{MZ}}{\text{RZ}}.$$

That being clear, and if then RS be translated parallel to itself toward MP, the point T' would simultaneously approach T, and one could diminish the line T'T as much as one pleased without distorting the proportionality. If, therefore, in solving the problem the quantity T'T were neglected, an error would certainly be introduced into the equation

$$\text{TP} = \text{MP}\,\frac{\text{MZ}}{\text{RZ}},$$

but it was an error that could be made as small as might be wished, and the equation therefore could be approached as closely as might be wished to equality.

Similarly, the equation

$$\frac{\text{MZ}}{\text{RZ}} = \frac{2y + \text{RZ}}{2r - 2x - \text{MZ}}$$

was perfectly exact whatever the location of R. And as RS is brought closer to MP, the error introduced by setting

$$\frac{\text{MZ}}{\text{RZ}} = \frac{y}{a-x}$$

might similarly be made as small as desired. From these equations, then, Carnot by simple substitution obtained the solution

$$\text{TP} = \frac{y^2}{r-x},$$

which we know from the geometry is rigorously correct—even though the two equations from which he has here drawn it, i. e.

$$TP = y\frac{MZ}{RZ} \text{ and } \frac{MZ}{RZ} = \frac{y}{r-x},$$

are in fact false, since the perpendicular distance between RS and MP was at no point assumed to be zero, nor even very small, but only equal to some arbitrary line that could be made as small as one willed.[32]

Such, then, was an example of the process of compensating for error that made the infinitesimal analysis rigorous. Later in the work Carnot extended it to the integral calculus. The procedure was not at all to be confused with a method of approximation (unless in its hypothetical genesis). It involved rather the use of infinitesimal quantities in order to find relationships between given quantities, and its results contained no errors at all, not even infinitesimal errors. Carnot went on to generalize the elements of the system in a series of definitions that distinguish between the types of quantity that figured in calculation. To the conventional division into determinate and indeterminate quantities, Carnot preferred a tripartite classification, distinguishing between quantities:

1) that were invariant and determined by the conditions of the problem; to this class belong parameters such as the radius, MC, in the problem of the tangent just worked out;

2) that, though variable by nature, acquired determinate values by reason of conventions or hypotheses introduced into the problem; to this class belonged what were ordinarily meant by variables, the coordinates, subtangents, or normals to a curve like the lines DP, MP, TP, and MT of the example;

3) that were always variable and indeterminate and to which definite dimensions were never assigned; to this class belonged all infinitely small and infinitely large quantities and also those involving the addition of a finite and an infinite or infinitesimal quantity. Instances in the sample problem were the lines DQ, NQ, TQ, T'P, T'Q, MZ, RZ—infinitesimals (MZ and RZ) or functions of infinitesimals.

Quantities of the first two classes Carnot defined as "designated" and those of the third as "non-designated." What characterized the last was not that they were of minute size, but that they could be varied at will. Yet at the same time non-designated quantities were not purely arbitrary. On the contrary, they were related to quantities of the second class by one system of equations just as the latter were to quantities of the first class by another, not unrelated system of equations. These systems containing only designated quantities Carnot called complete or perfect. Equations contain-

---

[32] *Op. cit.*, § 2-10, I, pp. 4-12.

ing terms in non-designated quantity by contrast he called incomplete or imperfect. What characterized them was the nature of those quantities: not that the error in the equation was insignificant or the quantity approximately zero, but that at the will of the calculator either could be made as small as he pleased. Another way of looking at the infinitesimal calculus, therefore, was to say that it consisted in transforming insoluble or difficult complete equations into manageable imperfect equations and then managing the calculation so as to eliminate all non-designated quantities from the result. Their very absence from the result was the evidence that it was rigorously correct.[33]

In the way that he developed these comparisons the congruence between Carnot's points of view in the calculus and in mechanics appears to particularly good advantage. Consider, he asked the reader, any general system of quantities, some constant and some variable, and suppose the problem was to find the relations between them. Let any specified state of this general system be regarded as a datum. Its quantities and any variables depending exclusively on them would then be the "designated" ones. If now the system be considered in some different state invoked for the purpose of determining more readily what relations did obtain between the designated quantities of the fixed system, this latter state would serve an auxiliary role and its quantities would be auxiliary quantities. If further the auxiliary state be approached to the fixed state so that all the auxiliary quantities approach more and more closely to their designated analogues, and if it is in our power to reduce these remaining differences as far as we please, then the differences would be what is meant by infinitesimal quantities. Since they were merely auxiliary anyway, these arbitrary quantities must not appear in the solution to the problem, which was to determine the relations between the designated quantities. Reciprocally, it was a proof that the result was correct if no arbitrary quantities did occur in it. Thus, the error willfully introduced in order to solve the problem had been eliminated. If it did persist it could only be infinitesimal. But that was impossible since the result contained no infinitesimals. Hence the procedures of the calculus had somehow eliminated it, and that was their peculiar genius.

Such, then, was an engineer's view of the calculus. Perhaps it will do no violence to paraphrase a later mathematician, Lewis Carroll, on words to the effect that the question is not what quantities mean, but who is master, we or they? In Carnot's sensibility, mathematics was a tool to the mind. Yet the attitude implied no disdain. Like a good workman he was respectful of his tools. He admired those who, like Lagrange, handled them with greater elegance and virtuosity than he did, but his admiration did not inhibit him from saying how he saw matters himself. A few months after the appearance of Carnot's *Réflexions*, Lagrange published his own account of the calcu-

[33] *Op. cit.*, § 12-18, I, pp. 14-21.

lus, which he acknowledged having composed in treatise form in consequence of the obligation to expound the subject before the students of the *Ecole polytechnique*.[34] A story—almost surely apocryphal—in Carnot's entourage had it that when Lagrange was preparing the text for the press, Prieur de la Côte-d'Or, encountering Lagrange, urged him to look up Carnot's book, and that Lagrange declined, saying enigmatically, "We can meet without seeing each other." When he later did read the *Réflexions* (so the tale goes), Lagrange wrote Carnot that if he had known the book, he would not have needed to undertake his own account.[35] Perhaps, though probably not—in any event, the parallelism of their interests is as notable as the difference in their styles both in mathematics and mechanics, a case of the high road and the low road.

Carnot never expressed anything like jealousy in the matter. To the later edition of the *Réflexions* as to his *Principes fondamentaux* he added remarks congratulating himself that his more homespun arguments had found confirmation in the analytical formulations of the foremost mathematician of their times. In that edition, indeed, he quoted in conclusion a paragraph from Lagrange of which he said, "There is my entire theory summarized with greater clarity and precision than I could have given it myself":

It seems to me that, since in the differential calculus as it is used, we consider and in effect calculate with quantities that are or are supposed to be infinitesimal, the correct metaphysics of this calculus consists in the circumstance that the error arising from this false supposition is retrieved or compensated by that which is created by the procedures of the calculus itself, in accordance with which only infinitesimal quantities of the same order are retained in differentiating. For example, in regarding a curve as a polygon with an infinite number of sides, each one infinitesimal and the prolongation of which is tangent to the curve, it is clear that an erroneous supposition has been made. But the error is corrected in calculation by the omission we there make of the infinitesimal quantities. The point is easy enough to see in particular examples although it is one of which it would perhaps be difficult to give a general demonstration.[36]

So much constituted the central thrust of the argument of Carnot's *Réflexions sur la métaphysique du calcul infinitésimal*. In the definitive, 1813 version what had been a brief summary of the rules of differentiation included in the 1797 edition has been expanded into an entire second chapter expounding the notations and practical operations of the calculus in textbook fashion. The juxtaposition with the doctrine of com-

[34] "Théorie des fonctions analitiques," *Journal de l'Ecole Polytechnique*, cahier 9 (an V, 1797).

[35] Hippolyte Carnot, *op. cit.*, I, 13-14.
[36] Quoted in *Réflexions*, § 37, I, pp. 39-40.

pensation of errors invests the book with that same combination of novelty in point of view with elementariness in scientific content that characterizes the later writing in the mechanics of machines and the *Géométrie de position*. The third and final chapter is historical and comparative, and purports to show how Carnot's position reconciles the two main traditions of the calculus, the Leibnizian method of infinitesimals and the Newtonian doctrine of the limiting ratio of nascent or vanishing quantities. It is clear from this discussion that he was widely read in the literature of the subject. He held the classical method of exhaustion to be the common ancestor of both traditions, the lineage coming down through Cavalieri's analysis of indivisibles and Descartes' method of indeterminates.

For himself he now professed to stand in the line of Leibniz and the justification of infinitesimals by the principle of continuity (an emphasis consistent with his habitual recourse to continuity of motion in mechanics). It might very well be, he admitted, that Leibniz had failed to define an infinitely small quantity before employing it, and that he even left it ambiguous whether he himself had taken the calculus to be absolutely rigorous or merely a method of approximation. Never mind whether Leibniz had had his doubts on these points, however. The prodigious success of the algorithm in his own hands and later in those of l'Hôpital, the Bernoulli brothers, and their successors reduced these objections to the dimensions of quibbles. Indeed, all that Carnot now claimed to be explaining, with the credentials of a century behind the calculus, was how Leibniz might properly have replied. He might have insisted that the method was perfectly rigorous once the infinitesimals were eliminated from the solution so that they contained only finite algebraic quantities. His critics had put themselves in a worse quandary by insisting that infinitely small quantities should logically be treated as having disappeared. For if so, all the terms in their equations would vanish leaving only a string of zeros with which to calculate relations.[37]

Not that Carnot was insensitive to the advantages of the Newtonian method of limiting ratios expounded in Section 1 of the *Principia*.[38] He was, indeed, more respectful of its algorithm, the fluxional calculus, and the eighteenth-century British mathematics in general than have been most historians of science, who have tended to repeat as history the contumely heaped upon it by Babbage, Peacock, and other British mathematical reformers of the early nineteenth century. Both in the *Réflexions* and in the *Géométrie de position* Carnot (and Chasles after him) made appreciative reference to the work of Maclaurin and of several later geometers, notably Robert Simpson, John Landen, and Matthew Stewart, appealing for the translation of their writings into French.[39] Carnot himself would have appreciated the point made

[37] *Ibid.*, § 30-32, I, pp. 29-33, and § 174, II, 73-74.

[38] *Principia mathematica*, Book I, Section 1.

[39] *Réflexions*, § 158, II, pp. 59-60; *Géométrie de position*, § 435, pp. 481-483; Chasles, *Aperçu historique ... des méthodes en géomé-*

recently by Youschkevitch that if the European mathematicians constructed the house of analysis in the eighteenth century, the British synthetic and critical tradition made greater progress in the study of the foundations[40]—an odd reversal (it may be remarked) of what is usually taken to be the characteristically British role of pragmatism over against continental preoccupation with deep principle.

In Carnot's view, Newton's method of first and last ratios of nascent and vanishing quantities had modified the ancient method of exhaustion by obviating the necessity for reducing every problem to a proof from the absurd. In this classic technique the properties of a problematic system would be determined through their correspondence to the known quantities of an auxiliary system that was closely enough related to permit approaching the one to the other. Newton's method had substituted the principle of continuity for reduction to the absurd,[41] and made an enormous gain in economy. In view of Carnot's own approach both to mechanics and to mathematics, it seems probable that it was through reflecting on the Newtonian doctrine of first and last ratios, its geometrical basis and its strong-minded voluntarism, that he first developed the justification he thought to bring to the complementary Leibnizian doctrine of infinitesimals. He professed to see no difference between them in point of rigor, although in fact he admitted the justice of the criticisms addressed to Leibniz on that score whereas of Newton's method he reported no reproach except a certain clumsiness, the consequence of its not being directly expressible in an algorithm. And it may well be that his account, perfectly correct what is more, of Newton's reasoning will serve as a clue to where he found his own:

> When any two quantities approach each other more and more closely in value so that their proportion or quotient differs less and less and as little as one wills from unity, these two quantities are said to have for *last ratio a ratio of equality*. In general, when it is supposed that diverse quantities approach respectively and simultaneously to other quantities considered as fixed, until they all simultaneously differ as little as one wills, the relations between the fixed quantities are the *last ratios* of those supposed to approach them in value, and those fixed quantities are themselves called the *limits* or *last values* of those which thus approach them.[42]

Carnot argued that the Newtonian difference between a quantity and its limit or final ratio was what Leibniz meant by an infinitesimal quantity, and that, therefore,

trie, 2nd ed. (Paris, 1875), pp. 170-186.

[40] In his introductory essay to the Russian translation of the *Réflexions*: Lazar Karno, *Razmyshlenia o metafizike ischislenia beskonechno-malykh*, trans. N. M. Solovev, with a critical introduction by A. P. Youschkevitch and a biographical sketch of Carnot by M. E. Podgorny (Moscow, 1933), pp. 16-18.

[41] *Réflexions*, § 112, II, pp. 10-11.

[42] *Ibid.*, § 129, II, pp. 33-34.

the two accounts of the calculus came down to the same thing. In fact, however, it will appear to the reader that it was he who wished to bring them to the same thing by his Newtonian definition of an infinitesimal as that which we can render as small as we will. He invoked the authority of d'Alembert, and indeed of Newton himself, to deny any distinction between the method of first and last ratios and that of limits.[43] It is surely significant, moreover (as Professor Youschkevitch points out in the essay that follows on the Berlin dissertation), that in the initial, 1785 version Carnot's reasoning draws heavily on the theory of limits. Why he should thereafter have progressively deemphasized it, no text permits us to say. His doing so was consistent, however, with a desire evident elsewhere in his work, particularly in the *Principes fondamentaux* of 1803, to accommodate his scientific style more closely to the prevailing analytic taste.

C. ANALYSIS AND SYNTHESIS

A concluding note appended to the final 1813 version of the *Réflexions* compares the compensation of error in the infinitesimal calculus to the theory of negative quantity that Carnot had developed in his geometric writings to account for algebraic operations. In both branches, finite and infinitesimal, though in somewhat different ways, analysis consisted essentially in introducing auxiliary quantities that realistically speaking were erroneous in order to facilitate the process of calculation. The difference was that the procedures of infinitesimal analysis themselves eliminated the errors in the course of computation, whereas in finite analysis they remain in the solution where we recognize them by comparison with a rational or physical reality and correct them accordingly.

The true character of analysis, its distinctive genius and the secret of the advantage it enjoyed over synthesis, lay in its capacity to employ negative, imaginary, and infinitesimal forms. In Carnot's view any other distinction between the two main modes of reasoning in mathematics was illusory. Conventionally it was said that synthesis moved from the known to the unknown whereas analysis supposed the solution known, but in terms that are unknown, the finding of which constitutes the problem. That widely adopted distinction Carnot considered to be false. If any truly subtle work of synthesis were perused, the reader would find that the essential procedure was the same as in analysis. The author would be reasoning on the unknown as if it had been found. The only real difference between synthesis and analysis was that, once the equation was formulated, and that was always a synthetic job, analysis permitted itself to employ inverse and imaginary quantities and physically unreal or metrically impossible operations, and synthesis did not.

[43] *Ibid.*, § 134, II, p. 37.

In the synthetic method the object of the reasoning could never be lost to view. It had to be kept constantly before the mind's eye, real and clear, together with whatever combinations it might enter into. If synthesis sometimes employed signs and symbols, it did so only as abbreviations or shorthand. Not so analysis, which disposed of all the resources of synthesis and also of others. It made combinations with objects that do not exist. It represented them by symbols and mingled real beings with imaginary ones. Then in the course of solving problems, it transformed equations systematically and in the course of the calculation eliminated the imaginary entities after they had served the purpose of auxiliaries. In the end, therefore, whatever was unintelligible was made to disappear, and there remained only what could equally have been discovered synthetically. But it had been obtained more easily and directly, almost one might say mechanically.[44]

Recognizable now are the elements of the Carnot style, like father, like son, in mathematical as in mechanical reasoning. Carnot adverted to something of the sort in a set of passages in the *Corrélation des figures*, where it comes down to the point that we make progress in knowledge by breaking complex problems into simple ones and analyzing the relations of systems that are unknown by comparison to those that are known.[45] Reduced to those terms the point might seem like banality itself, and only in the execution is it evident that the mode of handling problems inaugurated a combination of the engineering with the rational in a genre that was novel in its quality if not in all or perhaps any of its elements except reversibility in cyclic processes. The manner and the matter seem incompatible in a way—verbal in expression and yet exact in substance, redundant in argumentation and yet general in bearing. But the signet is the location of the reasoning in the ideal system and the control in the concrete system, the one relevant to determining the other by its approachability through a continuous change of state. It is from mechanics that we know the laws of the abstract machine, from hydrodynamics of the ideal heat engine, and from geometry of extended quantity. All are sciences of experience in a genetic sense, to be sure, and therefore operable in accordance with the more actual experience through which one knows that something is lost in impact and that a heat exchange must go one way. Perhaps it was in the mathematical reasoning, however, that Lazare Carnot most resolutely converted the imperfections of the actual into analytical assets. In the discussion of the calculus he treated inequalities as if they were equalities. He treated quantities that were approaching a goal as if they had reached it. What mattered was not the unavoidability of error, but that we should be masters of eliminating it once incurred. Thus could imprecise or "imperfect" equations furnish precise and rigorous results,

[44] *Ibid.*, note, § 20, II, pp. 103-104; Carnot incorporated a more extensive comparison of analysis to synthesis in the "Digression sur la nature des quantités dites négatives," n. 8 above.

[45] *Op. cit.*, § 45-51, pp. 26-30.

and inverse quantities real ones, not by some fortuitous cancellation of errors, but by their compensation or inversion, which really meant the calculator's power of compensating for them.[46]

[46] It is very interesting that his first discussion of the method should have occurred in the concluding passages of the theoretical portion of the 1780 prize memoir on the theory of machines. His hope clearly was to qualify the very abstract reasoning of his memoir for the contest by relating it to the effects of friction and binding. This, the "useful part" of the problem could be stated, "The forces applied to a given machine being known and such that neglecting friction and stiffness in cords there would be equilibrium, determine what must be added to the impelling forces to put the machine on the point of moving so that if these forces were increased, motion would begin." A rigorous solution being extremely difficult, the method that Carnot was proposing would deliberately introduce an erroneous supposition in order to achieve a solution exact enough to satisfy "any reasonable man." The supposition was that friction on a point is proportional to pressure, and the resulting error, a small one at worst, was then to be reduced "as far as one wishes" by successive applications of Carnot's empirically determined formulas for the relation of friction to pressure. See Appendix C, § 160, below.

LAZARE CARNOT AND
THE COMPETITION OF THE BERLIN ACADEMY IN 1786
ON THE MATHEMATICAL THEORY OF THE INFINITE

Lazare Carnot and the Competition of the Berlin Academy in 1786 on the Mathematical Theory of the Infinite

BY A. P. YOUSCHKEVITCH

LAZARE CARNOT'S *Réflexions sur la métaphysique du calcul infinitésimal* occupies a distinguished place among the numerous works on the foundations of infinitesimal analysis that appeared in the eighteenth century. Not only did the first edition of 1797 attract attention in France, but it was very rapidly translated—into Portuguese in 1798, German in 1800, English in 1800–01, Italian in 1803, and Russian in 1823. A revised and completed text published by Carnot in 1813 was reissued from time to time in France for more than a century, most recently in 1921. Its English translation came out in 1832. So enduring a success for a book of which the central idea—the theory of the so-called compensation of errors—did not satisfy either the majority of the author's contemporaries, or the generations that followed, was almost certainly not accidental. Carnot saw into the logical and psychological difficulties that inhered in the infinitesimal analysis of his period, and that beginners sometimes encounter even in our day. He gave a subtle and interesting comparison of all the infinitesimal methods known in the seventeenth and eighteenth centuries, and proposed a clear and rational interpretation of certain procedures and fundamental notions, one close to that which has become standard since the reform of analysis brought about initially by Bolzano, Cauchy, and Gauss. Finally, his book is written in an engaging manner and is still interesting to read. I remember that Professor L. K. Lakhtin, whose introductory course in analysis I took in the University of Moscow in the autumn of 1923, recommended the "Réflexions" to his students to broaden their horizons and advised them to translate it into Russian (Lakhtin probably did not know of the translation of 1823, which had become a bibliographical rarity). In 1933 a new Russian translation, in the preparation of which I had occasion to participate, was published and was reprinted in 1936, in the series "Classics of Natural Science."

In the preface of the first edition in 1797, Carnot wrote: "It was some years ago that the author of these reflections composed them in the form in which they are now being published."[1] This remark led the Italian historian of mathematics, G. Vivanti, to suggest in a précis, "Infinitesimalrechnung" contributed to Volume IV of M. Cantor's *Vorlesungen über Geschichte der Mathematik*, the possibility that we owe the

---

[1] *Oeuvres mathématiques du citoyen Carnot* (Basel, 1797), p. 128.

Réflexions of Carnot to the competition set by the Academy of Berlin for the year 1786 on the question of the theory of mathematical infinity.[2] Now not only is Vivanti's supposition of 1908 confirmed, but we are also in possession of and able to print the manuscript submitted by Carnot to the competition in Berlin.[3] That eventuality came about in the following manner. In unpublished correspondence of Carnot with J.K.F. Hauff, who was the translator of his *Réflexions* into German, Professor Charles C. Gillispie noticed a remark by Carnot himself according to which the work was a consequence of the essay he had submitted for the competition just mentioned. Mr. Gillispie told me of that letter in the course of a conversation when we happened to meet in Paris in May 1968. Soon thereafter I was able to stop over in the German Democratic Republic, and, thanks to the kind permission of the secretariat of the German Academy of Science in Berlin, it was possible for me to go over all the documents in its Central Archives that concern the competition of 1786. It required very little time to identify among the numerous manuscripts submitted for the competition that which without any doubt belonged to Carnot. The first pages of the manuscript confirmed its identity as did a distinctive figure that coincided with the first figure of the two printed editions of the *Réflexions*. So, too, did the handwriting, with which I had some slight familiarity from facsimiles of Carnot's signature. Joined to the manuscript was a letter addressed by the author from Arras, where Carnot was then living, to J.A.S. Formey, permanent secretary of the Academy of Berlin. The letter was dated 8 September 1785 and bore no signature. All the writings sent in for the competition had to remain anonymous until the prize should be announced, and were submitted under a device or motto together with a sealed note associating the device with the name of the author.

Photocopies made it possible for me to study in more detail in Moscow the entry that Carnot submitted for the contest, and it appears to be distinctly more interesting than one would have hoped in judging only from Carnot's own statement in the *Réflexions* to the effect that he was printing the book in a form written down some years previously. We do not know when the author of the *Réflexions* revised the manuscript of 1785, but the text of the edition of 1797 is definitely different from that of the manuscript both in form and content, although the theory of the compensation of errors is fundamental in both cases, and important passages are reproduced in the printed version without alteration. These differences are worth analysis, and that is the purpose of the present essay. But it will be necessary first to appreciate certain circumstances in which the Berlin contest of 1786 occurred.

[2] M. Cantor, *Vorlesungen über Geschichte der Mathematik*, Vol. IV (Leipzig, 1908), p. 647.

[3] Thus, the supposition of J. E. Hofmann, according to which Carnot could not have participated in the 1786 contest, is no longer tenable. Cf. J. E. Hofmann, *Geschichte der Mathematik*, Vol. III (Berlin, 1957), p. 68.

There is a very widespread opinion according to which eighteenth-century mathematics paid little attention to justification of the infinitesimal calculus, the period (in F. Klein's terms) having been a creative and uncritical one. That opinion is erroneous. On the contrary, the mathematicians of the age, including the very greatest among them, perceived numerous difficulties related to the notions of the infinitely small and great and to operations involving such quantities, and they attempted to resolve these arduous problems. It is true that infinitesimal procedures were at that time frequently employed in what from the point of view of later standards might often seem an unconsidered manner, but that appearance is to be explained in the first instance by the relative reliability of these procedures within the scope of most of the functions being studied, which normally were analytic, and not by the heedlessness of the investigators. All the same, eighteenth-century analysts were far from agreement on questions "of the metaphysics of the infinitesimal calculus"—agreement that, as it seemed for some time, was established in the middle of the last century only to be dissipated a few decades later, following discovery of the paradoxes of the theory of sets. Actually, throughout the entire eighteenth century, proponents of the different conceptions of the infinitesimal calculus disputed the issues in a continuous discussion that had been initiated most pointedly by the ingenious and frequently justified criticism of Berkeley's *The Analyst* in 1734. Not to enter into the details of these discussions, we necessarily limit ourselves here to the most summary of observations.[4]

In the investigations of mathematical analysis itself, as also in its application to geometry and physics, the fundamental operations involving infinitesimal quantities, infinitely small and infinitely great (for the two meanings of "infinite" were usually specified) were based on a principle that was sometimes stated explicitly and sometimes understood tacitly, i.e., the omission of all higher order terms from expressions and relationships containing infinitely small quantities of various orders. This approach permitted considering a curve as if it were a polygon composed of an infinite number of infinitely small sides. It further permitted considering as uniform motion a non-uniform motion on an infinitely small portion of path. In general it served the purpose nowadays called linearization. Newton had aspired to ground this practice in his theory of limits (*Principia Mathematica Philosophiae Naturalis*, 1687) and then in the definitive version of his method of fluxions, or velocities of change of flowing quantities, or fluents (*Tractatus de Quadratura Curvarum*, 1704). Most English mathematicians in the eighteenth century followed Newton. Maclaurin, on the other hand, preferred giving lengthy and meticulous demonstrations of the principal theorems of the calculus of fluxions by means of the ancient method of exhaustion, itself the initial and the still fairly cumbersome form of the method of limits.[5] Mean-

[4] Cf. Carl B. Boyer, *The History of the Calculus and its Conceptual Development* (New York, 1959).

[5] Cf. Florian Cajori, *A history of the Con-*

while, Leibniz's differential and integral calculus was dominant on the continent, its basic idea not that of the fluxion, but rather of the infinitely small differential of a variable quantity, corresponding to what in Newton was the moment, or the vanishing part of the fluent. The differential of a function, $y = f(x)$, was understood and defined to be the infinitely small increment, $dy = \Delta y$, but in practice dy was calculated as the principal part of the increment Δy linear with respect to Δx, i.e., as $f'(x)\,\Delta x$, and this divergence was one of the sources of paradoxes and logical difficulties.[6]

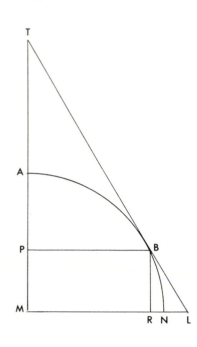

The principle of omitting infinitely small quantities of higher order was more widely known through the medium of the first published course in the differential calculus, L'Hôpital's *Analyse des infiniment petits* of 1696, where it was enunciated in the form of two postulates. The first postulate granted that two quantities may be taken indifferently for each other if their difference is infinitely small with respect to each of them. The second permitted a curved line to be considered as an assembly of an infinite number of infinitely small straight lines. Berkeley objected to these hypotheses in an argument that contained as one feature the idea of mutually compensating errors, which he developed through the example of the tangent TB to the parabola $y^2 = px$ at the point B, where $AP = x$ and $PB = y$.

Let us suppose that the point N of the parabola is infinitely close to B so that the lengths, $PM = dx$ and $RN = dy$, will then be the infinitely small increments or differentials of the corresponding abscissa and ordinate. By the second postulate the arc BN may then be assimilated to a straight line that prolongs the tangent TB through L. Since under this assumption the triangles TPB and BRN are similar, the subtangent TP is given by the equation

$$TP = y\,\frac{dx}{dy} \qquad\qquad (1)$$

Moreover, calculating dx on the basis of the first postulate from the equations

$$p\,dx = (y + dy)^2 - y^2 = 2y\,dy + (dy)^2$$

ceptions of Limits and Fluxions in Great Britain from Newton to Woodhouse (Chicago and London, 1919).

[6] Leibniz began his first article on differential calculus in 1684 with the definition of dy equivalent to our $f'(x)\,\Delta x$, where Δx is the arbitrary increment of the variable, but that definition was never applied either in his own work or in that of any mathematician of the eighteenth century. Cf. my article: A. P. Youschkevitch, "Gottfried Wilhelm Leibniz und die Grundlagen der Infinitesimalrechnung"; *Studia Leibnitiana Supplements*, vol. II, B. II (Wiesbaden, 1969).

THE MATHEMATICAL THEORY OF THE INFINITE

gives

$$p \, dx = 2y \, dy, \qquad (2)$$

so that

$$TP = \frac{2y^2}{p} = 2x. \qquad (3)$$

The authenticity of the final result was beyond doubt. The ancient Greeks had already established it without invoking the principles of omission of the infinitesimal calculus.

How could the paradoxical fact of drawing correct conclusions from false suppositions be explained? Berkeley's reply was that it came about through the reciprocal elimination of opposing errors of excess and defect committed in the course of the calculation. Equation (1) is false, since the triangle TPB is similar to the triangle BRL wherein L falls exactly on the prolongation of the tangent TB. Nevertheless, the equation

$$TP = \frac{y \, dx}{dy + z} \qquad (1')$$

is true, when $z = NL$. By the same reasoning, Equation (2) is false, and the equation

$$p \, dx = 2y \, dy + (dy)^2 \qquad (2')$$

is true, whence it follows that

$$TP = \frac{y \, dx}{\dfrac{p \, dx}{2y} - \dfrac{(dy)^2}{2y} + z} \qquad (3')$$

But Equation (3) coincides with Equation (3') seeing that $z = \frac{(dy)^2}{2y}$. Berkeley established that equivalence on the basis of the 33rd theorem of Book I of Apollonius's *Conics*, where it is demonstrated in a synthetic manner that if a straight line passes through the points B and T in such a way that $TA = AP$, it is tangent to the parabola AB. "Therefore," wrote Berkeley, "the two errors being equal and contrary destroy each other," and "by virtue of a twofold mistake you arrive, though not at science, yet at truth."[7]

In Berkeley's view, that errors were necessarily present deprived the infinitesimal calculus of the character of an authentic science even though they cancelled out, and he assumed that it should be possible to do without the infinitely small parts of lines or other quantities. In the latter half of the eighteenth century, the idea of compensation underwent further development on the continent, where the disputes created by *The Analyst* contributed, by and large, to invigorating investigation of the foundations of analysis. Euler arrived at the conviction that the only rigorous method of justifying the infinitesimal procedures generally allowed consisted in interpreting infinitely small terms and differentials in particular, as being absolute zeros or vanish-

[7] G. Berkeley, *The Analyst*, etc., § 21 (London, 1734).

ing quantities. Accordingly Euler worked out an original calculus of zeros in which $\frac{dy}{dx}$ was introduced as the limiting value of an indeterminate form $\frac{0}{0}$ that is rigorously reached by the ratio $\frac{\Delta y}{\Delta x}$ when Δx and Δy are reduced to absolutely nothing. But while applying the notion of limits to the very definition of $\frac{dy}{dx}$, Euler in no way developed the method of limits itself.[8] As for the compensation of errors, Euler mentioned it only to refute it on the spot in the preface to his *Institutiones calculi differentialis* of 1755, wherein he expounded his own conception.

At almost the same juncture d'Alembert in the articles "Différentiel" (1754) and "Limite" (1765) in the famous *Encyclopédie* declared himself a convinced proponent of Newton's method of limits freed from all mechanical considerations although not entirely arithmeticized. Keeping in mind the various views being advanced at the time on infinitesimal quantities, he asserted (just as Maclaurin did), that such terms were really used only as a kind of shorthand for the purpose of abbreviating the expressions. Such quantities are not actually encountered in analysis, which operates only with the limits of finite quantities. Strictly speaking, it is not isolated quantities that are differentiated but only the equations, and what is found thereby are the limiting values of the ratios of finite differences of the variables. In d'Alembert's eyes the procedure of omitting infinitely small terms was in substance a purely formal one, nothing more than the consequence of the passage to the limit. Nevertheless, d'Alembert never penetrated beyond the framework generally characteristic of the method of limits, even though he took the view that "the metaphysics of the differential calculus . . . is still more important and possibly harder to develop than even the very rules of this calculus."[9] He did give the definition of the limit of a monotone variable together with two general theorems: one on the uniqueness of the limit and the other on the limit of the products. He had no explicit definition of the differential, but it is obvious that by dy and dx he understood the arbitrary quantities of which the ratio is equal to the limit of the ratio of the corresponding finite increments, Δy and Δx, when Δx tends to zero. But the constitution of analysis on the basis of the method of limits still lay in the future.

The views of Euler and even more of d'Alembert attracted a number of partisans without, however, satisfying everyone. It was in a note published in volume II of *Miscellanea Taurinensia* (1760–1761) that Lagrange first enunciated as a general property of the infinitesimal calculus the circumstance that this calculus "in itself redresses the false hypotheses that are made in it."[10] He explained that faculty by

[8] Cf. A. P. Youschkevitch, "Euler und Lagrange über die Grundlagen der Analysis," *Sammelband zu Ehren des 250. Geburtstages Leonhard Eulers*, ed. K. Schröder (Berlin, 1959).

[9] *Encyclopédie méthodique ou par ordre des matières*, T. I. (Padua, 1787), p. 524 b (article "Différentiel").

[10] *Oeuvres de J. L. Lagrange*, vol. VII (Paris, 1877), p. 598 (article "Note sur la métaphysique du calcul infinitésimal").

extending to all curves the same considerations that Berkeley had exemplified in the problem of the tangent to the parabola. Lagrange's note did not go unnoticed. It was mentioned, for instance, in the posthumous publication of La Caille's course in mathematics, edited in 1770 by Marie, where, in addition, the idea that the non-rigorous procedures of the differential calculus lead to the truth only through the compensation of errors was attributed to Newton![11] Actually, however, Lagrange never attempted to prove the doctrine of compensation. About ten years later, in 1772, he put forward a new idea, which was to construct analysis independently "of all metaphysics and all theories of infinitely small or vanishing quantities."[12] He would do so by means of the notion of the derivative $u' = f'(x)$ of the function $u = f(x)$, defined as the coefficient of the second term in Taylor's series expansion of the given function, namely, $f(x + \xi) = f(x) + f'(x)\,\xi + \ldots$ and the properties of this series, which could, Lagrange thought, be generally established in purely algebraic fashion. It was only a quarter of a century later, in his theory of analytic functions of 1797, that Lagrange developed in detail this new conception of analysis. By then, other scientists as well were starting it independently. In the 1770's and 1780's Lagrange was busy rather with investigations in algebra, in the theory of numbers, and in mechanics. Nevertheless, the "metaphysics of the infinitesimal calculus," as he, like d'Alembert, called it, continued to interest him keenly, and it is thanks to him that the Mathematics Section of the Academy of Berlin, of which he was director from 1766 to 1787, proposed for the prize contest for the year 1786 the problem of the infinite in mathematics. The Academy adopted the proposition in its public meeting of 3 June 1784.

The announcement of the competition read as follows:

> The utility derived from mathematics, the esteem it is held in, and the honorable name of "exact science" *par excellence* justly given it, are all the due of the clarity of its principles, the rigor of its proofs, and the precision of its theorems.
>
> In order to ensure the perpetuation of these valuable advantages in this elegant part of knowledge, there is needed *a clear and precise theory of what is called Infinite in Mathematics.*
>
> It is widely known that advanced Geometry regularly employs the *infinitely great* and the *infinitely small*. The geometers of antiquity, however, and even the ancient analysts, took pains to avoid anything approaching the infinite, whereas certain eminent modern analysts admit that the phrase *infinite magnitude* is a contradiction in terms.
>
> The Academy, therefore, desires an explanation of how it is that so many correct theorems have been deduced from a contradictory supposition, together with

---

[11] N. L. de la Caille, *Leçons élémentaires de mathématiques* (Paris, 1770), p. 337.

[12] *Oeuvres de J. L. Lagrange*, vol. III (Paris, 1869), p. 443.

enunciation of a sure, a clear, in short a truly mathematical principle that may properly be substituted for that of the *infinite* without, however, rendering investigations carried out by its means overly difficult or overly lengthy. It is required that the subject be treated in all possible generality and with all possible rigor, clarity, and simplicity.[13]

The deadline for the competition was set for 1 January 1786, and the prize was to be a gold medal of weight equal to 50 ducats. The competition was open to all comers with the exception of regular members of the Academy.

Twenty-three memoirs in all were submitted for the contest, twenty-one of which are preserved in the Central Archives of the German Academy of Sciences in Berlin. Those formerly numbered 7 and 18 are missing. The members of the Mathematics Section under the chairmanship of Lagrange examined the memoirs, and its unanimous decision was confirmed by the Academy in its public meeting of 1 June 1786. The official finding on the outcome of the contest reads as follows:

> The Academy has received many essays on this subject. Their authors have all overlooked explaining *how so many correct theorems have been deduced from a contradictory supposition*, such as that of an infinite quantity. They have all, more or less, disregarded the qualities of *clarity, simplicity*, and above all *rigor* that were required. Most of them have not even perceived that the principle sought should not be restricted to the infinitesimal calculus, but should be extended to algebra and to geometry treated in the manner of the ancients.
>
> The feeling of the Academy, therefore, is that its query has not met with a full response.
>
> However, it has found that the entrant who comes closest to its intentions is the author of the French essay that has for motto:
> *The Infinite is the abyss in which our thoughts are engulfed.*[14]
> The Academy has, therefore, voted him the prize.[15]

The winner was the Swiss mathematician, Simon L'Huilier, who was living in Warsaw at the time. The Academy published his *Exposition élémentaire des principes des calculs supérieurs* that same year, 1786.[16] In addition, the Academy singled

[13] *Nouveaux Mémoires de l'Académie Royale des Sciences, Arts et Belles-Lettres,* 1784 (Berlin, 1786), pp. 12-13. Cf. the registers of the Academy of Berlin, I:IV, 32 fol. 373-374 (Zentrales Archiv der deutschen Akademie der Wissenschaften zu Berlin).

[14] The motto was taken from Bailly's *Histoire de l'astronomie moderne* (Paris, 1785).

[15] *Nouveaux Mémoires de l'Académie Royale des Sciences, Arts et Belles-Lettres,* 1786 (Berlin, 1788), p. 8. Cf. the registers of the Academy of Berlin, I:IV, 32, fol. 418 (Zentrales Archiv der deutschen Akademie der Wissenschaften zu Berlin).

[16] The manuscript of L'Huilier's memoir is not preserved in the archives, but no doubt it bore one of the two missing numbers of the original enumeration, either 7 or 18.

out for honorable mention[17] an entry with the motto, "Peritia fit mihi amor," the author of which is unknown. In all probability it was never published, and the sealed slips with the names of the entrants were destroyed at the end of the competition.

I have examined all the essays preserved from the competition, and it is clear that the judgment made of them by the Mathematics Section, or to be more precise by Lagrange,[18] was essentially just. It will be convenient to leave aside Carnot's memoir for the moment and come back to it further on, but none of the others contained even so much as an attempt to explain how it was that infinitesimal analysis established theorems that are correct after starting from suppositions that are false. Among those other memoirs, moreover, L'Huilier's undoubtedly stood out in its quality, although its fundamental idea was not in the slightest degree original. According to L'Huilier his essay represented "the development of ideas . . . that M. d'Alembert had only sketched and as it were proposed in the article 'Différentiel' in the *Encyclopédie* and in his *Mélanges*."[19] Above all, in the opening chapter of the *Exposition* L'Huilier somewhat improved the theory of limits itself by adding to the two theorems of d'Alembert already mentioned an important proposition on the limit of the ratio, $\lim \frac{x}{y} = \frac{\lim x}{\lim y}$, and by giving detailed proofs of all the theorems he employed, in certain cases with the assistance of the ancient method of exhaustion. He introduced for the first time in a printed text the symbol for a limit in the form "lim"; in that chapter he denoted, for example, by $\frac{dP}{dx}$ the $\lim \frac{\Delta P}{\Delta x}$; but the contribution of L'Huilier to the general theory of limits ended there. The superfluous distinction between the limits of quantities and the limits of ratios entailed useless theorems. At the same time we do not find in L'Huilier the theorem on the limit of the sum or difference of variables. The definition of the limit retained a restriction for the case of monotone variables, and the theorems were proved separately for increasing and for decreasing quantities. Moreover, in the Latin edition of the *Exposition*, which appeared in 1795, L'Huilier pointed out particularly that a ratio or a variable quantity could also approach a limit in a non-monotone manner, illustrating the possibility by the example of a decreasing geometric series with alternate signs.[20] It is curious that this circumstance had been omitted by d'Alembert as it was by many later authors, even at the opening of the nineteenth century.

In succeeding chapters L'Huilier gave a detailed and systematic explanation of the

---

[17] The official term was "accessit," which is to say "that which is closest."

[18] The other mathematicians in the Academy of Berlin at that time were Beguelin, Jean III Bernoulli, and Castillon, all much inferior to Lagrange both in talent and authority.

[19] S. L'Huilier, *Exposition élémentaire des principes des calculs supérieurs* (Berlin, 1786), p. 167.

[20] S. L'Huilier, *Principiorum calculi differentialis et integralis expositio elementaris* (Tübingen, 1795), p. 17. Cf. E. S. Schatounova, "Teoria predelov Simona L'Huilier," *Istorico-mathématitcheskié isslédovania*, XVII (Moscow, 1966), pp. 329-330.

elements of the differential and integral calculus with its application to geometry and in part to mechanics. For its time it really was a sound execution of the program sketched by d'Alembert. I should add that Chapter XI, a very full one, contains a criticism that is not without interest on contemporary ideas on the actual infinite as well as other conceptions of analysis. The notion of a potential infinitesimal quantity was by no means unknown to L'Huilier. He expounded it in the first chapter and even gave a theorem on the sum of a finite number of infinitely small terms. He said nothing, however, on the interdependence of the notions of the limit and the potential infinitesimal quantity, and he did not appreciate the intrinsic advantages of Leibniz's algorithm of the infinitesimal calculus.

The award of the prize to L'Huilier was in sum natural, which however cannot be said of the choice of subject given an Honorable Mention. The memoir thus cited (No. 21 in the original listing and 19 in the present sequence) is distinguished by no merit whatever, representing merely a brief summary of the principles of analysis based on the ideas of d'Alembert expounded in the manner of a beginner. It is true that the other memoirs, except for that of Carnot, were even weaker. It is all the more astonishing that the Academy's judgment held that *all* the participants in the contest had omitted to explain the very property of the infinitesimal analysis proposed as the paradox to resolve. This finding can be accounted for only on the supposition of some misunderstanding. For one author, namely Carnot, did devote his memoir primarily to that question. He stated it expressly in the first lines of the abstract placed at the beginning of his manuscript:

> At first it [infinitesimal analysis] was naturally bound to be regarded as a simple method of approximation. Later it was discovered that, in spite of the errors introduced in expressing the conditions of each problem, the results were, nevertheless, perfectly exact. The outcome proves that in reality these errors always must destroy each other through a necessary and infallible consequence of the operations of the calculus.[21]

It would have been perfectly possible to interpret and evaluate Carnot's demonstration in various ways, but to have passed it over in silence in the official judgment is incomprehensible, all the more that Lagrange himself took the compensation of errors to be the motive force in Leibniz's calculus, not only in 1760 but also later and indeed until the end of his life. He referred to it in the preface to the new edition of his *Théorie des fonctions analytiques*, which appeared in 1813, the year of his death. In regard to the demonstration that Carnot had proposed for the compensation of errors,

[21] L. Carnot, Dissertation sur la théorie de l'infini mathématique, below, p. 176 (Zentrales Archiv der deutschen Akademie der Wissenschaften zu Berlin. Sign. AAW: 1261-62. Preisschrift N 5 für das Jahr 1786).

Lagrange certainly considered both in 1786 and later that it had not succeeded. Here is what Lagrange wrote of the compensation of errors in the preface just mentioned: "The point can easily be made by means of examples, but it would perhaps be difficult to give a general demonstration."[22] That is why there are no grounds for believing the anecdote according to which Lagrange, after reading Carnot's *Réflexions*, wrote the author saying that if he had known his book earlier, he would not have had to develop his own theory.[23] Carnot, moreover, never mentioned that supposed letter in § 37 of the second edition of the *Réflexions*, where he cited in support of his own theory the entire passage from the *Théorie des fonctions analytiques*, including the sentence just quoted. Yet in § 167 he did seek to strengthen his point of view on another question with a reference to conversations with Lagrange.

Let us now look at Carnot's manuscript, which bears the title "Dissertation sur la théorie de l'infini mathématique," and which was submitted under the motto: "Only he who knows how to avoid the abuse of words can judge sanely of things."

The manuscript of the Dissertation consists of ninety pages written in Carnot's hand. Besides the introductory remarks and a preliminary abstract or "Sommaire" there are one hundred paragraphs divided into thirteen sections with marginal headings. The first edition of the *Réflexions* in 1797, a volume briefer than the manuscript by almost twenty per cent, consists of sixty-eight paragraphs divided into seventeen sections. The 1813 edition is a volume almost three times the length of its predecessor and contains 174 paragraphs. It was divided into three main chapters and twenty sections. In addition, it had a considerable supplement on the theory of negative quantity, a question already touched upon in the manuscript "Dissertation." The differences between the several variants derive from modifications of the structure as well as the content of Carnot's work.

The opening paragraphs of the manuscript (§ 1–36) introduce fundamental notions, sometimes in the form of definitions and sometimes of descriptions more or less rigorous in character and occasionally rather vague. They concern first, systems of "designated quantities" (our $x, y, z \ldots$, $a, b, c \ldots$, and their functions) and systems of "auxiliary quantities," which in the conditions of the problem tend to approach the "designated quantities" in an indefinite and continuous manner;[24] second, a limit; third, the infinitely great and infinitely small; fourth, "imperfect equations," i.e., an approximate equality that becomes exact in the limit; fifth, differentials of different orders and integral; sixth, variation; seventh and last, "differential moment" (the limit of the ratio of the infinitesimal increment of a function to the infinitesimal

[22] J. L. Lagrange, *Théorie des fonctions analytiques*, Nouvelle édition (Paris, 1813), p. 3.
[23] Hippolyte Carnot, *Mémoires sur Carnot*, vol. I (Paris, 1861), pp. 13-14.
[24] Carnot wrote: "by insensible degrees."

increment of the independent variable). The notion of differential moment, i.e., of a derivative, assumed great significance and found application in what followed.

Next, § 37 and 38 take up the links between infinitesimal analysis and the procedures for calculating various magnitudes by approximation. Here a lengthy remark underscores the algorithmic advantages of Leibniz's calculus over the method of limits applied by Newton in the *Philosophiae naturalis principia mathematica*, and it is affirmed that the discovery of the calculus belonged only to Leibniz, for the reason that "Newton did not reduce the method of first and last ratios [i.e., limits] to an algorithm by his calculus of fluxions until long after the discovery of the differential calculus, or at least he did not publish it earlier, and men can be judged only by their works."[25] That remark was not included in the printed text, but nearly the same opinion was advanced, although more discreetly, in § 164–165 of the second (1813) edition of the *Réflexions*.

Next, in § 39–50, the compensation of errors is explained for the example of constructing the subtangent to the circle, $y^2 = 2ax - x^2$, according to the method of Leibniz, and the equations corresponding to (1) and (2) in Berkeley's example were adduced as "imperfect." Thence Carnot turned to his fundamental problem, building up the theory of imperfect equations in a way that would account in full generality for the compensation of errors characteristic of Leibniz's calculus.

As we have already seen, the idea of the compensation of errors was not itself a novelty. It is highly probable that if Carnot did not know Berkeley's *Analyst*, at least he had read the note already mentioned that Lagrange published in *Miscellanea Taurinensia* (1760–1761). What was new, however, was the scheme of proving that errors inevitably do compensate each other in the procedures of the calculus. That project Carnot attempted to bring off in a series of three theorems. According to Theorem I (§ 51) if in an exact or imperfect equation there be substituted for one variable another differing infinitely little, the given equation is thereby transformed either into an exact or at least an imperfect equation. That assertion followed from the definition of an imperfect equation, of which the parts were (as we would say) continuous functions. Writing the given equation in the form, A = B, and the equation obtained by the substitution, $x' = x + z$, where z is infinitely small, in the form, A′ = B′, Carnot showed that if the limit of the ratio $\frac{A}{B}$ equals unity, so does the limit of the ratio $\frac{A'}{B'}$. In Theorem II (§ 53) Carnot showed that an equation containing only designated quantities could not be imperfect. From these two propositions, he derived the fundamental Theorem III (§ 55): every equation (a) obtained by means of such transformations of exact or imperfect equations, during which they remained at least

[25] L. Carnot, Dissertation, App. A, p. 201.

imperfect; and (b) containing only designated quantities, is necessarily and rigorously exact.

Carnot placed the theory of imperfect equations at the basis of the interpretation of Leibniz's calculus that considered infinitely small quantities to be true variables distinct from zero. But another approach to analysis existed in which infinitely small quantities were taken to be equal to zero. In § 58–77 is discussed the "calculus of vanishing quantities," or of zeros, compared with that of Leibniz. (Throughout, the name of Euler was not mentioned.) From Carnot's point of view the two conceptions were equally legitimate, and as he says in § 72, infinitesimal analysis is a composite of both. The method of vanishing quantities differed from the ordinary method of infinitely small quantities insofar as in the first case all the differential relations were exact, and it is unnecessary to neglect any quantities whatever. On the other hand, vanishing quantities were merely mental constructs or algebraic forms similar to imaginaries, but this standing in no way prevented either the one or the other from exhibiting determinate mathematical properties (§ 67). In this connection, it is important to keep in mind that in Carnot's view only positive values are true quantities. He considered not only imaginary numbers but also zero and negative numbers to be useful fictions.

In § 65 Carnot stated his Theorem IV that he described as fundamental for the calculus of vanishing quantities. It is in fact merely a paraphrase of Theorem III, with which he proceeded to combine it in order to obtain "the fundamental principle of the calculus of the Infinite," namely Theorem V stated in § 77:

> In an equation whose two sides differ infinitely little . . . [in both senses of the word], there may be substituted for one of the quantities figuring in it another differing infinitely little from the first, without thereby introducing any error into the designated result, and the same will be true of all propositions that can be expressed by similar equations.[26]

The choice of which of these two substantially equivalent conceptions to apply in practice was a function of the character of the corresponding problems. To employ ordinary non-vanishing infinitely small quantities was distinctly more natural and convenient in applications of the integral calculus to the measurement of geometrical magnitudes (§ 83). A brief note here extends the doctrine of the compensation of errors also to the method of indivisibles. The correctness of the results achieved by that approach is to be explained in that, for instance, the absurdity introduced by supposing a body to be composed of two-dimensional elements is eliminated by the further absurdity introduced in supposing the number of these elements to be abso-

[26] L. Carnot, Dissertation, App. A, p. 233.

lutely infinite. In this connection, Carnot remarked that difficulties of this sort inhere not only in infinitesimal analysis but also in ordinary analysis, which operates with "purely fictitious quantities" of the type of negative numbers, "which strictly speaking can represent true quantities only when they are preceded by positive quantities of greater magnitude than they,"[27] a thought that Carnot developed in detail in his *Géométrie de position* of 1803 and later in the supplements to the second edition of the *Réflexions.*

Having characterized in this fashion the two modes of the infinitesimal calculus, Carnot devoted § 84–92 to the method of limits (we shall return to a detailed discussion of this part of the Dissertation), and § 93–97 to the method of indeterminate coefficients. He then went on to exhibit the intimate relation that existed between these two methods and the infinitesimal calculus, and in addition established that the procedures applied in them can be rendered altogether similar to those of Leibnizian analysis as generally understood.

Although the view just mentioned was fully developed only at the end of Carnot's manuscript, as was also the case in the published editions of the *Réflexions*, it nevertheless held a fundamental place in his thinking. When justly comprehended, all the methods under consideration—the two variants of the infinitesimal calculus, as well as the methods of limits, of fluxions, and of indeterminate coefficients (to which were added in the second edition of the *Réflexions* the method of exhaustion, that of indivisibles, and the theory of analytic functions or derivatives)—rested on common principles. Summarizing his investigation in the conclusion to his manuscript, Carnot wrote: "In a word, is it not always the same ideas that we have to convey, the same relations that we have to express, and do all these different methods really differ otherwise than in the mode of conveying these ideas and expressing these relations?"[28] Leibniz's infinitesimal calculus stands out among the different methods by reason of the perfection of its algorithm and its wealth of applications, but no other infinitesimal method is to be excluded. In every particular case it is better to employ the approach most suited to it (§ 98).

The last two paragraphs of the "Conclusion" contain general considerations on the progress of mathematics (§ 99–100). Drawing on examples from arithmetic, algebra, and analysis, and specially on the primordial notions of number and quantity, Carnot showed that the distinctive characteristic of the development of the mathematical sciences is the tendency to arrive at conceptions and lay down theories of greater and greater generality, for "to generalize is almost always to simplify."[29] After enlarging on that observation for several pages, Carnot ended his manuscript with the words: "But this is too lengthy a discussion of metaphysics in a work the purpose of which

[27] L. Carnot, Dissertation, App. A, p. 239.   [29] *Ibid.*, p. 257.
[28] *Ibid.*, App. A, p. 253.

is to reduce to clear and luminous principles notions that can at first appear to be vague and imprecise."[30]

Turning now to a discussion of Carnot's theory of limits, we must first of all emphasize that it plays a highly important and we may even say a determining role in the Dissertation. Carnot's purpose was to rationalize the infinitesimal analysis, but the argument rested on the properties of limits. To that end it was necessary to introduce a whole series of new definitions and propositions on the general theory of limits, and along those lines Carnot achieved results not inferior and in some ways even superior to those of L'Huilier in his prize-winning work.

Above all, infinitely great and infinitely small quantities were defined with the aid of the notion of a limit. In the very first paragraph Carnot drew the reader's attention to the circumstance that the exact notion of the infinitely small is very simple, following immediately from that of a limit, and that the latter is so clear that mathematicians usually think its definition unnecessary.[31] Specifically, an infinitely small quantity is the difference of two quantities having the same limit. Moreover, the fundamental definition is given a little further on, in § 12: "Among infinitesimal quantities, those of which the limit or last value is zero are called *infinitely small*."[32] A note inserted at this point then demonstrates the equivalence of these two definitions (though not, it is true, in an altogether satisfactory manner). The same note further emphasizes with special care that the infinitely small are variable quantities of which the values are determinate and finite. As for "infinitely small quantity," the term is badly chosen and misleads beginners in the subject. It would be more appropriate to speak of the "indefinitely small," and indeed Carnot did employ that expression at many junctures.[33] Still, in § 12 Carnot went on to define an infinitely great quantity as one that has 1/0 for its limit.[34] He explained that this last symbol in no way represents a true quantity as does the symbol o. What is to be understood as infinitely great is a variable quantity the inverse of an infinitely small one (as Carnot observed in § 20 of the first edition of the *Réflexions*). In § 15 of the manuscript the definitions just given are accompanied by a distinction between the "sensible" or "assignable" infinite and the "absolute" or "metaphysical" infinite. When it is a ques-

[30] *Ibid.*, p. 262.

[31] The precise definition of the limit was not given in the "Dissertation," but this notion was described with sufficient detail and clarity in § 1, 4, and 5. It was not assumed that the variable quantity be monotone.

[32] L. Carnot, Dissertation, p. 181.

[33] It is worth noticing that when L'Huilier was considering the limits of ratios of infinitely small and great quantities, he also recommended the use of the novel terms "infinible-

ment petit" and "infinible" to express their property of being rendered as small or as great as one wished. Cf. L'Huilier, *op. cit.*, p. 147.

[34] D'Alembert had already written that "the *infinite* as it is considered in analysis is strictly the *limit* of the finite, that is to say, the *terminus towards which the finite always tends without ever reaching it.*" Cf. *Oeuvres philosophiques, historiques, et littéraires de d'Alembert*, t. II (Paris, an XIII—1805), p. 346.

tion of the infinitely small, the former case then involves an "indefinitely small" variable, from the values of which the limiting or last value, i.e., o, is excluded. The latter case involves a "vanishing quantity," that is to say, the limiting value itself, or nullity. Vanishing quantities were always to be considered along with the indefinitely small quantities of which they were the limits.

Carnot put the idea of limits at the basis not only of the definition of infinitesimal quantity, as we have seen, but also of imperfect equations and differential moments. In § 36, where he first mentioned differential moments, Carnot also introduced the sign for a limit in the form $\mathcal{L}$ (L'Huilier was then writing lim) that is also met with in the two editions of the *Réflexions*. Inasmuch as the differential is identified with an infinitely small increment, the differential moment was represented by symbols of the type $\mathcal{L} \frac{dy}{dx}$. With the further purpose of exhibiting the analogy with differentials, Carnot represented that same quantity also by the symbol Dy, already used by Johann I Bernoulli and towards the end of the eighteenth century by Arbogast. The demonstrations of the theorems on imperfect equations likewise depended on applying the notion of limits, as did the calculus of vanishing quantities.

While thus applying the idea and properties of limits to rationalizing Leibniz's analysis, Carnot nevertheless thought it would be impossible to replace the infinitesimal calculus by the method of limits as then understood, seeing that it ruled out operations employing isolated infinitely small quantities. In § 84 of the Dissertation we read: "Although the method of compensation of errors and the calculus of vanishing quantities are entirely founded on the properties of what we have called limits, they nevertheless differ from what is normally called *method of limits* or *first and last ratios* in that in the latter method neither the quantities that we have called infinitesimals nor even their ratios are allowed to enter separately, but only the limits or last values of these ratios."[35] Had it not been for this defect, the method of limits would have been preferable to the infinitesimal calculus since, contrary to the calculus of zeros, it operated only with real quantities, and contrary to the method of compensation of errors, it followed a direct and luminous route.

In Carnot's view it was, nevertheless, possible to eliminate the defect just noticed in the method of limits. He laid down the preliminaries to that end in a series of three general theorems on the properties of limits (§ 86–88). According to Theorem VI, which in all probability no one had ever before enunciated, the limit of a designated quantity is equal to this quantity (i.e., in particular, lim $a = a$). As for Theorem VII, it asserted the uniqueness of the limit, whereas Theorem VIII, which Carnot stated for the first time simultaneously with L'Huilier, laid down that ". . . the limit of the ratio of two quantities is equal to the ratio of their limits," or in Carnot's notation,

[35] L. Carnot, Dissertation, App. A, p. 239.

$\mathcal{L}\frac{Y}{Z}=\frac{\mathcal{L}Y}{\mathcal{L}Z}$.[36] The last proposition extended to the case in which $\mathcal{L}\,Y=\mathcal{L}\,Z=0$, i.e., the limit of the ratio of two indefinitely small quantities is equal to the ratio of the corresponding vanishing quantities. That result derived from the manner in which the symbol $\frac{0}{0}$ was conceived in the calculus of zeros. (It must be noted that we could not regard as demonstrations the reasonings by which Theorems VI through VIII were argued.)

In order to separate the infinitesimals dy and dz in finite expressions of the form $\mathcal{L}\frac{dy}{dz}$, the variables y and z were considered as functions of one and the same variable, let us say of x. In accordance, then, with Theorem VII, the expression $\mathcal{L}\frac{dy}{dz}$ could be represented by the form $\mathcal{L}\frac{dy/dx}{dz/dx}=\frac{\mathcal{L}dy/dx}{\mathcal{L}dz/dx}$ and also in the form $\frac{Dy}{Dz}$. In the latter fraction, the numerator and denominator are fully separate and the characteristic D is subjected to the same algorithm as the characteristic d (§ 90–91). "Thus, by this means not only does the method of limits become as simple as the infinitesimal calculus without depriving it of any of its rigor, but also the procedures of the two methods are made completely identical, since the only thing necessary for that is to substitute for the infinitely small quantities their moments of evanescence,[37] which are finite quantities whose ratios are the same as those of the corresponding vanishing quantities."[38] It must be added that in Newton's method of fluxions, what we regard as the derivatives of any variable with respect to another, say $\frac{dy}{dz}$, are always expressed in the form of the ratio of the fluxions $\dot{y}/\dot{z}$, where y and z depend on "time," t, as a universal variable parameter, and that is why in this calculus, the fluxions $\dot{y}$ and $\dot{z}$ play the same role as our differentials dy and dz. In the second edition of the *Réflexions*, § 142–143 advance this same observation. As A. N. Kolmogorov remarks, the equivalence of the procedures of the calculus of fluxions to the definition of the differential of a function as the product of the derivative by the arbitrary constant increment of the variable Δt, is altogether natural: if this last is set equal to unity, then $dx=\dot{x}\Delta t=\dot{x}$, and $dy=\dot{y}\Delta t=\dot{y}$.[39]

It was likewise completely clear to Carnot that in the method of limits all imperfect equations were transformed into perfectly exact equations. In the first edition of the *Réflexions* he showed that to be true by means of two examples. The first of them (*Réflexions*, § XXXVIII) was nothing other than the problem of the subtangent to the curve, $y=f(x)$.[40] Here, imperfect equations of the form (1) and (2) were replaced

---

[36] *Ibid.*, p. 242.

[37] Carnot also called the differential moment (moment différentiel), or "the moment of evanescence" (moment d'évanouissement).

[38] L. Carnot, Dissertation, App. A, p. 244.

[39] A. N. Kolmogorov, "Newton i sovremen-

noe matematicheskoe myshlenie," *Moskovski Ouniversitet pamiati Isaaca Newtona* (Moscow, 1946), p. 40.

[40] Carnot constructed the tangent to the circle $y^2=2ax-x^2$ mentioned above.

by the limiting relations for the subtangent

$$s_t = y \ \lim_{\Delta x \to 0} \frac{\Delta y}{\Delta x} \tag{4}$$

and for

$$\lim_{\Delta x \to 0} \frac{\Delta x}{\Delta y} = \frac{1}{f'(x)} = \frac{1}{y'} \tag{5}$$

[I am writing these expressions in modern notation.] Carnot chose the second example (*Réflexions,* § LVI) from the rules of the differential calculus and showed that to the imperfect equation, $d \cdot xy = x\,dy + y\,dx$ [his notation], in the method of limits corresponded the exact equation, $\lim \left(\frac{d \cdot xy}{dy}\right) = y \times \lim \left(\frac{dx}{dy}\right) + x$.[41] He retained the former of these examples in the second edition of the *Réflexions* (§ 135).

Probably the most interesting feature of the Dissertation is Carnot's aspiration to put the theory of limits at the basis of the infinitesimal calculus and to invest the former with all the algorithmic advantages presented by the latter. In this respect as in several others—for example, the primary emphasis accorded the reciprocal relations of the notions of the limit and the infinitely small—Carnot partially anticipated the reform of analysis that Cauchy later undertook with such marked success. For Cauchy also set himself the goal of combining the rigor pertaining to the theory of limits with the simplicity gained through direct operations with infinitesimals. In order to achieve that end, however, a much more profound revision of the foundations of the calculus would be requisite. Although Carnot established the formal identity of the rules of operation with the symbols Dy and dy, he still did not construct the formal apparatus of the infinitesimal calculus and of differentials on the basis of the theory of limits. He did disengage from each other the finite quantities Dy and Dz (proportional to dy and dz in modern usage), but not the infinitely small differentials dy and dz (in his usage). Since Carnot continued in the traditional manner to understand by the differential of a function its infinitesimal increment, the differential expressions that he failed to reduce to a form containing only derivatives remained imperfect. Among other advantages, Cauchy's new definition of the differential eliminated that difficulty at the outset, rendered all the rules for differentiation exact instead of approximate, and made superfluous imperfect equations.

It must be added that Carnot's theory of imperfect equations, though it yielded nothing in point of rigor to other conceptions of eighteenth-century analysis, offered no advantages in that respect. It will suffice to remark that the applicable scope of the central theorem, Theorem III, appears to have been altogether undetermined for the reason that there were no means of learning what classes of quantities and functions

[41] Here Carnot wrote the limit in the same way that L'Huilier did.

would lead to imperfect equations and what the permissible types of transformation of such equations might have been. In order to answer such questions, it would have been necessary first of all to lay down a much more fully developed theory of limits, and then to establish the fundamental properties of continuous functions, etc. Carnot himself had not even enunciated the notion of a continuous function in the modern sense, although he was moving toward it in the definition of imperfect equations.

Nevertheless, however inadequate from the point of view of classical nineteenth-century analysis may have been the foundations of the infinitesimal calculus proposed by Carnot in 1785, his Dissertation was a remarkable piece of work. It remains only to compare it briefly to the published editions of the *Réflexions*.

As has been said, Carnot's fundamental conception remained the same during the preparation of the *Réflexions,* and entire passages reappear with little or no change. The structure of the work was revised, however, certain theorems and definitions were improved, and several sections were dropped or introduced.

The *Réflexions* of 1797 opened with preliminary remarks on the purpose of the work and on the ties between analysis and the method of approximation (§ 37–38 of the manuscript, except for the remark there on Leibniz's priority, which was dropped). Then the problem of the tangent to the circle was considered (§ 39–50 of the manuscript), and only after that did Carnot give his definitions of fundamental notions and his three theorems on imperfect equations (§ 1–36 and 51–56 of the manuscript). Such a sequence invested the exposition with greater elegance; moreover, some statements were formulated with greater precision. As earlier, the notion of limits was at the basis of the definition of infinitesimal quantities and imperfect equations, but the definition of differential moment was now eliminated. Next, Leibniz's calculus was compared to other methods, but in a different order now: first of all to the method of indeterminate coefficients (§ 93–97 of the manuscript), then to the method of limits (§ 84–91 of the manuscript), and finally to the calculus of vanishing quantities (§ 57–77 of the manuscript). The explanations of the latter two methods were considerably reduced. Most notably, the entire argument about an algorithm for the method of limits and the theorem on the limit of the ratio of two quantities were both eliminated. On the other hand, as has already been mentioned, the solution of the problem of the tangent to the circle by the method of limits was added. The book ends with a succinct exposition of the rules of differentiation and integration and also of certain other problems not found in the manuscript. The concluding observations in the manuscript on the progress of mathematics were entirely omitted from the published work.

In an overall comparison, the published *Réflexions* of 1797 differed most notably from the manuscript of 1785 in paying a much less sustained attention to the theory of limits, although the basic idea of limits was still fundamental to the argument.

This tendency to deemphasis of limit-theory was even more marked in the second edition of the *Réflexions* of 1813. There in Chapter I, which is devoted to the general principles of infinitesimal analysis, infinitesimal quantities and imperfect equations are defined without reference to the notion of limits, which was not there utilized explicitly. The second chapter contains an exposition of infinitesimal analysis that is much more detailed than in the first edition, and the third also gives a circumstantial comparison of the methods that may be substituted for analysis. It was, obviously, this second edition of the *Réflexions* that has been most widely disseminated, whereas the manuscript of the Dissertation was known only to a small group of contemporaneous members of the Mathematics Section of the Berlin Academy. Consequently, a number of the remarkable ideas that Carnot failed to include in the final version of his work have continued to be unknown or very little known.

It remains a difficult task to evaluate the extent of Carnot's influence on the immediate development of the "metaphysics of the infinitesimal calculus." We do not know how well Bolzano, Gauss, Cauchy, and the other founders of classical nineteenth-century analysis knew his work or what role his ideas played in the formation of their conceptions. Only one thing is certain: the first edition of the *Réflexions* was already being widely read in many countries, and it is extremely improbable that the leading mathematicians at the beginning of the last century could have failed to notice it. As evidence there may be adduced the high opinion expressed in the second edition of Lacroix's course of analysis, a very popular textbook, where among other things it is said:

> M. Carnot then shows how imperfect equations become rigorous at the end of the calculation, and by what mark their legitimacy may be recognized. . . . The work of M. Carnot is not to be judged by the little I have said of it, and the merit of his memoir does not consist only in his characteristic way of envisaging the differential calculus, but still more in the comparison that he makes of the different points of view from which the calculus has been presented.[42]

[42] S. F. Lacroix, *Traité du calcul différentiel et du calcul intégral*, 2d ed., vol. I (Paris, 1810), p. XVII.

Appendix A

The text of Carnot's "Dissertation sur la théorie de l'infini mathématique; ouvrage destiné à concourir au prix qu'a proposé L'Académie Royale des Sciences, Arts, et Belles Lettres de Berlin pour l'année 1786," is here reproduced by kind permission of the *Deutsche Akademie der Wissenschaften zu Berlin*, where the original is conserved in the Central Archives, Sign AAW, 1261-1262. To accompany this edition Professor A. P. Youschkevitch has prepared a series of notes which the reader will find following Appendix A. A transcription of the footnote in § 10 is included. The figure to which Carnot refers in § 39, p. 203 below, is identical with that which he reproduced in the published *Réflexions*. It appears above on p. 135, the only difference being that the lettering of the printed figure is in capitals.

à Monsieur

Monsieur formey

à arras le 8 7bre 1785.

Monsieur.

L'académie ayant proposé pour
sujet du prix de mathématique qu'elle
doit adjuger en 1786, la théorie de
l'infini. je vous serai infiniment obligé
si vous voulez bien présenter au
concours la pièce ci-jointe.

j'ai laissé au bureau des diligences
12 livres argent de france pour
affranchir le paquet; cette somme étant
à peu-près ce que coûteroit ici un
pareil envoi.

Dissertation

sur la théorie de l'infini mathématique,
ouvrage destiné à concourir au prix,
qu'a proposé l'académie Royale des Sciences,
arts et belles-lettres de berlin,
pour l'année 1786.

celui-là seul peut juger sainement des choses,
qui sait éviter l'abus des mots.

L'académie Royale des Sciences, arts
et belles-lettres de berlin, propose pour sujet du
prix de mathématiques, qu'elle doit adjuger
en 1786, la question suivante.

elle demande une théorie claire et précise
de ce qu'on appelle infini en mathématiques.

l'académie désire, qu'on indique un principe
sûr à substituer à l'infini, sans rendre trop
difficiles ou trop longues, les recherches qu'on
abrége par ce moyen.

il n'est aucune découverte qui ait produit
dans les sciences mathématiques, une révolution
aussi heureuse et aussi prompte, que celle de

l'analyse infinitésimale; aucune n'a fourni des moyens plus simples ni plus efficaces, pour pénétrer dans la connoissance des loix de la nature. en décomposant, pour ainsi dire, les corps et toutes les quantités, jusque dans leurs éléments, elle semble en avoir indiqué la structure intérieure et l'organisation. mais comme tout ce qui est extrême, échape aux sens et à l'imagination. on n'a jamais pu se former qu'une idée imparfaite de ces éléments, espèces d'êtres singuliers, qui tantôt jouent le rôle de véritables quantités, tantôt doivent être traités comme absolument nuls et semblent par leurs propriétés équivoques, tenir le milieu, entre la grandeur et le zéro, entre l'existence et le néant.* heureusement, cette difficulté

---

* je parle ici conformément aux idées vagues que les commençants se font pour l'ordinaire des quantités infinitésimales. mais dans le vrai, rien n'est plus simple, que la notion exacte de l'infini. elle tient immédiatement à celle des limites et des premières et dernières raisons, notions sur lesquelles on n'a jamais élevé de difficulté. et que les géomètres tiennent pour si claires, qu'ils ne se donnent pas même communément la peine de les définir. en effet, qu'est-ce qu'une quantité infiniment petite? ce n'est autre chose, que la différence de deux grandeurs qui ont pour limites une même troisième grandeur. (par ce terme de grandeur, j'entends ici, une quantité effective. c'est-à-dire, qui ne soit, ni 0, ni $\frac{0}{0}$) voila en admettant la notion des limites, une réponse claire, exacte et précise. toutes les fois, que deux quantités quelconques ont pour dernière raison, une raison d'égalité; on dit qu'elles diffèrent infiniment peu, que leur différence est infiniment petite relativement à chacune d'elles, ou que le quotient de cette différence par chacune de ces grandeurs, est une quantité infiniment petite. or rien de moins vague que cela, ni de plus facile à concevoir.

n'a point nui aux progrès de la découverte; il est certaines idées primitives qui laissent toujours quelque nuage dans l'esprit, mais dont les premières conséquences une fois tirées, ouvrent un champ vaste et facile à parcourir. Celle de l'infini a

lorsqu'on sait ce que c'est que première et dernière raison on m'objectera peut-être, qu'il m'a plu, à moi, de définir ainsi les quantités infiniment petites, mais que cette définition ne convient point à l'idée qu'on s'en fait ordinairement; puisqu'il est très-possible qu'une grandeur variable, quoiqu'elle ait o pour limite, soit cependant succeptible de valeurs diverses plus grandes que telle ou telle quantité donnée, aulieu qu'il est de fait, qu'on se représente toujours les quantités infiniment petites, sinon comme absolument nulles, aumoins comme plus petites que toutes les quantités données. je réponds à cela, que si on se les représente de cette manière, dans la pratique du calcul, c'est parcequ'on le veut bien et uniquement pour soulager son imagination; mais que rien n'y oblige, qu'on peut prendre les differentielles des quantités aussi grandes et plus grandes que les quantités même, dont elles sont les differentielles, sans que cela change la moindre chose au procedé du calcul; qu'il en resultera seulement un peu plus de difficulté à saisir les dernières raisons de ces quantités ou les limites de leurs rapports, limites dont la considération est d'un très-grand secours; qu'enfin l'exactitude du calcul dont il s'agit, n'est pas fondée, comme semble l'indiquer cette denomination d'infiniment petite, sur ce que les quantités qu'on y néglige ou qu'on y traite comme nulles soient réellement inassignables comme le disent quelques personnes (cette expression ne signifie rien dutout, car il est évident que toute quantité, si elle existe, peut être assignée) mais bien, sur ce que les erreurs auxquelles on donne lieu en négligeant ces quantités se détruisent d'elles-mêmes, par une compensation qui est une suite nécessaire et infaillible des operations du calcul. or cette compensation auroit lieu également, quelques valeurs, soit petites soit grandes, qu'on attribuat aux quantités appelées infiniment petites. nous verrons dans la suite quelle est l'origine de cette denomination, qui ne fait rien à la chose, mais qui trompe les commençants; c'est pourquoi le nom de quantités indéfiniment petites, me paroitroit plus convenable.

toujours paru telle, et plusieurs géomètres en ont fait le plus heureux usage, qui n'en avoient peut-être point approfondi la notion. cependant les philosophes s'efforceront toujours de remonter aux principes, et l'on doit tout espérer de leurs efforts, excités surtout, par l'intérêt qu'annonce prendre à cette recherche une société célèbre. je ne me flatte point d'avoir réussi dans cette entreprise difficile, et je me croirai bien récompensé de mon travail, si j'ai pu jeter quelques degrés de lumière, sur un objet si intéressant.

Sommaire.

Définitions exactes et explication des principaux termes relatifs à la théorie de l'infini. origine que peut avoir eue l'analyse infinitésimale. on a dû naturellement la regarder d'abord, comme une simple méthode d'approximation. on a découvert ensuite, que malgré les erreurs commises dans l'expression des conditions de chaque problème, les résultats étoient néanmoins de la plus parfaite exactitude. preuve qu'en effet ces erreurs doivent se détruire toujours, par une suite nécessaire et infaillible des opérations du calcul. il y a deux manières d'envisager la question de l'infini; la première consiste à attribuer aux variables qu'on nomme infiniment petites, des valeurs déterminées qui soient effectives. c'est-à-dire, qui ne soient point égales à zéro, et alors, l'analyse infinitésimale doit être regardée comme un calcul d'erreurs compensées: la seconde consiste à attribuer aux quantités infiniment petites, des valeurs absolument nulles; et alors l'analyse infinitésimale est un calcul exact dans tous ses procédés, mais il a pour sujet des quantités évanouissantes, qui sont des êtres de raison. comparaison de la méthode des erreurs compensées avec celle des évanouissantes. elles conduisent précisément aux mêmes opérations et aux mêmes résultats. avantages et inconvénients qui sont propres à chacune d'elles. réunion de ces deux méthodes en un même calcul, qui est l'analyse infinitésimale proprement dite. explication de ce qu'on nomme méthode des limites.

ou des premières et dernières raisons. elle est la même au fond, que la méthode ordinaire du calcul infinitésimal, mais les procédés en sont plus difficiles. moyen de rendre la méthode des limites aussi simple que le calcul infinitésimal ordinaire. l'analyse infinitésimale n'est autre chose, qu'une application ou si l'on veut, une extension de la méthode des indéterminées. celle-ci peut suppléer à l'autre, sans augmenter sensiblement la difficulté du calcul. Conclusion.

définitions exactes et explication des principaux termes relatifs à la théorie de l'infini.

I imaginons un système quelconque de quantités, ~~[texte barré]~~ et qu'il soit question de trouver les rapports et relations qui existent entre elles.

pour y parvenir, concevons un nouveau système de quantités différent du premier, mais tel, qu'en le faisant changer par degrés insensibles suivant une loi quelconque, on puisse l'en faire approcher d'aussi près qu'on voudra, ou qu'il finisse même, par se confondre absolument avec lui.

Supposons maintenant, qu'au lieu de comparer directement ensemble les quantités du premier système, pour obtenir les relations désirées, on commence par chercher celles qui existent entre les quantités de ce premier système et celles du second, qu'ensuite, on élimine ces dernières du calcul; les résultats qu'on obtiendra par cette élimination, ne contenant plus que les quantités du premier système, exprimeront évidemment, les rapports et relations, qu'on s'étoit d'abord proposé de trouver.

or telle est en général, la marche du calcul

qu'on nomme analyse infinitésimale.*

2. d'abord, le premier système étant fixe ou indépendant de l'état où l'on suppose que se trouve le second, je l'appelerai Système désigné le second au contraire, n'étant en quelque sorte, qu'un moyen de Comparaison imaginé pour faciliter la recherche proposée je l'appelerai Système auxiliaire.

3. de même toutes les quantités qui composent le premier système ou qui en font des fonctions quelconques, c'est-à-dire, qui dépendent de ces premières quantités et d'aucune autre s'appelleront quantités désignées, toutes les autres au contraire, c'est-à-dire, toutes celles qui ne dépendroient point du tout, ou qui dépendroient seulement en partie de ces quantités désignées, je les nommerai quantités auxiliaires.

de même encore, toute fonction toute équation tout résultat quelconque en un mot, qui ne contiendra que des quantités désignées se nommera résultat désigné et au contraire, on le nommera résultat auxiliaire lorsqu'il ne dépendra pas entièrement de ces quantités désignées.

* ceux qui connoissent l'analyse infinitésimale sentiront bien la vérité de cette assertion. Car, qu'on ait à trouver les relations qui existent entre des grandeurs proposées a, b, c &c x, y, z &c les unes constantes les autres variables: on commencera suivant les principes de ce calcul, par concevoir un autre système de grandeurs qui puisse être supposé aussi peu différent qu'on voudra du premier et composé des quantités a, b, c &c $(x+dx) (y+dy), (z+dz),$ &c ainsi que de toutes les autres quantités qui comme $dx, dy, dz,$ &c sont fonctions de ces premières; puis ayant trouvé les relations qui existent entre celles du premier système et celles du second, on éliminera ces dernières, soit par les règles de l'algèbre ordinaire soit par les moyens abrégés que fournit l'art de négliger à propos les quantités qui n'influent point sur le résultat cherché. alors on a évidemment la relation désirée telle qu'on la vouloit avoir.

4. le système désigné sera aussi nommé limite du système auxiliaire; parce qu'en effet, c'est comme le terme ou la limite du changement auquel celui-ci est supposé assujetti, le système désigné n'étant autre chose que le dernier état du système auxiliaire, celui auquel on le rapporte, et dont il approche de plus en plus.

5. par la même raison, j'appelerai limite ou dernière valeur d'une quantité, le terme dont cette quantité approche de plus en plus, à mesure que le système auxiliaire approche davantage du système désigné.

6. la limite du rapport de deux quantités ou la dernière valeur du quotient de l'une divisée par l'autre, s'appelle première raison ou dernière raison de ces quantités.*

7. ainsi en général, on nomme dernières valeurs et dernières raisons des quantités, celles qui sont en effet, les dernières de celles qu'assigne à ces quantités et à leurs rapports, la loi de continuité, lorsque l'on conçoit que le système auxiliaire va en s'approchant continuellement du système désigné.

___

* on emploie indifféremment l'une ou l'autre de ces expressions, parce qu'on peut également supposer, ou comme nous l'avons fait (1) que le système auxiliaire s'approche continuellement du système désigné, desorte que celui-ci, soit considéré comme le dernier état du système auxiliaire ou qu'au contraire il s'en éloigne continuellement; desorte que le système désigné soit regardé comme le premier état du système auxiliaire or le nom de dernière raison se rapporte à la première manière de concevoir la chose, et celui de première raison se rapporte à la seconde.

8. de même on appelera limite d'une fonction d'une équation ou d'un résultat quelconque, le dernier état de ce résultat; c'est-à-dire, le dernier de ceux que détermine la loi de continuité, lorsque le système auxiliaire va en s'approchant par degrés insensibles du système désigné.

9. ainsi par exemple, si deux quantités avoient chacune 0 pour limite; leur dernière raison, quoique égale à $\frac{0}{0}$ ne seroit pas pour cela une quantité arbitraire: car parmi toutes les valeurs qu'on pourroit attribuer à cette fraction, il y en a évidemment une seule donnée par la loi de continuité; c'est-à-dire, qui soit le terme dont approche de plus en plus le rapport des quantités en question, à mesure qu'elles décroissent, ou que le système auxiliaire approche davantage du système désigné: or c'est ce terme ou cette limite, qu'on nomme dernière raison des quantités proposées.

10. à chacune des quantités désignées, répond donc une quantité auxiliaire dont elle est la limite: or si malgré le changement du système général de ces quantités auxiliaires, il s'en trouve quelques-unes qui ne changent pas, c'est-à-dire, qui restent constament égales à leurs limites ou quantités désignées qui y répondent; on les nommera quantités constantes: celles au contraire, qui changent à mesure que les deux systèmes se rapprochent, se nommeront variables auxiliaires*

* par exemple si l'on se propose de trouver les rapports qui existent entre a, b, c, &c x, y, z, &c et que pour y parvenir, on ait recours à un système auxiliaire composé des quantités a, b, c, &c (x + dx), (y + dy), &c celles a, b, c, qu'on suppose les mêmes dans le système auxiliaire que dans le système désigné seront les constantes celles (x + dx) (y + dy) &c qui changent par le rapprochement des deux systèmes seront les variables auxiliaires, et leurs limites x, y, z, &c seront les variables désignées.

et enfin, les quantités désignées qui répondent à ces variables auxiliaires ou qui en font les limites, se nommeront variables désignées. par l'expression générique de variables, nous entendrons également, soit les variables désignées soit les variables auxiliaires, à moins que le sens de la phrase ne dénote particulièrement l'une de ces espèces plutôt que l'autre.

11. toute quantité dont la dernière valeur est une quantité quelconque effective, c'est-à-dire dont la limite n'est ni o ni $\frac{1}{o}$, se nomme quantité finie; toutes celles au contraire, qui n'ont point de limites effectives, c'est-à-dire dont les dernières valeurs font o ou $\frac{1}{o}$ se nomment quantités infinitésimales.

12. parmi les quantités infinitésimales, celles dont la limite ou dernière valeur est o, se nomment infiniment petites*; celles au contraire dont la limite ou dernière valeur est $\frac{1}{o}$ se nomment infinies ou infiniment grandes.

* nous avons dit (note 1) qu'une quantité infiniment petite n'est autre chose que la différence de deux autres quantités qui ont pour limite commune une troisième grandeur quelconque effective, ce qui revient au même que la définition présente. car soit x une quantité qui ait o pour dernière valeur, et X une autre quantité qui ait pour limite une grandeur quelconque effective Y. il est clair qu'on a $x = (X + x) - X$, c'est-à-dire que x est la différence des quantités $X + x$ et X, lesquelles ont évidemment pour limites la même grandeur Y. on trouvera (42) la preuve que les quantités dont les géomètres font usage sous la dénomination de quantités infiniment petites, sont en effet telles que nous venons de les définir. ce nom porteroit cependant à croire, que les grandeurs auxquelles on le donne, doivent non seulement avoir o pour limite; mais que de plus elles ne peuvent avoir elles-mêmes que des valeurs très-petites, c'est-à-dire, qu'on ne seroit pas maître de leur attribuer des valeurs égales à telle ou telle grandeur donnée. mais cela n'est pas: ce sont des quantités variables auxquelles, de même qu'à toute autre variable on peut attribuer diverses valeurs déterminées: ainsi l'expression d'infiniment petite est assez impropre, et ne contribue pas peu, à faire prendre une idée fausse de ces quantités. nous verrons (17) quelle est l'origine de cette dénomination, à laquelle il conviendroit peut-être, de substituer celle d'indéfiniment petite.

dire suivant l'usage vulgaire que l'infini est ce qui n'a point de bornes, ce qui est sans limite ou ce dont la limite n'existe pas; c'est donc en donner une idée, tout à la fois, fort simple et fort exacte, puisqu'en effet toutes les quantités infinitésimales, soit infinies soit infiniment petites n'ont point de vraies limites, ou plutôt, ont pour limites, les unes 0, les autres $\frac{1}{0}$ qui ne sont point de vraies quantités.

13. les quantités infinitésimales ne sont donc pas des êtres chimériques, mais de simples quantités variables, caractérisées par la nature de leurs limites, et qui supposent par conséquent toujours l'existence de deux systèmes de quantités dont l'un est pris pour terme de comparaison ou limite de l'autre. lors donc, qu'on emploie les expressions de limite, quantité infinie ou infiniment petite, dernière valeur ou dernière raison et autres qui en dépendent, on suppose tacitement la comparaison de deux systèmes de cette nature sans qu'il soit besoin de le dire expressément.

14. on nomme calcul infinitésimal, l'art d'employer les quantités infinitésimales d'un ou de plusieurs systèmes auxiliaires, pour découvrir les rapports et relations quelconques, des quantités finies qui composent un système désigné.

15. les quantités infinitésimales étant de simples variables dont la limite est 0 ou $\frac{1}{0}$, on peut leur attribuer successivement comme à toutes les autres

variables, diverses valeurs déterminées, et parmi ces valeurs on doit compter la dernière de toutes, qui est o pour les quantités infiniment petites et $\frac{1}{o}$ pour les quantités infiniment grandes: je distinguerai donc l'infini mathématique en deux espèces, savoir, l'infini sensible ou assignable, et l'infini absolu ou métaphysique, qui n'est autre chose que la limite du premier.

Si donc on attribue à une quantité infiniment petite une valeur quelconque déterminée, qui ne soit point égale à zéro, cette valeur sera ce que j'appelle quantité infiniment petite sensible ou assignable et que je désignerai aussi par le nom d'indéfiniment petite; aulieu que si cette valeur est la dernière de toutes, c'est-à-dire, si elle est absolument nulle, elle sera alors ce que j'appelle quantité infiniment petite absolue et que je désignerai aussi par le nom de quantité évanouissante.

Les quantités que je nomme évanouissantes, ne sont donc autre chose que des quantités absolument nulles auxquelles on donne cette dénomination particulière d'évanouissantes, pour avertir que de tous les rapports et relations dont elles sont susceptibles, on ne veut considérer que ceux qui leur sont assignés par la loi de continuité, lorsque l'on conçoit le système auxiliaire s'approchant par degrés insensibles du système désigné. C'est ce que de grands géomètres ont exprimé en disant, que les évanouissantes sont des quantités considérées, non avant de s'évanouir, non après qu'elles sont évanouies, mais à l'instant même qu'elles s'évanouissent.

ainsi une quantité évanouissante ou infiniment petite absolue, n'est pas à proprement parler, ce qu'on nomme en général une quantité infiniment petite, mais seulement la dernière valeur de cette quantité; ce n'est dis-je, qu'une valeur déterminée qu'on peut attribuer comme toute autre à cette quantité variable qu'on nomme en général infiniment petite.

de même, si l'on attribue à ce que je nomme en général quantité infiniment grande une valeur quelconque déterminée qui ne soit point $\frac{1}{0}$, cette valeur sera ce que j'appelle infiniment grande assignable mais si l'on suppose que cette valeur soit la dernière de toutes, c'est-à-dire, égale à $\frac{1}{0}$, elle sera ce que j'appelle infiniment grande absolue ou métaphisique.

16. j'emploirai ces distinctions lorsque je voudrai spécifier la nature de l'infini dont j'aurai à parler, mais lorsque je me servirai simplement des termes généraux d'infini, d'infiniment grand, d'infiniment petit, ce que je dirai devra également s'appliquer à l'infini sensible et à l'infini absolu à moins que le sens de la phrase, ne dénote particulièrement l'un de ces infinis plutôt que l'autre.

17. l'objet de la question étant toujours de parvenir à la connoissance des rapports qui existent, non entre les diverses quantités qui composent le système auxiliaire mais seulement entre les quantités désignées qui en sont les limites ou dernières valeurs; on sent que pour saisir avec facilité ces dernières valeurs et les dernières raisons de ces quantités, il faut rapprocher autant qu'il est possible par l'imagination, le système auxiliaire du système désigné. or, à mesure que ces deux systèmes se rapprochent les quantités dont la limite est 0 deviennent de plus en plus petites, et celles dont la limite est $\frac{1}{0}$ deviennent de plus en plus grandes: il faut donc pour saisir avec facilité leurs limites et leurs dernières raisons, attribuer aux premières les plus petites valeurs que faire se pourra dans la pratique; et aux autres, au contraire, les plus grandes qu'il sera possible.

de là viennent les dénominations de quantités infiniment petites et de quantités infiniment grandes qu'on leur attribue; le nom d'infiniment petites qu'on donne aux premières, ayant pour objet de faire entendre que dans l'usage on a coutume de leur attribuer en effet, des valeurs assez petites, pour que l'imagination saisisse facilement leurs dernières raisons au moyen de la loi de continuité; et le nom d'infiniment grandes qu'on donne aux autres, indiquant que dans la pratique, on leur attribue en effet, des valeurs assez grandes pour que l'on puisse aisément saisir leurs dernières raisons et relations quelconques par la même loi de continuité. ainsi ce n'est pas seulement, à cause qu'on peut attribuer aux quantités dont la limite est 0 ou $\frac{0}{0}$, des valeurs déterminées aussi petites ou aussi grandes qu'on le veut; ce n'est point dis-je, à cause de cela qu'on les nomme infinitésimales, puisque les autres quantités indéterminées, peuvent être également supposées aussi grandes ou aussi petite qu'on le veut; mais c'est qu'on est invité par la comparaison des deux systêmes désigné et auxiliaire, à les rapprocher autant qu'il est possible pour soulager son imagination, et par conséquent à attribuer en effet aux unes des valeurs aussi petites qu'on le peut dans la pratique. et aux autres au contraire, des valeurs aussi grandes, qu'il est possible de le faire. Cependant, comme rien n'oblige absolument* de leur attribuer une valeur déterminée plutôt qu'une autre, il s'en suit, qu'on explique bien par là, l'origine de leur nom, mais que ce n'est point précisément ce qui constitue le caractère de ces quantités, lequel consiste uniquement comme on l'a dit (12) dans la nature de leurs limites.

* absolument parlant, rien n'empêche par exemple, dans une figure de géométrie de représenter les différentielles des abscisses et ordonnées, par des lignes plus longues que les abscisses et ordonnées elles-mêmes, et si on ne le fait pas ordinairement, il est aisé de voir, que c'est uniquement, afin que l'imagination ait moins à travailler pour saisir la dernière raison de ces différentielles. mais dira-t-on si l'on suppose les différentielles plus grandes que telle ou telle grandeur donnée serat-on ~~en~~ en droit de les négliger en comparaison de ces quantités données ? je réponds qu'on aura la mêmes droit dans les mêmes circonstances; parce que les erreurs auxquelles on donnera lieu par là, se compenseront d'elles-mêmes comme je le prouverai (55).

18. deux quantités sont dites différer infiniment peu ou être infiniment peu différentes l'une de l'autre lorsque leur dernière raison est une raison d'égalité.

en général, on dit que deux systèmes de quantités sont infiniment peu différents l'un de l'autre ou diffèrent infiniment peu, lorsque chacune des quantités qui composent l'un de ces systèmes diffère infiniment peu de la quantité qui lui correspond dans l'autre.

19. toute équation exacte a visiblement pour limite une autre équation exacte; mais la réciproque n'est pas toujours vrai. Car il peut arriver que la limite d'une équation fausse, soit cependant une équation vraie. par exemple, si A et B diffèrent infiniment peu l'équation A = B sera une équation fausse, mais sa limite sera néanmoins une équation exacte, et la dernière raison de A à B sera une raison d'égalité. cela posé pour distinguer ces sortes d'équations, dont on fait grand usage dans l'analyse infinitésimale ces équations exactes et de celles qui sont absolument fausses, je les nommerai équations imparfaites.

c'est-à-dire, que j'appelle imparfaite toute équation fausse dont les deux membres se trouvent avoir pour dernière raison une raison d'égalité.

l'erreur qui a lieu dans de semblables équations diminue donc à mesure que le système auxiliaire approche du système désigné c'est-à-dire, que les

les deux membres approchent d'autant plus du rapport d'égalité, qu'on attribue des valeurs moindres aux quantités auxiliaires que j'ai nommées indéfiniment petites desorte qu'une équation imparfaite diffère pour ainsi dire, aussi peu qu'on veut d'une équation vraie, et qu'on est maître d'en atténuer l'erreur autant qu'on le juge à propos.

20. en général je nommerai imparfaite toute proportion hypothèse ou proposition quelconque, dont l'expression ou traduction algébrique pourra être mise sous la forme d'une équation imparfaite. Ces sortes de propositions, ne doivent par conséquent être entendues qu'avec certaines modifications; mais ces modifications sont toujours faciles à saisir puisque les propositions dont il s'agit, tiennent toujours si l'on peut s'exprimer ainsi, d'aussi près qu'on veut à la vérité.

21. dans tout calcul infinitésimal, on prend toujours au moins tacitement, une quantité infiniment petite pour servir de terme de comparaison à toutes les autres quantités infinitésimales, j'appelle cette quantité ou terme de comparaison, base infinitésimale.

22. toute quantité dont le rapport à la base infinitésimale est une quantité ~~infiniment petite~~ finie (II) s'appelle infiniment petite du premier ordre.

toute quantité dont le rapport à la base infinitésimale est une quantité infiniment petite du premier ordre, je nomme infiniment petite du second ordre.

Toute quantité dont le rapport à la base infinitésimale est une quantité infiniment petite du second ordre, je nomme infiniment petite du troisième ordre et ainsi de suite.

C'est-à-dire donc, qu'en général si l'on nomme x la base infinitésimale, toute quantité égale à Ax^n A étant une quantité finie, sera infiniment petite de l'ordre n.

23. Si l'on supposoit $n=o$, Ax^n deviendroit Ax^o ou A. c'est-à-dire, que toute quantité infiniment petite de l'ordre o, n'est autre chose qu'une quantité finie.

24. toute quantité finie, divisée par une quantité infiniment petite a évidement pour limite $\frac{1}{o}$, et par conséquent (12) est une quantité infinie Cela posé,

toute quantité finie, divisée par la base infinitésimale sera nommée infiniment grande du premier ordre.

toute quantité infiniment grande du premier ordre divisée par la base infinitésimale, sera nommée, infiniment grande du second ordre, et ainsi de suite.

ainsi en général, si l'on nomme x la base infinitésimale, toute quantité égale a Ax^{-n} A étant une quantité finie sera infiniment grande de l'ordre n.

25. Si l'on suppose $n=o$ Ax^{-n} deviendra Ax^o ou A.

c'est-à-dire, que toute quantité infiniment grande de l'ordre o, est comme la quantité infiniment petite du même ordre, une quantité finie.

26. il suit visiblement de la subordination qu'on vient d'établir entre les quantités infinitésimales, que c'est la même chose de dire d'une quantité qu'elle est infiniment petite de l'ordre n, ou infiniment grande de l'ordre -n; et que réciproquement, c'est la même chose de dire d'une quantité qu'elle est infinie de l'ordre n, ou infiniment petite de l'ordre -n.

il s'en suit aussi évidemment, que le produit de deux quantités infiniment petites du premier ordre est infiniment petit du second ordre; qu'en général le produit d'une quantité infiniment petite de l'ordre m par une quantité infiniment petite de l'ordre n est une quantité infiniment petite de l'ordre m+n, que leur quotient, est au contraire, une quantité infiniment petite de l'ordre m-n. et que pareillement le produit ou le quotient de deux quantités infinies l'une de l'ordre m, l'autre de l'ordre n, est une quantité infinie de l'ordre m+n s'il est question du produit, et de l'ordre m-n, s'il est question du quotient.

il est aisé d'extraire encore de ces définitions plusieurs conséquences dans le détail desquelles je n'entreverai pas parce que je n'entreprends point un traité d'analyse infinitésimale, et que mon but est simplement, d'en faire connoître l'esprit et les premiers principes.

27. imaginons maintenant, un système quelconque de quantités diverses, les unes constantes, les autres variables, et liées entre elles par une loi quelconque donnée, ainsi que seroient par exemple, les paramètres abscisses et ordonnées d'une courbe. Soient x, y, z, &c les variables de ce système, que je suppose changer par degrés insensibles, suivant la loi prescrite par les relations données entre les constantes et les variables qui le composent. Cela posé, considérons ce système, dans deux états infiniment peu différents l'un de l'autre ; supposons c'est-à-dire, que les nouvelles variables diffèrent infiniment peu des premières, et prennons ce second état du système, pour système auxiliaire, tandis que le premier sera le système désigné.

cela étant, si l'on prend la différence d'une quelconque des quantités de ce système auxiliaire à la quantité correspondante du système désigné, la limite ou dernière valeur de cette différence sera évidemment égale à zéro. donc (12) ce sera une quantité infiniment petite: c'est pourquoi, je la désignerai dans le discours, par l'expression diminutive de différentielle de cette quantité désignée, et je la représenterai dans le calcul par la quantité même dont elle est la différentielle précédée de la lettre ou caractéristique d. c'est-à-dire, que si x, y, z, &c sont les quantités du système désigné, x', y', z', &c les quantités correspondantes, du Système auxiliaire ; les quantités $x'-x$, $y'-y$, $z'-z$, &c seront les différentielles respectives de x, y, z, &c et on les représentera dans le calcul, par dx, dy, dz, &c demanière qu'on aura $dx = x'-x$, $dy = y'-y$, $dz = z'-z$, &c.

28. les quantités x, y, z, &c, qui ont pour différentielles dx, dy, dz, &c sont nommées sommes ou intégrales de ces différentielles, et on représente ces intégrales dans le calcul, par leurs différentielles précédées de la lettre $\int$. ainsi $\int dx$ représente l'intégrale ou la somme de dx, $\int dy$ l'intégrale de dy, &c. la différentielle se nomme aussi élement ou fluxion. et l'intégrale se nomme aussi fluente. ainsi dx est l'élement, la fluxion ou la différentielle de x; et réciproquement x ou $\int dx$, est la somme, l'intégrale ou la fluente* de dx.

29. l'opération par laquelle on assigne la différentielle d'une quantité, se nomme différentiation, et réciproquement l'opération par laquelle on assigne l'intégrale ou la somme d'une différentielle, se nomme intégration ou sommation.

différentier une quantité ou une intégrale c'est donc assigner la différentielle, fluxion ou élement. et réciproquement, intégrer ou sommer une différentielle

* ces dénominations viennent, des différents points de vue, sous lesquels on peut envisager la question de l'infini nous avons vu (27) d'ou vient le nom de différentielle celui de fluxion est propre à exprimer que le changement qui s'opère dans le Système auxiliaire est assujetti à la loi de continuité, et l'on verra (83) d'ou vient celui d'élement qui est opposé à celui de somme ; celui de différentiel est opposé à celui d'intégrale et celui de fluxion à celui de fluente.

c'est en assigner la somme fluente ou intégrale.

30. l'art de différentier les diverses quantités proposées, ou d'assigner leurs différentielles, je nomme Calcul différentiel ou Calcul des fluxions ; et réciproquement l'art d'intégrer les quantités différentielles ou d'en trouver la somme j'appelle Calcul intégral ou des fluentes.

31. le systême auxiliaire composé des quantités x', y', z' &c correspondantes à celles du premier systême et (3) de toutes celles qui comme $x'-x$ ou dx, $y'-y$ ou dy, &c en sont des fonctions quelconques; le systême dis-je étant assujetti dans son changement à la loi de continuité il est évident, que nous pouvons concevoir le systême désigné dans un troisième état infiniment peu différent du second; ou considerer le troisième état comme un second systême auxiliaire composé de quantités x'', y'', z'', &c $x''-x'$, $y''-y'$, $z''-z'$, &c correspondantes à celles x', y', z', &c $x'-x$, $y'-y$, $z'-z$, &c qui composent le premier et qui different infiniment peu chacune de la correspondante. Cela posé, demême, que la différence de chacune des quantités du premier systême auxiliaire à sa correspondante dans le systême désigné, a été nommée différentielle de cette dernière. demême aussi, la différence de chacune des quantités qui composent le second systême auxiliaire à sa quantité correspondante dans le premier sera nommée différentielle de cette dernière. et je la distinguerai

dans le calcul, par la même caractéristique d mise en avant : c'est-à-dire que $x''-x'$, $y''-y'$, $z''-z'$, &c, $(x''-x)-(x'-x)$, $(y''-y)-(y'-y)$, $(z''-z')-(z'-z)$ &c seront les différentielles de x', y', z', &c, $(x'-x)$ ou dx, $(y'-y)$ ou dy $(z'-z)$ ou dz &c et on les écrira ainsi, dx', dy', dz' &c $d(x'-x)$ ou ddx (parceque $x'-x=dx$), $d(y'-y)$ ou ddy, $d(z'-z)$ ou ddz, &c : et pour rapporter tout autant qu'il est possible au système désigné, je nommerai ces nouvelles quantités ddx, ddy, ddz, &c secondes différences de x, y, z, &c ou leurs différentielles du second ordre.

pareillement, le second système auxiliaire composé des quantités x'', y'', z'', &c dx', dy', dz', &c ddx, ddy, ddz, &c étant assujetti dans son changement à la loi de continuité, on pourra concevoir un troisième système auxiliaire composé de quantités analogues infiniment peu différentes de celles qui composent le second système auxiliaire. nous nommerons donc encore les différences respectives des quantités qui composent ce troisième système auxiliaire à celles du second; nous les nommerons dis-je différentielles de celles-ci, et nous les exprimerons dans le calcul, par les quantités dont elles font différentielle précédées de la caractéristique d ainsi $x'''-x''$, $y'''-y''$ &c $dx''-dx'$ $dy''-dy'$, &c, $ddx'-ddx$, $ddy'-ddy$, &c seront les différentielles de x'', y'', &c dx', dy', &c ddx, ddy, &c et ces différentielles seront exprimées dans le calcul par dx'', dy'', &c ddx', ddy', &c $dddx$, $dddy$, &c enfin, pour rapporter encore ces nouvelles quantités $dddx$, $dddy$, &c au système désigné; je les nommerai troisièmes différences des quantités x, y, &c ou leurs différentielles du troisième ordre.

par les mêmes raisons, on pourra imaginer un quatrième, un cinquième, enfin un nombre quelconque de systèmes auxiliaires qui seront autant d'états consécutifs du système désigné, et composés chacun de quantités infiniment peu différentes, des quantités correspondantes du système précédent. ce qui donnera à chaque fois, des différentielles d'un nouvel ordre.

32. on comprend sous le nom de calculs différentiel et intégral, ou de calcul des fluxions et fluentes, non seulement l'art de rechercher les différences premières et de les intégrer, mais aussi, tout ce qui regarde la recherche des différentielles des ordres supérieurs et de revenir de celles-ci aux quantités qui les produisent par leur différentiation.

la partie de ce calcul qui enseigne à rechercher les différentielles des ordres supérieurs se nomme calcul différentio-différentiel.

33. toutes ces quantités différentielles forment évidemment une liaison entre le système désigné et les divers systèmes auxiliaires dont nous avons parlé: d'où il est aisé de sentir que l'art d'intégrer et de différentier

{ 195 }

24

doit entrer pour beaucoup dans la théorie de l'infini, et en effet, il y joue un si grand rôle, que l'on prend souvent la partie pour le tout, en comprenant sous le nom de calcul différentiel et intégral toute la théorie de l'infini.

34. les divers systèmes auxiliaires dont nous venons de parler ne sont autre chose par hypothèse, que le système même proposé, considéré dans plusieurs états consécutifs dépendants des rapports prescrits entre les quantités qui le composent, et par lesquels est déterminée la loi de continuité suivant laquelle se fait le changement.

concevons maintenant, un autre système auxiliaire, mais qui ne soit point assujetti à la loi donnée qui lie les différentielles dont nous venons de parler, avec le système désigné. supposons c'est-à-dire, que ce nouveau système auxiliaire soit ~~composé~~ différent du premier, et suive dans son changement, une autre loi quelconque. cela posé, si l'on prend comme ci-dessus, la différence de chacune des quantités de ce nouveau système, à la quantité correspondante du système désigné. cette différence sera une quantité infiniment petite que l'on distingue des différentielles par une dénomination particulière. on les nomme variations; desorte, qu'il y a cette différence entre variations et différentielles, savoir

{ 196 }

que le système auxiliaire qui donne celles-ci, est assujetti dans son changement, à une loi prescrite et donnée; aulieu que celui qui donne ce qu'on nomme variations change suivant une autre loi quelconque.

pour exprimer ces variations dans le calcul, on emploie une caractéristique particulière, autre que celle dont on se sert pour les différentielles et qui cependant lui est analogue; parce qu'en effet, la variation d'une quantité n'est à proprement parler que sa différentielle même, prise dans un autre système auxiliaire : ainsi par exemple, si l'on désigne la différentielle de x par dx on pourra désigner sa variation par δx.

on peut encore imaginer en quel nombre on voudra, d'autres systèmes auxiliaires, assujettis dans leurs changements et dans leurs relations réciproques à des loix absolument arbitraires, ou en partie prescrites par les conditions du problème et en partie arbitraires; et alors, si l'on prend les différences respectives des quantités qui composent le système auxiliaire, aux quantités correspondantes du système désigné on aura des variations de différents genres. par exemple, si l'on conçoit que le système auxiliaire soit composé de deux parties dont l'une lui soit commune avec le système désigné et parconséquent constante (10) et dont l'autre lui soit commune avec le système auxiliaire qui donne les différentielles et dont la loi est prescrite; les variations se nomment dans ce cas différentielles partielles.

or demême; qu'il y a des différentielles secondes, troisièmes &c on pourra prendre des variations de différents ordres, des différentielles de variations, des variations de différentielles &c et toutes ces quantités pourront se trouver mêlées dans un même calcul dont le calcul différentiel ordinaire, est le cas particulier où il n'y a qu'un seul système auxiliaire, et où ce système est assujetti à une loi donnée. or l'analyse de l'infini en général, est l'art d'employer le ministère ou l'entremise de tous ces systèmes et quantités auxiliaires, pour découvrir les rapports et relations quelconques, des quantités qui composent le système désigné.

35. nous avons distingué (15) deux sortes d'infinis, savoir, l'infini sensible et l'infini absolu; on est donc maître d'attribuer aux différentielles et variations, des valeurs qui soient des quantités effectives, ou des valeurs absolument nulles: dans ce dernier cas, on les nommera différentielles et variations absolues ou évanouissantes: dans l'autre au contraire, on les nommera différentielles et variations vraies ou sensibles : mais si l'onne désigne pas expressément la nature de celles dont on parle; ce qu'on en dira, sera applicable aux deux espèces de différentielles et de variations; à moins, que le sens de la phrase ne montre, qu'il ne peut être question que de l'une des deux.

36. pour réunir ici, toutes les définitions dont j'aurai besoin, je vais encore expliquer, ce que j'entends par ~~moment évanouissant, ou d'évanouissement,~~ moment différentiel, moment de variation, et moment intégral ou d'intégrale. j'avertis d'abord, que je désigne la limite ou dernière valeur d'une quantité quelconque par cette même quantité précédée de la caractéristique L; ainsi par exemple la limite de $\frac{dx}{dy}$ ou la dernière raison de dx à dy sera exprimée par $L\frac{dx}{dy}$; de plus nous avons dit (21) que dans l'analyse infinitésimale, on prend toujours, au moins tacitement une quantité infiniment petite pour servir de base ou terme de comparaison. Cela posé.

je nommerai moment différentiel d'une quantité quelconque, la dernière raison de sa différentielle avec la base infinitésimale; c'est-à-dire, que si dx par exemple, est prise pour base, et que dy soit la différentielle de y; la dernière raison de dy à dx, ou $L\frac{dy}{dx}$, est ce que je nomme moment différentiel de y, et je le désignerai dans le calcul de cette manière Dy: ainsi nous aurons $Dy = L\frac{dy}{dx}$, dx étant prise pour base infinitésimale*.

réciproquement y sera nommée moment intégral de Dy, et je le désignerai ainsi $\int Dy$.

par la même raison, si l'on différentie Dy, qui est une quantité finie, qu'on divise cette différentielle par la base infinitésimale dx, et qu'on prenne la dernière

_____
* en général je nommerai moment d'évanouissement d'une quantité quelconque infiniment petite, la dernière raison de cette quantité à celle qui est prise pour base: ainsi Dy est le moment d'évanouissement ou moment évanouissant de dy.

~~la dernière~~ valeur de ce rapport; ce sera le moment différentiel de la quantité finie Dy; ainsi ce moment sera $\mathcal{L}\frac{Dy}{dx}$ ou DDy qui est aussi une quantité finie; et réciproquement, Dy sera le moment intégral de DDy.

chaque quantité aura de même ses moments de variation de différents ordres, qu'on désignera par des caractéristiques analogues. l'usage de ces expressions sera expliqué dans la suite.

je prie le lecteur de donner une attention particulière aux définitions précédentes; j'ai tâché de leur donner toute l'exactitude et la clarté possibles, de manière qu'elles puissent servir, à des personnes qui n'auroient aucune idée de cette théorie. je vais maintenant rechercher, quelle a pu être la marche des inventeurs dans sa découverte.

origine que peut avoir eu l'analyse infinitésimale.

37. la difficulté qu'on rencontre souvent à exprimer exactement par des équations, les différentes conditions d'un problème et à resoudre ces équations, a pu faire naître les premières idées du calcul infinitésimal*. lorsqu'il est trop difficile en effet, de

---

* il ne s'agit pas ici de la marche qu'on a suivie réellement, dans la découverte, mais de celle qu'on auroit pu suivre assez naturellement. je n'entreprendrai pas non plus de discuter si c'est neuton qui est l'inventeur de l'analyse infinitésimale ou si c'est lebnitz: je ne puis cependant m'empêcher de dire, que je suis intimement persuadé que c'est à lebnitz seul, qu'on doit cette brillante découverte. neuton a trouvé la méthode des premières et dernières raisons, mais qu'il y a loin de la

trouver la solution exacte d'une question; il est naturel de chercher aumoins, à en approcher le plus qu'il est possible, en négligeant les quantités qui embarassent les combinaisons; si l'on prévoit, que ces quantités négligées, ne peuvent à cause de leur peu de valeur, produire qu'une erreur legere, dans le résultat du calcul. c'est ainsi par exemple, que ne pouvant découvrir qu'avec peine les propriétés des courbes on aura imaginé de les regarder, comme des polygones d'un grand nombre de côtés. en effet, si l'on conçoit un polygone régulier inscrit dans un cercle, il est visible, que les deux figures, quoique toujours différentes, et

subtile synthèse employée par ce grand homme dans le livre des principes et qui exige tant de sagacité dans le calculateur; à l'algorithme si simple et si facile à mettre en pratique du calcul proposé par lebnitz! ces deux méthodes sont les mêmes au fond, j'en conviens; mais au fond, ne sont-elles pas aussi l'une et l'autre, les mêmes, que la méthode d'exhaustion. et dira-t-on pour cela, que les anciens, avoient connaissance de l'analyse infinitésimale? Si parceque'on découvre une vérité, on prétend être l'auteur de toutes les conséquences qui en dérivent, quelque éloignées qu'elles soient de leur origine, il faudra donc dire que les bernouillis, les eulers, les d'alembert, les delagrange les neuton même, n'ont rien inventé; parceque les premiers principes des sciences, étoient connus avant eux. pour moi je pense que le calcul infinitésimal tient de beaucoup plus près à la méthode des indéterminées de descartes, qu'à la méthode des premières et dernières raisons de neuton, et cependant, descartes s'il vivoit encore, auroit honte de se donner pour l'inventeur du calcul différentiel. on ne dispute point à neuton le système de l'attraction, malgré ce qu'en avoient dit les anciens, et malgré ce qu'avoient écrit kepler sur le mouvement des planetes et huguens sur la théorie des forces centrales; pourquoi lui attribueroit-on une gloire dont il n'a pas besoin, et qu'un autre a méritée. en un mot ce qui caractérise l'invention du calcul infinitésimal, est la forme analytique, parceque c'est ce qui le distingue d'une manière tranchante de la méthode d'exhaustion; or neuton n'a réduit la méthode des premières et dernières raisons en algorithme par son calcul des fluxions, que longtems après la découverte du calcul différentiel, où du moins, il ne la pas publié plutôt, et l'on ne peut juger des hommes que par leurs ouvrages. il n'est donc pas plus l'inventeur du calcul infinitésimal que les anciens, le sont de l'algebre ordinaire; quoiqu'ils ayent fait plusieurs raisonnements vraiment analytiques, c'est-a-dire, ou ils comparent les quantités inconnues comme si elles étoient données, ce qui est la base du calcul algébrique.

incapables de jamais devenir identiques, se ressemblent cependant de plus en plus, à mesure que le nombre des côtés du polygone augmente; que leurs périmètres, leurs surfaces, les solides formés par leurs révolutions autour d'un axe donné, les lignes analogues menées au dedans ou au dehors de ces figures, les angles formés par ces lignes &c sont, sinon respectivement égaux, au moins, d'autant plus approchants de l'égalité que le nombre des côtés devient plus grand: d'où il suit qu'en supposant ce nombre très-grand en effet, on pourra sans erreur sensible, attribuer au cercle circonscrit, les propriétés qu'on aura trouvées appartenir au polygone inscrit.

en outre, chacun des côtés diminue évidemment de grandeur, à mesure que le nombre de ces côtés augmente, et par conséquent, si l'on suppose, que le polygone soit réellement composé d'un très grand nombre de côtés, on pourra dire aussi que chacun d'eux est réellement très-petit.

38. Cela posé, s'il se trouvoit par hazard dans le cours du calcul, une circonstance particulière, où l'on pût simplifier beaucoup les opérations en négligeant par exemple, un de ces petits côtés, par comparaison à une ligne donnée; c'est-à-dire, en employant dans le calcul cette ligne donnée aulieu d'une quantité qui seroit égale à la somme faite

de cette ligne et du petit côté en question; il est clair, qu'on pourroit le faire sans inconvénient: car l'erreur qui en résulteroit, ne pourroit être qu'extrêmement petite, et ne mériteroit pas, qu'on se mit en peine, pour en connoître la valeur.

39. par exemple, soit proposé de mener une tangente au point m de la circonférence mbdC. (voyez la figure)

Soient, c le centre du cercle, dcb l'axe, supposons l'abscisse $dp = x$, l'ordonnée correspondante et perpendiculaire $mp = y$, et soit tp la soutangente cherchée.

pour la trouver, considérons le cercle comme un polygone d'un très-grand nombre de côtés, soit mn un de ces côtés, prolongeons le jusqu'à l'axe, ce sera la tangente en question puisque cette ligne ne pénètre pas dans l'intérieur du polygone; abaissons de plus la perpendiculaire mo, sur nq parallèle à mp, et nommons a le rayon du cercle. cela posé, nous aurons évidement, $mo : no :: tp : mp$. ou $\frac{mo}{no} = \frac{tp}{mp}$.

d'une autre part, l'équation de la courbe étant pour le point m, $yy = 2ax - xx$, elle sera pour le point n $(y + no)^2 = 2a(x + mo) - (x + mo)^2$. ôtant de cette équation, la première trouvée pour le point m et réduisant, on a $\frac{mo}{no} = \frac{2y + no}{2a - 2x - mo}$. égalant donc, cette valeur de $\frac{mo}{no}$ à celle trouvée ci dessus, et multipliant par mp ou y, il vient $tp = \frac{y(2y + no)}{2a - 2x - mo}$.

si donc mo et no étoient connues, on auroit la valeur

cherchée de tp; or ces quantités mo, no, sont très-
petites, puisqu'elles sont moindres chacune, que le
côté mn, qui par hypothèse est lui même très-
petit. donc (38) on peut négliger sans erreur
sensible ces quantités, par comparaison aux quantités
y et $2x - 2a$ auxquelles elles sont ajoutées. donc l'équation
se réduit à $tp = \frac{y^2}{a-x}$. ce qu'il falloit trouver.

40. Si ce résultat n'est pas absolument exact il
est au moins évident que dans la pratique, il peut
passer pour tel, puisque les quantités mo, no, sont
extrêmement petites: mais quelqu'un qui n'auroit
aucune idée de la doctrine des infinis, seroit
peut-être fort étonné, si on lui disoit que l'équation
$tp = \frac{y^2}{a-x}$, non seulement approche beaucoup du vrai,
mais est réellement de la plus parfaite exactitude.
c'est cependant, une chose dont il est aisé de s'assurer
en cherchant tp d'après le principe que la tangente
est perpendiculaire à l'extrémité du rayon. car il est
visible que les triangles semblables cpm, mpt, donnent
cp:mp::mp:tp. d'où l'on tire $tp = \frac{mp^2}{cp}$ comme ci dessus.

on a dû naturellement
la regarder d'abord
comme une simple
méthode d'approxi=
mation.

41. pour second exemple, supposons qu'il soit question
de trouver la surface d'un cercle donné.

Considérons encore cette courbe comme un
polygone régulier d'un grand nombre de côtés. l'aire
d'un polygone régulier quelconque, est égale au
produit de son périmètre, par la moitié de la
perpendiculaire menée du centre sur l'un des côtés.

donc le cercle étant considéré comme un polygone
d'un grand nombre de côtés; sa surface doit être égale
au produit de sa circonférence par la moitié du rayon.
proposition, qui n'est pas moins exacte, que le résultat
trouvé ci-dessus.

42. quelque vagues et peu précises que puissent donc
paroître d'abord, ces expressions de très-grand et de
très-petit ou autres équivalantes, on voit par les
deux exemples précédents, que ce n'est cependant
pas sans utilité, qu'on les emploie dans les combinais^{ns}
mathématiques: tâchons donc, s'il est possible, de fixer
exactement, le sens qu'on doit attacher à ces expressions.

or il suffit pour cela, d'observer attentivement, la
marche de l'esprit, dans l'usage qu'on en fait; car on
remarquera aisément 1° une comparaison tacite
entre deux systèmes de quantités dont l'un qui est
le système désigné, est composé du cercle et des tangente
sous-tangente, abscisse et ordonnée correspondantes au
point proposé; l'autre qui est le système auxiliaire
est composé du polygone inscrit qui est supposé avoir
un de ses angles au point m, lequel correspond au
cercle; du prolongement de son côté mn adjacent au
point proposé, lequel correspond à la tangente, des
lignes dq, nq qui correspondent à l'abscisse et à
l'ordonnée, et enfin, de toutes les quantités qui, comme
mo, no dépendent, en partie, de ce système auxiliaire

2º que plus le système auxiliaire approche du système désigné, plus les quantités mn, mo, no, que nous avons désignées par l'expression de très-petites, diminuent de manière, que leur limite ou dernière valeur est o. 3º qu'enfin, libres d'attribuer à ces quantités mn, mo, no, des valeurs plus ou moins grandes, nous avons voulu les supposer assez petites, pour que l'imagination pût saisir avec facilité leurs dernières raisons ou les limites de leurs rapports; parceque ces limites se trouvant liées avec les quantités désignées dont nous voulions avoir la relation, on a senti, que c'étoit un moyen simple d'y parvenir.

mais il est évident (12 et 17) que dans ces propriétés, consiste précisément, le caractère des quantités que nous avons nommées infiniment petites* dont nous avons donné une exacte définition. on doit donc cesser de regarder ces expressions de très-grand et de très-petit employées ci-dessus, comme des termes vagues et sans

_____

* il n'est pas nécessaire, lorsqu'on entreprend la solution d'un problème, de faire l'énumération des quantités qui composent les systèmes désigné et auxiliaire. il n'est pas même nécessaire d'avertir qu'on suppose l'existence et la comparaison de deux systèmes de cette nature: il suffit (13) que la supposition en soit implicitement renfermée dans quelque autre hypothèse ou proposition quelconque. en un mot, dès que le sens de la phrase ou quelque expression du discours suppose soit explicitement soit tacitement, qu'on a recours à des quantités auxiliaires, pour découvrir les rapports qui existent entre les parties d'un système désigné, c'est assez pour que le calcul appartienne à ce qu'on nomme analyse infinitésimale. ainsi dans le cas présent, la seule supposition de la ligne nq, menée à une distance arbitraire de mp, suffit, pour que le problème proposé, appartienne à l'analyse infinitésimale parceque cette supposition renferme implicitement la comparaison des deux systèmes dont nous avons parlé ci-dessus et dont l'un est pris pour terme de comparaison ou limite de l'autre.

précision: on peut même remarquer, que les quantités x, y étant liées par la nature de la courbe, desorte, qu'elles varient suivant une loi prescrite et donnée, à mesure que le système auxiliaire approche du système désigné: on peut dis-je remarquer que les quantités mo, no, sont précisément (27) ce que nous avons nommé différentielles de x, y; c'est-à-dire, que l'on a $mo = dx$ et $no = dy$.

dans le cas présent, ces différentielles dx, dy, sont (35) de celles que j'ai nommées <u>vraies ou sensibles</u>, parceque nous leur avons attribué des valeurs effectives; mais on pourroit aussi, en envisageant la question sous un autre point de vue, les supposer absolument égales à zéro comme on le verra plus bas; et alors, elles seroient de celles que j'ai nommées différentielles absolues ou évanouissante

43. par les deux exemples rapportés ci-dessus, il est aisé de connoître, combien l'usage de ces sortes de quantités est utile pour faciliter la solution des diverses questions qui peuvent être proposées; car leur notion une fois admise, toutes les courbes pourront aussi bien que le cercle, être considérées comme des polygones d'un grand nombre ou d'une infinité de côtés. toutes les surfaces, pourront être partagées en une multitude de bandes ou de zones. tous les corps en corpuscules toutes les quantités en un mot, pourront être décomposées en particules de même espèce qu'elles. de là naissent, beaucoup de nouveaux rapports et de nouvelles combinaisons: et l'on peut juger par cet apperçu des ressources que doit fournir dans le calcul l'introduction de ces quantités

élémentaires ou infinitésimales.

44. mais l'avantage qu'elles procurent, est bien plus considérable encore, qu'on n'avoit d'abord eu lieu de l'espérer. car il suit des exemples rapportés, que ce qui n'avoit été regardé en premier lieu, que comme une simple méthode d'approximation, conduit au moins en certains cas, à des résultats parfaitement exacts. il seroit donc intéressant, de savoir distinguer ceux où cela arrive, d'y ramener les autres autant qu'il est possible, et de changer ainsi cette méthode d'approximation, en un calcul absolument exact et rigoureux. or tel est l'objet, de l'analyse infinitésimale.

45. voyons donc d'abord, comment dans l'équation $tp = \frac{y(2y + no)}{2a - 2x - mo}$. trouvée (39), il a pu se faire, qu'en négligeant mo et no, on n'ait point altéré la justesse du résultat; ou plutôt, comment il est devenu exact par la suppression de ces quantités, et pourquoi, il ne l'étoit pas auparavant.

or on peut rendre fort simplement raison de ce qui est arrivé dans la solution du problème traité ci dessus, en remarquant, que l'hypothèse d'où l'on est parti étant fausse, puisqu'il est absolument impossible qu'un cercle puisse être jamais considéré comme un vrai polygone quel que puisse être le nombre de ses côtés; il a dû résulter de cette hypothèse une erreur dans l'équation $tp = \frac{y(2y + no)}{2a - 2x - mo}$ et que le résultat

on a découvert ensuite, que malgré les erreurs commises dans l'expression des conditions de chaque problème les résultats étoient néanmoins de la plus parfaite exactitude.

$tp = \frac{y^2}{a-x}$ étant certainement exact, comme on le prouve par la comparaison des deux triangles Cpm, mpt, on a pu négliger mo et no, et même on a du le faire, pour rectifier le calcul, et détruire l'erreur à laquelle avoit donné lieu la fausse hypothèse, d'où l'on étoit parti. négliger les quantités de cette nature, est donc non-seulement permis en pareil cas; mais il le faut et c'est la seule manière d'exprimer exactement les conditions du problème.

46. le résultat $tp = \frac{y^2}{a-x}$, n'a donc été obtenu, que par une compensation d'erreurs, et cette compensation, peut être rendue plus sensible encore, en traitant l'exemple rapporté ci-dessus d'une manière un peu différente, c'est-à-dire, en considérant le cercle comme une véritable courbe, et non pas comme un polygone.

pour cela, soient menées rs à une distance quelconque de mp, et la sécante rt: nous aurons évidemment, tp : mp :: mz : rz. et partant $tp + t't = mp \cdot \frac{mz}{rz}$. cela posé, si nous imaginons que rs se meuve parallèlement à elle-même, en s'approchant continuellement de mp; il est visible que le point t' s'approchera en même tems de plus en plus du point t, et qu'on pourra par conséquent rendre la ligne t't aussi petite qu'on voudra, sans que la proportion établie ci-dessus, cesse pour cela d'avoir lieu. si donc, je néglige cette quantité t't dans l'équation que je viens de trouver, il en résultera à la vérité une erreur dans l'équation $tp = \frac{mp \cdot mz}{rz}$, à laquelle la première sera alors réduite; mais cette erreur pourra être rendue aussi legère qu'on le jugera à propos: c'est-à-dire, que le rapport des deux membres, approchera autant qu'on voudra, du rapport d'égalité.

pareillement, nous avons (39) $\frac{mz}{rz} = \frac{2y + rz}{2a - 2x - mz}$, or plus rs approche de mp plus mz et rz sont petites, et par conséquent si je les néglige dans le second membre de cette équation, il y aura une erreur dans celle $\frac{mz}{rz} = \frac{y}{a-x}$ à laquelle elle sera réduite alors; mais cette erreur pourra comme la première, être supposée aussi petite qu'on voudra.

Cela etant, sans avoir égard, à des erreurs que je serai toujours maître d'atténuer autant que je le jugerai à propos, je traite les deux équations tp = mp. $\frac{mz}{rz}$ et $\frac{mz}{rz} = \frac{y}{a-x}$, comme si elles étoient parfaitement exactes, substituant donc, dans la dernière, la valeur de $\frac{mz}{rz}$ tirée de l'autre, j'ai pour résultat tp = $\frac{y^2}{a-x}$ comme ci dessus.

Ce résultat est parfaitement juste, puisqu'il est conforme à celui qu'on a obtenu, par la comparaison des triangles Cpm, mpt, et cependant, les équations tp = mp. $\frac{mz}{rz}$, et $\frac{mz}{rz} = \frac{y}{a-x}$ d'où il a été tiré sont certainement fausses toutes deux, puisque la distance de rs à mp, n'a point été supposée nulle, mais bien égale à une ligne quelconque arbitraire* il faut par conséquent de toute nécessité que les erreurs se soient compensées mutuellement par la comparaison des deux équations erronées.

47. voila donc le fait des erreurs compensées bien acquis et bien prouvé, il s'agit maintenant de l'expliquer, de rechercher quel est le signe auquel on reconnoit que la

* cela prouve bien ce que nous avons dit dans la note première et dans celle l'article 17 c'est-à-dire qu'on n'est point tenu d'attribuer aux quantités dont les géometres font usage sous le nom d'infiniment petites, des valeurs moindres que telle ou telle grandeur donnée, que si on le fait c'est uniquement pour aider l'imagination et que la compensation d'erreurs n'auroit pas moins lieu quand même on leur attribueroit d'autres valeurs quelconques.

compensation a lieu, et les moyens de la produire en chaque circonstance particulière.

or il suffit pour cela de remarquer, que les erreurs commises dans les équations $tp = mp \cdot \frac{mz}{rz}$ et $\frac{mz}{rz} = \frac{y}{a-x}$, pouvant être rendues aussi petites qu'on veut, celle qui auroit lieu, s'il s'en trouvoit une dans l'équation résultante $tp = \frac{y^2}{a-x}$ pourroit également être rendue aussi petite qu'on voudroit. or cela n'est pas, puisque le point m, par où doit passer la tangente étant donné, il ne se trouve aucune des quantités $a, x, y, tp,$ de cette équation qui soit arbitraire: donc il ne peut y avoir en effet, aucune erreur dans cette équation.

48. il suit de là évidemment, que la compensation des erreurs qui se trouvoient dans les équations $tp = \frac{mp \cdot mz}{rz}$ et $\frac{mz}{rz} = \frac{y}{a-x}$ se manifeste dans le résultat par l'absence des quantités $mz, rz,$ qui causoient les erreurs. ainsi la méthode qu'on vient d'employer consiste en deux points qui sont 1° introduire dans le calcul certaines quantités indéterminées au moyen desquelles on puisse faciliter l'expression des conditions du problème, en se permettant s'il le faut quelques erreurs, pourvu que les erreurs soient toujours de nature à ce qu'on puisse constamment se réserver la faculté de les atténuer autant qu'on voudra, en diminuant de plus en plus la valeur des indéterminées introduites. 2° éliminer les indéterminées, et par ce moyen, opérer la compensation de ces erreurs.

49. l'inventeur de cette méthode, a donc pu être conduit à sa découverte par un raisonnement bien simple, si à la

place d'une quantité quelconque donnée soit immédiatement soit par les conditions du problème a-t-il pu dire, j'emploie dans le calcul une autre quantité qui ne soit point égale à la première il en résultera une erreur: mais si la différence des deux quantités employées l'une pour l'autre dépend de ma volonté, et que je sois maître de la rendre aussi petite que je le voudrai; cette erreur ne sera point dangereuse, puisque quand même je ne pourrois pas la détruire entièrement, il dépendra toujours de moi, de la modérer autant que je le jugerai à propos, en diminuant de plus en plus, cette quantité arbitraire qui en est la cause. je pourrois même commettre à la fois plusieurs erreurs semblables, sans qu'il en arrivât pour cela aucun inconvénient, puisque je serai toujours maître du degré de précision que je voudrai donner à mes résultats. il y a plus encore, c'est qu'il pourroit arriver par hazard que ces erreurs se compensassent mutuellement, et que j'obtinsse par là des résultats parfaitement exacts; de sorte que j'aurois alors plus que je n'avois osé l'espérer. Car je n'avois compté que sur une approximation, qu'à la vérité, je pouvois pousser aussi loin que je le voulois; aulieu que si je trouvois le moyen d'obtenir la compensation désirée mes résultats, seroient rigoureusement exacts. mais comment opérer cette compensation, et dans tous les cas? C'est ce qu'un peu de réflexion aura pu faire découvrir: en effet,

supposons aura pu dire l'inventeur, que la compensation
désirée ait effectivement lieu, et voyons par quel signe, elle
doit se manifester dans le résultat du calcul. or ce qui
doit naturellement arriver, c'est que les erreurs ayant
disparu, les quantités qui causoient ces erreurs auront
disparu de même: car ces quantités ayant par hypothèse,
des valeurs arbitraires, elles ne doivent plus entrer,
au moins nécessairement, dans des formules ou résultats
qui ne le font pas, et qui étant devenus exacts par
supposition dépendent uniquement, non de la volonté
mais de la nature des choses dont on s'étoit proposé de
trouver la relation exprimée par ces résultats: donc
le signe qui annonce, que la compensation désirée a
lieu est l'absence des quantités arbitraires qui produisoient
les erreurs, et partant, il ne s'agit pour opérer cette
compensation que d'éliminer ces quantités arbitraires.
50. pour fixer davantage les idées et donner aux
principes qui en dérivent, le degré de précision et de
généralité qui leur convient, je remarquerai que les
équations $tp = mp \cdot \frac{mz}{rz}$ et $\frac{mz}{rz} = \frac{y}{a-x}$ font précisément
de celles que j'ai nommées (19) équations imparfaites
c'est-à-dire que ces équations quoique fausses ont
pour limites des équations exactes puisqu'il est évident
d'une part, que la dernière raison de mz à rz est $\frac{tp}{mp}$
et que d'une autre part, cette même dernière raison
a pour valeur $\frac{y}{a-x}$. d'où il suit (18) que $\frac{tp}{mp}$ et $\frac{y}{a-x}$

diffèrent chacune infiniment peu, de $\frac{mz}{rz}$.

on voit par là, que la compensation d'erreurs dont il a été question ci dessus, est fondée sur la théorie de ce que nous appelons équations imparfaites: voici donc en peu de mots quelle est cette théorie.

théoreme I.

51. si dans une equation soit exacte soit imparfaite, à la place de l'une quelconque des quantités qui y entrent, on en substitue une autre qui en diffère infiniment peu, c'est-à-dire dont le rapport à la premiere, ait pour limite ou dernière valeur, l'unité. l'équation qui en résultera, sera nécessairement, ou exacte, ou aumoins, de celles que j'ai nommées imparfaites c'est-à-dire qu'elle ne peut jamais devenir par cette transformation, une équation absolument fausse.

en effet, supposons que l'équation proposée soit $A=B$ et que par la substitution dont il s'agit elle devienne $A'=B'$ on aura donc par hypothèse $\mathcal{L}\frac{A}{B}=1$, et il s'agit (19) de prouver qu'on doit avoir de même $\mathcal{L}\frac{A'}{B'}=1$.

Soit x' la quantité substituée à x et que l'on ait $x'-x=z$ ou $x'=x(1+\frac{z}{x})$; donc la substitution de x' à la place de x, se réduit à multiplier x partout où elle se rencontre dans les termes de l'équation $A=B$ de la multiplier dis-je, par $1+\frac{z}{x}$, c'est-à-dire, que par cette multiplication A et B, se trouveront changées en A' et B'. mais puisque x et x' diffèrent infiniment peu, leur rapport qui est $1+\frac{z}{x}$ a pour dernière valeur l'unité; c'est-à-dire que $\mathcal{L}(1+\frac{z}{x})=1$ donc $\mathcal{L}\frac{A'}{B'}$ ne diffère pas de $\mathcal{L}\frac{A}{B}$: mais $\mathcal{L}\frac{A}{B}=\frac{A}{B}=1$ donc $\mathcal{L}\frac{A'}{B'}=1$, donc $A'=B'$ est une équation imparfaite, ce qu'il falloit prouver.

Corollaire.

52. il en sera demême si dans un raisonnement, dans une proposition, ou dans une hypothèse quelconque, on emploie une quantité pour une autre qui en différeroit infiniment peu; c'est-à-dire, que les propositions ou résultats quelconques, auxquels on seroit conduit par ces raisonnements, ne seroient point faux absolument, mais de ceux que j'ai nommés imparfaits ou susceptibles d'être traduits par des équations imparfaites.

Théorème II.

53. toute équation qui ne contient que des quantités désignées ne peut être une équation imparfaite, et ne peut y être ramenée, ni par transposition des termes d'un membre dans l'autre, ni par multiplication ou division de ces termes par une même quantité, ni enfin par élévation de ces membres à des puissances d'un même ordre, ou extraction de racines de mêmes degrés.

en effet, soit $A = B$ l'équation proposée; si cette équation étoit imparfaite, on n'auroit pas exactement $(19)\, A = B$ ni par conséquent, $\frac{A}{B} = 1$. Cependant on auroit par hypothèse $\mathcal{L}\frac{A}{B} = 1$. or cela ne se peut. car si l'on a $\mathcal{L}\frac{A}{B} = 1$ comme A et B, ne contiennent par supposition que des quantités désignées, elles resteront les mêmes, malgré le changement du système auxiliaire; donc $\frac{A}{B}$ ne différera point de $\mathcal{L}\frac{A}{B}$ donc il n'est pas possible, qu'on ait comme on l'a supposé $\mathcal{L}\frac{A}{B} = 1$ sans avoir en même tems $\frac{A}{B} = 1$ ou $A = B$ équation qui étant exacte, ne peut devenir fausse ni même imparfaite, par les transformations de l'algèbre.

Corollaire.

54. de même, Si dans une proposition ou résultat quelconque il n'est question que de quantités désignées, elle ne pourra jamais être une proposition imparfaite, et ne se traduira jamais, que par des équations, ou absolument exactes, ou absolument fausses.

théoreme III.

55. toute équation que l'on obtiendra par des transformations semblables à celle qui est indiquée dans le théorème I ou autres quelconques; pourvu qu'elle conserve le caractère des équations imparfaites, c'est-à-dire, pourvu que ses deux membres ayent toujours pour dernière raison, une raison d'égalité, ou que du moins, on puisse l'y ramener par quelques-unes des transformations indiquées dans le théorème II; toute équation dis-je de cette nature, et de laquelle on sera parvenu à éliminer les quantités auxiliaires, sera une équation nécessairement et rigoureusement exacte.

Car par le théorème I ce ne peut être une équation absolument fausse, et par le théorème II, ce ne peut être une équation imparfaite. donc elle est nécessairement et rigoureusement exacte.

Corollaire.

56. il en est de même évidemment des conséquences et résultats quelconques déduits d'autres propositions ou résultats imparfaits, pourvu que dans ces conséquences et résultats, il soit uniquement question de quantités désignées. en un mot, on doit dire de toutes ces propositions, la même chose, que des équations qui en peuvent être regardées comme la traduction algébrique.

Scholie.

57. on voit par la théorie précédente, que si ayant exprimé par des équations imparfaites, les conditions d'un problème; on parvient ensuite à en éliminer toutes les quantités auxiliaires

Soit par des transformations algébriques, soit par celles qui sont indiquées dans le théorème 1. il faudra nécessairement qu'il se soit opéré dans le cours du calcul une compensation d'erreurs. ainsi en attribuant comme nous l'avons fait jusqu'ici, aux quantités infinitésimales des valeurs effectives, qui ne soient point o ni ⁰, on voit que la théorie de l'infini n'est autre chose qu'un calcul d'erreurs compensées, et que l'avantage de ce calcul consiste en ce que les conditions d'une question étant souvent fort difficiles à exprimer exactement par des équations rigoureuses, tandis qu'il seroit facile de le faire par des équations imparfaites; il donne le moyen de tirer de ces équations imparfaites, les mêmes résultats et des rapports tout aussi certains, que si les équations primitives eussent été véritablement de la plus parfaite exactitude. et cela, par la simple élimination des quantités dont la présence occasionnoit les erreurs.

     la raison de cela est simple: qu'on ait à découvrir les relations qui existent entre plusieurs quantités quelconques désignées; s'il est difficile de trouver directement des équations qui expriment ces relations. il est naturel de recourir à quelque quantité intermédiaire, qui leur serve de terme de comparaison; par ce moyen, on pourra obtenir sinon les équations même cherchées, au moins d'autres équations où les quantités désignées, se trouveront mêlées avec les quantités auxiliaires: il ne sera donc plus question que d'éliminer celles-ci. mais si les valeurs de ces quantités auxiliaires sont arbitraires, il est aisé de sentir que chaque équation où elles se trouveront mêlées avec les quantités désignées, pourra se décomposer en deux autres, l'une entre les quantités désignées, l'autre entre quantités arbitraires; à peuprès de même qu'une équation qui contient des quantités réeles et des quantités imaginaires peut se décomposer en deux l'une entre quantités réeles, l'autre entre quantités imaginaires: or comme on n'a besoin, que de l'équation qui existe entre les quantités désignées il est clair, qu'on peut sans inconvénient dans celle où elles se trouvent mêlées avec les arbitraires négliger les quantités qui embarrassent le calcul, lorsque les erreurs qui doivent en résulter, ne peuvent tomber que sur l'équation entre arbitraires qu'elle renferme. or c'est ce que l'on fait en observant ce principe que, lorsque deux quantités quelconques, ont pour dernière raison une raison d'égalité, on peut employer l'une pour l'autre dans le calcul sans qu'il en resulte pour cela aucune erreur dans le résultat désigné. et c'est dans l'art de faire un bon usage de ce principe qui revient au même que le théorème III, c'est dis-je dans cet art que consiste principalement, le calcul des erreurs compensées.

il y a deux manieres d'envisager la question de l'infini: la premiere consiste à attribuer aux variables qu'on nomme infiniment petites des valeurs déterminées qui soient effectives, c'est-a-dire, qui ne soient point égales à zéro et alors l'analyse infinitésimale doit être regardée comme un calcul d'erreurs compensées; la seconde consiste à attribuer aux quantités infiniment petites, des valeurs absolu=

58. voila à quoi se réduit l'esprit de l'analyse infinitésimale lorsque l'on veut attribuer aux indéterminées qu'on nomme infiniment petites des valeurs effectives, mais il est encore une autre maniere d'envisager cette analyse, et cela se fait en attribuant aux quantités infiniment petites des valeurs absolument nulles: et c'est ce qu'on nomme calcul des évanouissantes.

dans le calcul des erreurs compensées, nous avons

ment nulles et alors l'analyse infinitési- male est un calcul exact dans tous ses procédés, mais qui a pour sujet des quanti- tés évanouissantes, lesquelles sont des zéro de raison.

vu, que pour simplifier les opérations, on se permet quelquefois des erreurs; mais ces erreurs se réduisent toujours, a traiter comme égales ou employer l'une pour l'autre des quantités qui diffèrent infiniment peu; c'est-à-dire, dont la dernière raison est une raison d'égalité d'abord (39), c'étoit à force de ~~rectifier~~ les erreurs, qu'on s'étoit proposé de rectifier le calcul, ensuite (55) c'est par une compensation produite entre ces erreurs, qu'on a cherché le moyen de les faire évanouir: dans la méthode enfin, que nous allons exposer, c'est en supposant absolument nulles les quantités qui les occasionnent; qu'on fait disparoître ces quantités et avec elles, les erreurs qu'elles avoient introduites.

59. pour expliquer comment cela doit arriver, nous allons appliquer ce calcul à l'exemple déjà traité ci-dessus.

nous avons trouvé (46) $\frac{tp+t't}{y} = \frac{mz}{rz}$ et $\frac{mz}{rz} = \frac{zy+rz}{2a-2x-mz}$ équations parfaitement exactes, l'une et l'autre, quelques soient les valeurs de mz et de rz: égalant donc, les deux valeurs de $\frac{mz}{rz}$ j'ai $\frac{tp+t't}{y} = \frac{zy+rz}{2a-2x-mz}$, équation exacte, qui doit avoir lieu quelle que soit la distance qu'on voudra mettre entre les lignes rs et mp: supposons donc cette distance égale à zéro, nous aurons t't=0 mz=0, rz=0 et partant l'équation trouvée se réduira à $tp = \frac{y^2}{a-x}$ comme précédemment.

60. en supposant nulle comme nous venons de le faire la distance de rs à mp la fraction $\frac{mz}{rz}$ devient $\frac{0}{0}$ et les équations primitives de l'article précédent se réduisent à $\frac{tp}{y} = \frac{0}{0}$ et $\frac{0}{0} = \frac{zy}{2a-2x}$, on voit donc par là, combien il peut

être utile, de considérer les rapports qui existent entre les quantités évanouissantes: car il est souvent difficile de comparer directement, deux quantités telles que $\frac{tp}{...}$ et $\frac{y}{a-x}$, tandis qu'il est aisé de trouver le rapport de chacune d'elles en particulier à une même troisième grandeur qui se présente sous la forme de $\frac{0}{0}$. or c'est l'art de comparer ainsi des quantités évanouissantes entre elles et avec d'autres, et d'en tirer les relations qui existent entre des quantités proposées, qui fait l'objet du calcul de l'infini envisagé sous le second point de vue, dont il est ici question.

61. mais il se présente d'abord ici une difficulté. il est clair dira-t-on, qu'on peut attribuer aux quantités nulles autant de valeurs différentes qu'on juge à propos. car une quantité double, triple &c de o est toujours o; c'est-à-dire que la fraction $\frac{0}{0}$ est une quantité indéterminée; comment peut-on donc en tirer les rapports déterminés qui existent entre des quantités désignées, et dire que deux quantités déterminées sont égales entre elles parce qu'elles sont égales chacune à $\frac{0}{0}$? à cela je réponds, que par là même, que $\frac{0}{0}$ est une quantité indéterminée, si l'on conçoit que plusieurs quantités décroissent ensemble, et se réduisent toutes en même tems à zéro, les rapports qu'elles auront ~~les qu'elles l'ayyonaiste~~ en s'évanouissant, deviendront arbitraires et pourront par conséquent, être déterminées par quelle condition l'on voudra. cela posé prennons pour cette condition, qu'ayant pris l'une quelconque d'entre elles pour base ou terme de comparaison, les autres ayent avec cette première des rapports tels que chacun d'eux soit assigné. ~~p...............~~

par la loi de continuité à laquelle est assujetti le système auxiliaire dans son changement. alors les quantités nulles déterminées par cette condition, seront précisément ce qu'on entend, par ce terme d'évanouissantes: d'où il suit évidemment que cette hypothèse une fois faite, les rapports de ces quantités considérées deux à deux, ne seront plus arbitraires, mais ne seront autre chose, que les dernières raisons des quantités indéfiniment petites qui leur correspondent.

62. dans le cas traité ci-dessus par exemple, tant que mz ne s'évanouit pas, $\frac{mz}{rz}$ est plus grande que $\frac{tp}{mp}$; ces deux quantités ne deviennent égales, qu'au moment où mz se réduit à zéro. il est vrai qu'alors $\frac{mz}{rz}$ est aussi bien égale à toute autre quantité, qu'à $\frac{tp}{y}$, puisque $\frac{0}{0}$ est une quantité arbitraire; mais parmi ces diverses quantités, $\frac{tp}{y}$ est la seule qui soit assujettie à la loi de continuité et déterminée par elle: car si l'on construisoit une courbe dont l'abscisse fut mz et l'ordonnée proportionnelle à $\frac{mz}{rz}$, celle qui répondroit à l'abscisse nulle, seroit représentée par $\frac{tp}{y}$, et non pas par une quantité arbitraire, or c'est ce qui distingue les quantités que je nomme évanouissantes de celles qui sont simplement nulles.

63. comme égales à zéro, ces quantités doivent se négliger dans le calcul, lorsqu'elles se trouvent ajoutées à quelques quantités effectives ou en être retranchées; mais comme on vient de le voir elles n'en ont pas moins

entre elles des rapports très-intéressants à connoître. Ces rapports sont déterminés par la loi de continuité à laquelle le système auxiliaire est assujetti dans son changement: or pour saisir aisément cette loi de continuité, il est ~~oblig~~ aisé de sentir, qu'on est obligé de considérer les quantités en question, à quelque distance du terme où elles s'évanouissent entièrement, sinon, elles n'offriroient que le rapport indéfini de 0 à 0. cependant cette distance n'est point déterminée, parcequ'elle n'a point d'autre objet que de faire juger plus facilement, quels sont les rapports qui existent entre les quantités évanouissantes auxquelles elles répondent: Ce sont les rapports qu'on a en vue, et non pas ceux qui existent entre les quantités qui ne sont pas encore parvenues au terme de leur anéantissement: celles-ci étant de celles que j'ai nommées (15) indéfiniment petites, ne sont point destinées à entrer elles-mêmes dans le calcul dont il s'agit en ce moment, mais employées seulement, pour aider l'imagination, et indiquer la loi de continuité, qui détermine les rapports et relations quelconques des quantités évanouissantes auxquelles elles répondent.

64. ainsi dans la proportion $mz : rz :: tp : mp$. les quantités représentées par mz et rz, sont bien supposées absolument égales à zéro, mais comme c'est de leur rapport qu'on a besoin, il faut pour appercevoir son égalité avec $\frac{tp}{mp}$ considérer les quantités indéfiniment petites qui répondent

à ces quantités nulles, non afin de les introduire elles-mêmes dans le calcul, mais afin d'y faire entrer sous leurs noms mz, rz, les quantités évanouissantes qui en sont les dernières valeurs.

Les quantités mz, rz, représentent donc ici des quantités nulles, et on les emploie sous ces formes différentes mz, rz, plutôt que sous la forme commune de 0, parceque si on les employoit sous cette dernière forme 0, on ne pourroit plus distinguer dans les opérations où elles se trouveroient mêlées leurs diverses origines, c'est-à-dire quelles sont les diverses quantités indéfiniment petites qui leur répondent: or la considération aumoins mentale de celles-ci est nécessaire, pour saisir la loi de continuité qui détermine les rapports cherchés des quantités évanouissantes qui en sont les limites et par conséquent, il est essentiel de ne pas les perdre de vue.

De ces remarques sur les quantités évanouissantes, suit évidement cette proposition.

théorème IV.

65. Si dans une équation dont les deux membres ont une raison d'égalité, on substitue à l'une quelconque des quantités (évanouissante ou non) qui y entre une autre quantité qui ait aussi avec la première une raison d'égalité: la nouvelle équation qui en résultera, sera une équation exacte, et ses deux membres auront encore entre eux une raison d'égalité.

Il en sera demême aussi, de toutes les propositions, hypothèses ou résultats quelconques, qui pourront être traduits par de semblables équations.

66. Cette proposition évidente qui est le principe fondamental du calcul des évanouissantes, est analogue à celui qui permet (51) dans le calcul des erreurs

Comparaison de la méthode des erreurs compensées avec celle des évanouissantes.

compensées, d'employer ~~d'employer~~ dans les équations imparfaites une quantité quelconque à la place d'une autre qui en diffère infiniment peu : car si l'on suppose, que a et b soient des quantités indéfiniment petites, dont la dernière raison soit une raison d'égalité et que a' et b' soient les quantités évanouissantes qui leur répondent on pourra suivant le théorème III, employer dans le calcul a et b l'un pour l'autre ; et suivant le théorème IV ce seront a' et b' qu'on pourra substituer l'une à l'autre. il n'y a donc de différence, qu'en ce que dans le calcul des évanouissantes, il s'agit de l'infini absolu, aulieu qu'il s'agit dans l'autre de l'infini sensible mais les propriétés qu'on vient de rapporter, prescrivent comme on le voit les mêmes règles pour l'emploi des quantités infinitésimales sensibles, que pour celles qui sont absolues.

67. on voit par là, que les quantités indéfiniment petites qui font le sujet du calcul des erreurs compensées, ne diffèrent des quantités ordinaires que par leur limite qui est o, tandis que les quantités évanouissantes étant elles-mêmes o, sont des êtres de raison. mais cela n'empêche pas que ces évanouissantes n'ayent des propriétés mathématiques, et qu'on ne puisse les comparer tout aussi bien, que les quantités imaginaires qui n'existent pas non plus. Car il est tout aussi vrai de dire par exemple, que $2.o = 2.o + 3.o.o$, qu'il est vrai de dire que $\sqrt{-a} = \sqrt{-b} \cdot \sqrt{\frac{a}{b}}$. or personne ne révoque en doute, l'exactitude des résultats qu'on obtient par le calcul des imaginaires, quoique elles ne soient que des formes algébriques, ainsi, on n'a pas plus

de raison pour donner l'exclusion aux quantités évanouissantes. qu'importe en effet, que ces quantités soient ou non des êtres chimériques, si leurs rapports ne le sont pas; et que ces rapports soient la seule chose qui nous intéresse.

vantages et incon-
éniens qui sont
ropres à chacune
elles.

68. la différence donc, qui se trouve entre le calcul des erreurs compensées et celui des évanouissantes, consiste en ce que dans celui-ci, les propositions, équations et résultats quelconques, sont toujours exacts et rigoureux mais se rapportent à des quantités qui sont des êtres de raison: ils expriment des relations qui existent entre quantités qui n'existent pas elles-mêmes: tandis que dans l'autre méthode, au contraire, les propositions équations et résultats, ont bien pour sujet de véritables quantités; mais les propositions, équations et résultats sont faux, ou plutôt ils sont imparfaits: c'est-à-dire que sans être absolument exacts, on peut cependant eu égard aux circonstances, les employer comme tels sans qu'il doive en résulter aucune erreur, dans les conséquences qu'on se propose d'en déduire.*

69. pour rendre encore plus sensible le parallele des deux méthodes dont on vient d'expliquer les principes je vais les appliquer l'une et l'autre à un nouvel exemple.

Supposons donc, qu'il s'agisse de trouver le maximum

* il paroit que neuton considéroit les différentielles ou fluxions comme des quantités évanouissantes, qui ne sont autre chose que les indivisibles de Cavalleri, et que lebnitz au contraire, les considéroit comme des quantités effectives. ainsi leurs métaphisiques à cet égard etoient fort différentes; et il n'y auroit rien d'étonnant à ce qu'ils eussent pu découvrir cette analyse chacun de son côté.

de la fonction $2ax-xx$; c'est-à-dire, que a étant une constante donnée, et x une variable quelconque, il faille trouver, quelle est la valeur qu'on doit attribuer à x pour que $2ax-xx$ soit la plus grande possible.

Soit z la valeur cherchée de x, $2az-zz$, sera donc plus grande, que si l'on mettoit pour z une autre valeur quelconque; c'est-à-dire, qu'elle diminuera toujours, soit qu'on augmente ou qu'on diminue la quantité z.

Cela posé, soient deux quantités infiniment petites u, u', il suit de ce qu'on vient de dire, que soit qu'on substitue $z+u$ à la place de z, dans la quantité $2az-zz$, soit qu'on y substitue $z-u'$, le résultat sera toujours moindre que $2az-zz$: c'est-à-dire, qu'on a $2a(z+u)-(z+u)^2 < 2az-zz$, et $2a(z-u')-(z-u')^2 < 2az-zz$. donc u et u' étant des quantités arbitraires, on peut les supposer telles, que ces deux quantités moindres chacune que $2az-zz$ soient égales entre elles. Supposons le donc en effet, et nous aurons, $2a(z+u)-(z+u)^2 = 2a(z-u')-(z-u')^2$, ou réduisant $(2a-2z-(u-u')(u+u'))=0$ ou enfin $2a-u=2z-u'$.

Maintenant nous pouvons supposer à volonté, que les quantités u et u' soient infiniment petites absolues ou seulement indéfiniment petites: dans le premier cas u étant évanouissante, on peut la négliger, et l'équation se réduit à $2a=2z-u'$, dans laquelle on pourra par la même raison négliger u' et qui par conséquent se réduira à $a=z$.

Dans le second cas, u étant indéfiniment petite

si on la néglige, l'équation restante, $2a = 2z - u$, sera
évidement ce que j'ai nommé (19) équation imparfaite,
et comme u est aussi indéfiniment petite, on pourra
la négliger encore, sans que l'équation devienne fausse
pour cela (51) donc l'équation $a = z$ à laquelle se
réduira alors la précédente, ne sera pas une équation
fausse: or elle ne contient que des quantités désignées;
donc (55) ce sera une équation exacte. donc on
aura par cette méthode $z = a$ comme par la première
d'où l'on voit que ces deux méthodes quoique différentes
quant aux principes, conduisent précisément aux
mêmes procédés et aux mêmes résultats.

70. il est aisé de voir par tout ce qui vient d'être dit,
que pour transformer une équation imparfaite
proposée, en une autre qui soit exacte, il n'y a qu'à
supposer égales à zéro les quantités indéfiniment petites;
c'est-à-dire, substituer à ces quantités les évanouissantes
qui leur répondent. mais alors l'équation contiendra
des quantités nulles qu'il faudra éliminer pour tirer
de cette équation des relations déterminées. réciproquement,
lorsqu'on aura une équation exacte, qui contiendra des
évanouissantes et qu'on voudra la transformer en
une autre qui ne contienne plus que des quantités
effectives; il n'y aura qu'à substituer à ces évanouissantes
les quantités indéfiniment petites qui leur répondent.
mais alors l'équation deviendra imparfaite et pour

faire disparoître l'erreur, il faudra éliminer les quantités indéfiniment petites.

71. La forme des équations et le but qu'on se propose en les combinant, sont donc les mêmes dans les deux cas, et ces deux méthodes doivent par conséquent exiger les mêmes procédés et aboutir aux mêmes résultats: toute la différence consiste dans la manière d'envisager les quantités dont on y fait usage ou plutôt les signes qui les représentent, c'est-à-dire, que ce sont toujours les mêmes idées à rendre, les mêmes combinaisons à faire les mêmes relations à exprimer; mais les termes qu'on emploie dans l'une pour rendre ces idées et pour exprimer ces relations, ne sont pas les mêmes, que ceux qu'on emploie dans l'autre.

réunion de ces deux méthodes en un même calcul qui est l'analyse infinitésimale proprement dite.

72. Ce qu'on appelle analyse infinitésimale, est à proprement parler un composé des deux méthodes précédentes, celle des quantités évanouissantes renfermant la théorie de l'infini absolu, et celle des erreurs compensées, la théorie de l'infini sensible.

En effet, la méthode des évanouissantes, n'est autre chose comme on l'a vu plus haut, que le calcul rigoureusement exact de l'infini absolu ou métaphisique. celle des erreurs compensées, n'est autre chose qu'une fausse position par laquelle on attribue aux quantités infinitésimales sensibles et appréciables les mêmes rapports qu'à leurs limites ou quantités infinitésimales absolues qui y répondent, parceque cette fausse supposition ne peut entraîner à aucune erreur dans le résultat qu'on a en vue. or le calcul infinitésimal proprement dit, n'est autre chose que la réunion de ces deux méthodes; c'est-à-dire, un calcul, ou sous la dénomination commune de quantités infiniment petites, l'on fait usage suivant l'occurrence des cas, soit des quantités évanouissantes soit des quantités indéfiniment petites, en employant les unes préférablement aux autres, suivant qu'on trouve plus ou moins de facilité à les faire entrer, soit dans les raisonnements, soit dans les opérations du calcul. Cette méthode réunit donc les avantages particuliers à chacune des deux autres, en donnant comme celle des erreurs compensées la facilité de fixer son imagination sur des quantités vraies et sensibles. et comme celles des évanouissantes. le moyen d'exprimer rigoureusement lorsqu'on le veut, les rapports et relations qui existent entre les quantités proposées. on evite ainsi, et le deffaut d'exactitude qu'on pourroit ce semble (quoique sans fondement réel) repprocher à la méthode des erreurs compensées; et la difficulté que présente le calcul des évanouissantes par la nature des quantités qu'on y considère; lesquelles étant des êtres de raison, n'offrent à l'esprit que des rapports vagues, qu'on ne parvient à saisir et fixer, qu'en ayant recours mentalement, aux quantités indéfiniment petites qui y répondent.

73. dans le calcul infinitésimal on emploie (16. 35, 72) le seul terme générique d'infini, pour désigner, soit l'infini absolu soit l'infini relatif, et c'est au lecteur à juger par le sens de la phrase quel est celui dont il est question. mais cela ne peut faire aucune équivoque parce qu'on voit bien

si cette phrase peut se rapporter à des quantités effectives ou si elle ne peut avoir de rapport qu'à des quantités absolument nulles: dans le premier cas il est clair qu'il ne peut être question que de l'infini sensible, et dans le second, de l'infini absolu. par exemple, lorsqu'on dit que l'assymptote d'un hyperbole rencontre sa branche à une distance infinie, il est bien clair, qu'il ne peut être question pour cette distance, d'une quantité réele, puisque dans le vrai, l'assymptote ne rencontre jamais l'hyperbole. il s'agit donc alors évidement, de l'infini absolu; lequel n'existant pas fait voir qu'effectivement l'assymptote ne rencontre jamais l'hyperbole. mais si l'on demande que nous imaginions un polygone d'une infinité de côtes, inscrit dans un cercle; comme on ne peut imaginer que ce qui est possible; il est visible que nous ne pouvons satisfaire à cette demande sans que ces côtes ne soient de vraies lignes droites, car il est absolument impossible qu'un polygone soit identique avec un cercle; donc il s'agit alors de l'infini sensible, et l'on suppose tacitement en faisant cette demande, que les côtes du polygones soient ce que nous appelons indéfiniment petits.

74. mais si comme cela arrive très souvent, ce qu'on a à dire, sur les quantités infinitésimales, peut s'appliquer également, soit à des quantités évanouissantes, soit à des quantités effectives, il faudra regarder alors la proposition comme équivalante à deux autres dont l'une se rapporteroit à l'infini absolu, et l'autre à l'infini sensible. si donc on traduit algébriquement cette proposition ce sera

ou par une équation vraie entre quantités non existantes
ou par une équation imparfaite entre quantités effectives*
mais ces équations seront de même forme, ou plutôt, ne
seront que la même équation expliquée de deux manières
différentes et qui cependant reviennent au même. par exemple
dans la recherche du maximum trouvé (69) les quantités
u, u', sont infiniment petites, et l'on peut supposer à volonté,
qu'il s'agit de l'infini absolu ou de l'infini sensible. dans le
premier cas, l'équation trouvée, $2a = 2z - u'$ est exacte, mais
contient une quantité u' qui n'existe pas: dans le second,
cette équation n'est qu'imparfaite, mais elle a lieu entre
quantités existantes et réelles, et la forme de l'équation est
la même dans l'un et dans l'autre cas.

75. cette double signification d'une même expression
appliquée tantôt à zéro tantôt à une quantité effective
ou plutôt tout ensemble à l'une et à l'autre, est je crois
la véritable cause de l'embarras qu'on trouve à définir
exactement l'infini mathématique; les uns voulant qu'on
regarde toujours les quantités infiniment petites comme
absolument nulles, et les autres au contraire, voulant qu'on
les regarde toujours comme de véritables quantités: tandis
que dans l'usage habituel qu'on en fait en mathématiques,
il est visible que l'expression d'infini présente à la fois

---

*il n'est même absolument parlant, aucune proposition concernant l'infini, qui ne soit
dans ce cas, c'est à dire, qui ne puisse être regardée comme équivalante à deux autres dont
l'une exacte, se rapporte à l'infini absolu, et l'autre imparfaite se rapporte à l'infini sensible.
par exemple, la proposition que nous venons de rapporter sur l'hyperbole équivaut à ces
deux ci l'assymptote rencontre l'hyperbole à une distance absolument infinie, ce qui est
exactement vrai puisque cette distance absolument infinie n'existe pas. et l'assymptote rencontre
l'hyperbole à une distance indéfiniment grande ce qui est faux mais peut être regardé
comme vrai sans qu'il en résulte aucune erreur

deux acceptions fort différentes, et que c'est le sens de la phrase qui naturellement indique celui dont il s'agit, ou plutôt, que le sens de cette phrase est toujours double et équivalant à deux autres propositions l'une relative à l'infini absolu, l'autre à l'infini sensible: voilà pourquoi, lorsqu'on veut restreindre la signification du nom d'infiniment petites, aux seules quantités évanouissantes ou aux seules quantités indéfiniment petites, on ne peut plus leur appliquer des propositions très-simples d'ailleurs, ce qui fait bientôt sentir l'insuffisance de la définition qu'on a adoptée. je crois donc avoir donné ici, la véritable idée que chacun se fait de l'infini mathématique, souvent même, sans s'en appercevoir.

76. une remarque importante que nous avons déja faite, (70) c'est que les quantités infinitésimales de quelle espèce qu'elles soient, ne peuvent jamais entrer dans le calcul que comme auxiliaires, et seulement pour en faciliter les combinaisons; mais qu'elles doivent toujours se trouver exclues du résultat désigné. on sait en effet, qu'un calcul ou il entre des quantités infiniment petites n'est censé fini et l'on ne compte sur l'exactitude du résultat, que du moment où toutes ces quantités sont éliminées. Car en supposant qu'on regarde ces quantités comme des évanouissantes, il faut qu'elles soient toutes éliminées pour que le résultat offre quelque chose de déterminé et des rapports dont on puisse faire usage: et si

on les regarde comme de vraies quantités, on ne peut
être assuré (55) que la compensation des erreurs est
effectuée, qu'après l'élimination de toutes les quantités
auxiliaires qui les avoient occasionnées. on peut donc dire
en quelque sorte, que le calcul infinitésimal est un
calcul non fini ou qui n'est pas encore achevé: parce qu'en
effet, dès qu'on est parvenu à quelque résultat désigné,
il cesse d'être infinitésimal, et ressemble en tout au calcul
algébrique ordinaire.

le principe fondamental du calcul de l'infini est
renfermé dans les deux théorèmes III et IV; ainsi on peut
le réduire à la proposition suivante.

théorème V.

77. Si l'on a une équation dont les deux membres diffèrent
infiniment peu, (soit qu'il s'agisse de l'infini sensible ou de l'infini absolu) on pourra
substituer à la place de l'une quelconque des quantités qui y entrent,
une autre quantité infiniment peu différente de la première, sans qu'il
puisse y avoir pour cela aucune erreur dans le résultat désigné: et il en sera
de même, de toutes les propositions qui pourront se traduire par de semblables
équations.

78. d'après la notion que nous avons donnée (29) du
calcul différentiel, la règle générale pour trouver la
différentielle d'une fonction quelconque de variables
$x, y, z, \&c$ consiste à mettre dans cette fonction $x + dx$,
$y + dy$, $z + dz$, $\&c$ à la place de $x, y, z, \&c$ et à retrancher
du résultat la fonction donnée. ainsi, en supposant que φ
soit cette fonction, et qu'elle devienne φ' par la substitution
de $x + dx$, $y + dy$, $z + dz$, $\&c$ à la place de $x, y, z, \&c$ on aura $d\varphi = \varphi' - \varphi$.

mais comme dans le résultat qu'on obtiendra par ce moyen, il peut se trouver des termes qui soient infiniment petits relativement aux autres, on pourra suivant le théorème précédent les négliger; c'est-à-dire employer dans le calcul à la place de la vraie différentielle le résultat qu'on obtiendra, en y traitant comme nuls, les termes dont les rapports aux autres termes seront des quantités infiniment petites.

par ce moyen on pourra simplifier l'expression de ces différentielles, et il en résultera seulement, qu'au lieu des quantités exactes qu'on auroit eues pour exprimer les conditions du problème, on aura, si l'on suppose l'infini appreciable, des équations imparfaites qui conduiront au même but, par l'élimination des quantités auxiliaires qu'on aura introduites dans le calcul; et si l'on suppose que l'infini soit absolu, les équations ne cesseront pas d'être exactes, mais elles ne conduiront au terme désiré, que par l'élimination de toutes ces quantités évanouissantes.

par exemple, soit $\varphi = ax + by^2 + cz^3$. nous aurons donc $\varphi' = ax + adx + by^2 + 2bydy + b(dy)^2 + cz^3 + 3cz^2dz + 3cz(dz)^2 + c(dz)^3$. donc $\varphi' - \varphi$ ou $d\varphi = adx + 2bydy + b(dy)^2 + 3cz^2dz + 3cz(dz)^2 + c(dz)^3$. mais (25). $b(dy)^2$, $3cz(dz)^2$, $c(dz)^3$, sont des quantités infiniment petites d'un ordre supérieur à dx, $2bydy$, $3cz^2dz$, auxquelles elles sont ajoutées c'est-à-dire. qu'elles sont infiniment petites à leur égard donc (77) on peut les négliger ce qui donnera $d\varphi = adx + 2bydy + 3cz^2dz$. expression plus

simple que la différentielle rigoureuse, et qui cependant
employée à sa place dans le calcul, doit conduire
précisément au même résultat désigné.

on verra de même que la différentielle de l'équation
$yy = 2ax - xx$ est l'équation imparfaite $2ydy = 2adx - 2xdx$.
mais mon objet, n'est point d'expliquer ici les règles
particulières du calcul différentiel, il me suffit d'en avoir
indiqué le principe.

79. les règles de l'intégration se tirent de celles de la
différentiation. par exemple, si l'on se trouve avoir à
intégrer une équation de cette forme $2ydy = 2adx - 2xdx$,
on se rappelera qu'on a obtenu cette équation en différentiant
l'équation $yy = 2ax - xx$: donc réciproquement, celle-ci sera
l'intégrale de l'autre. en un mot, il s'agit de revenir de
la différentielle à l'intégrale qui l'a donnée; or c'est à
quoi l'on parvient, en observant la loi suivant laquelle
s'opère le passage de l'intégrale à la différentielle, et
employant cette observation, pour établir des règles par
lesquelles on puisse revenir, de la différentielle à l'intégrale.

80. nous avons déjà fait remarquer (33) combien l'art
de différentier et d'intégrer doit être utile dans le calcul
infinitésimal. pour donner une idée de l'usage qu'on en
fait, supposons qu'ayant à trouver les relations qui
existent entre plusieurs quantités variables x, y, z &c. liées
entre elles par une loi quelconque donnée, et que n'ayant
pu se procurer toutes les équations finies dont on auroit
besoin pour cela, on soit parvenu néanmoins à l'aide

{ 235 }

d'un système auxiliaire, à en trouver d'autres, où les mêmes quantités soient mêlées avec leurs différentielles: il est clair que si l'on peut réussir à éliminer ces différentielles, soit par les règles ordinaires de l'algèbre, soit par des intégrations qui les fassent disparoître; on obtiendra les équations finies, qu'on n'avoit pu avoir directement.

par exemple, qu'on ait l'équation finie $yy = 2ax - xx$ et l'équation différentielle $zdy = ydx$; et qu'il soit question de trouver la relation de x à z.

je différentie la première, et j'ai (78) $2ydy = 2adx - 2xdx$. je tire de cette équation, la valeur de dy que je substitue dans l'autre équation différentielle; après quoi réduisant, j'ai l'équation finie $z = \dfrac{yy}{a-x}$ ou $z = \dfrac{2ax-xx}{a-x}$ ce qu'il falloit trouver.

de même, si je cherche la relation de y à x, et que n'ayant pu la trouver directement; j'aie cependant découvert que l'équation imparfaite $2ydy = 2adx - 2xdx$, existe entre ces quantités et leurs différentielles; je n'aurai qu'à intégrer cette équation et il viendra (79) $yy = 2ax - xx$, qui exprime la relation désirée.

81. l'utilité des calculs différentiel et intégral se manifeste dans toutes les parties des mathématiques, elle est particulièrement sensible dans la théorie des lignes courbes.

pour faciliter la recherche de leurs propriétés on a imaginé de les regarder comme des figures rectilignes d'un nombre infini de côtés, or dans le calcul des erreurs compensées, cette supposition conduit à des équations imparfaites dont les erreurs se rectifient d'elles-mêmes, lorsqu'on parvient à éliminer les quantités infinitésimales introduites dans le calcul. aucontraire dans la méthode des évanouissantes, la figure rectiligne n'est qu'un terme de comparaison, qu'on emploie pour appercevoir plus facilement au moyen de sa ressemblance avec la courbe quelles sont les propriétés et affections de cette courbe elle-même ; on ne s'y arrête pas précisément aux rapports que fourniroit cette fausse hypothèse, mais à ceux qui y correspondent dans la courbe même et que la loi de continuité indique d'une manière d'autant plus sensible, qu'on suppose plus grand, le nombre des côtés du polygone, parceque son analogie avec la courbe en devient plus frapante, à mesure que le nombre de côtés augmente. dans le calcul infinitésimal proprement dit enfin, on suit à volonté, soit la marche des erreurs compensées soit celle des évanouissantes, suivant qu'on trouve plus de facilité à faire usage de l'une plutôt que de l'autre : mais alors, on désigne par l'expression générique d'infiniment petites ces quantités évanouissantes et les quantités indéfiniment petites qui leur répondent et c'est le sens de la phrase qui détermine quelles sont celles dont il s'agit en chaque cas particulier.

82. dans tous les cas, le polygone infinitésimal doit donc être regardé comme un système auxiliaire, dont les diverses parties ne sont introduites dans le calcul, que pour faciliter la recherche des propriétés du système désigné qui est la courbe proposées; et par conséquent, ce n'est que par l'élimination de ces diverses quantités auxiliaires, qu'on pourra

obtenir les rapports cherchés.

83. j'ai dit que les quantités appelées infiniment petites peuvent être à volonté, supposées évanouissantes ou indéfiniment petites; il s'agit seulement, de raisonner conséquemment à l'hypothèse qu'on aura adoptée: il est des cas où il est plus simple de considérer ces quantités comme absolument nulles, d'autres où cela est à peu près indifférent, d'autres enfin, où il est plus naturel et plus commode de supposer que ce soient des quantités effectives. Ce dernier cas arrive principalement, lorsqu'on emploie le calcul intégral à rechercher la valeur d'une étendue superficielle ou autre. Car lorsqu'on veut trouver la valeur d'une surface ou d'un solide dont la forme est irrégulière, ce qui se présente de plus naturel, est de partager cette surface ou ce solide, en plusieurs autres d'une forme plus simple, et plus faciles à évaluer; puis d'ajouter ensemble toutes ces quantités partielles ou d'en prendre la somme. De là, vient qu'on dit également sommer une quantité différentielle ou l'intégrer; et dans ce cas, on se sert encore souvent du terme d'élément pour signifier une différentielle: desorte, qu'aulieu de dire, qu'on intègre la différentielle de telle ou telle quantité; on s'exprime en disant, qu'on cherche la somme des éléments qui composent cette quantité: or cela suppose naturellement, qu'on ne regarde pas ces éléments comme absolument nuls; et en effet, il seroit absurde de dire, qu'une ligne est composée de points mathématiques, qu'une surface est composée de lignes ou qu'un solide est composé

de surfaces. il est donc plus naturel en ce cas (quoique cela ne soit pas absolument nécessaire)* de regarder les différentielles comme de vraies quantités, assez petites seulement, pour qu'on puisse aisément saisir les limites ou dernières valeurs de leurs rapports; et c'est ainsi, que dans chaque cas particulier, on se détermine à employer l'un ou l'autre infini, suivant qu'on le juge plus facile, relativement à l'objet qu'on a en vue.

84. quoique la méthode des erreurs compensées et le calcul des évanouissantes soient fondés entièrement sur les propriétés de ce que nous avons appelé limites cependant ils diffèrent de ce qu'on nomme proprement méthode des limites ou des premières et dernières raisons en ce que dans celle-ci, on ne fait point entrer séparement les quantités que nous avons nommées infinitésimales ni même leurs rapports, mais seulement, les limites ou dernières valeurs de ces rapports. tellement, que dx, dy, dz, &c. par exemple, étant des quantités infiniment petites, on ne les admettra point séparément dans le

Explication de ce qu'on entend proprement par méthode des limites ou des 1res et dernières raisons.

---

* je dis que cela n'est pas absolument nécessaire, parce qu'il est permis de regarder un prisme par exemple comme composé d'éléments superficiels ou sans aucune épaisseur, c'est-à-dire, dont la solidité soit absolument nulle; pourvu qu'on suppose en même tems, que le nombre de ces éléments superficiels est absolument infini. ce seront alors deux propositions absurdes chacune en particulier, mais dont le conséquent sera néanmoins exact; l'absurdité de l'une, détruisant l'effet de ce qui est absurde dans l'autre. au reste, cette difficulté n'est point particulière à l'analyse infinitésimale. l'analyse ordinaire est pleine de ces comparaisons de quantités purement fictives, d'où sortent des vérités très lumineuses; puisque sans compter les imaginaires qui ne sont que des formes algébriques, on rencontre à chaque pas des quantités négatives qui à le bien prendre ne peuvent représenter de véritables quantités, que lorsqu'elles sont précédées de quantités positives plus grandes qu'elles.

calcul mais on y fera seulement entrer les quantités $\int \frac{dx}{dy}$, $\int \frac{dz}{dx}$, $\int \frac{dy}{dz}$, &c. qui étant des grandeurs finies, font de cette méthode, moins un calcul particulier, qu'une simple application du calcul algébrique ordinaire.

il s'agit donc en se bornant à introduire ce genre de quantités finies dans l'algèbre ordinaire, il s'agit dis-je, de suppléer aux moyens que fournit l'analyse infinitésimale pour découvrir les propriétés, rapports et relations quelconques, des grandeurs qui composent un système proposé, et voilà, ce qu'on nomme proprement méthode des limites.

85. Si cette méthode étoit aussi facile à mettre en usage que le calcul infinitésimal ordinaire, il est certain qu'elle lui seroit préférable; car elle auroit l'avantage sur le calcul des évanouissantes ou de l'infini absolu, de n'offrir jamais à l'imagination que des quantités vraies et sensibles; au lieu que l'infini absolu est un être de raison, qu'on ne peut considérer en lui-même, mais seulement dans ses rapports avec d'autres quantités de même nature et aussi chimériques que lui. D'un autre côté, la marche toujours sûre et rigoureusement exacte de la méthode des limites, lui donneroit sur celle des erreurs compensées ou de l'infini appréciable, l'avantage de conduire au même résultat, par une route directe et toujours lumineuse, au lieu que celle-ci ne conduit en quelque sorte au vrai, qu'après avoir

fait pour ainsi dire parcourir tout le pays des erreurs.

mais la méthode des limites est sujette à une difficulté qui n'a pas lieu dans l'analyse infinitésimale. c'est qu'on ne peut y séparer comme dans celle-ci, les quantités infiniment petites l'une de l'autre; il faut toujours qu'elles s'y trouvent deux à deux; ce qui empêche, de faire subir aux équations ou elles se rencontrent, toutes les transformations qui serviroient à les éliminer, comme on doit le faire (70 et 76) et cette difficulté se fait bien moins sentir encore, dans les opérations même du calcul, que dans les proposition et raisonnements qui préparent ou supplent à ces opérations.

il est cependant un moyen d'éluder cet inconvénient nous l'expliquerons, mais il est nécessaire pour cela, de faire connoître auparavant quelques propriétés des limites.

théorème VI.

86. toute quantité désignée est nécessairement égale à sa limite.

Cela est clair, par la définition même des quantités désignées.

théorème VII.

87. Si deux quantités sont chacune limite d'une même troisième quantité elles sont nécessairement égales entre elles.

Car si l'on suppose que Y et Z soient chacune en particulier, limites de la même quantité X on aura par hypothèse Y = dernière valeur de X, et Z = dernière valeur de X; or X ne peut avoir évidement qu'une

Elle est la même au fond que la méthode ordinaire du calcul infinitésimale mais les procédés en sont plus difficiles.

seule dernière valeur, celle dont elle approche de plus en plus, à mesure que le système auxiliaire approche du système désigné, et qui est déterminée par la loi de continuité à laquelle ce système auxiliaire est assujetti dans son changement. donc Y est nécessairement égale à Z.

Théorème VIII

88. la dernière raison de deux quantités quelconques est toujours égale au rapport de leurs dernières valeurs. ou ce qui est la même chose, la limite du rapport de deux quantités est égale au rapport de leurs limites.

en effet, $\mathcal{L}\frac{Y}{Z}$ est la dernière valeur de la fraction $\frac{Y}{Z}$. or d'un autre côté, pour avoir cette dernière valeur de $\frac{Y}{Z}$ il faut évidement diviser la dernière valeur de Y par la dernière valeur de Z. donc $\mathcal{L}\frac{Y}{Z}$ est la même chose que $\frac{\mathcal{L}Y}{\mathcal{L}Z}$.

ainsi par exemple, la limite du rapport de deux quantités indéfiniment petites, est égale au rapport des quantités évanouissantes qui leur répondent.

il nous reste à montrer, comment on peut donner à la méthode des limites, la même simplicité qu'a l'analyse infinitésimale proprement dite.

89. on sait que c'est l'usage en mathématiques de rapporter toutes les espèces de grandeurs à des unités homogènes; lorsqu'on dit par exemple, que la surface d'un parallélograme est égale au produit de sa base par sa hauteur, on entend par cette surface,

moyen de rendre la méthode des limites aussi simple que le calcul infinitésimal ordinaire.

non pas l'aire même du parallélograme, mais son rapport à la quantité arbitraire prise pour unité des surfaces. de même, par les base et hauteur, de ce parallélograme, on entend, non pas ces lignes elles-mêmes, mais les rapports de ces lignes à l'unité linéaire. cela posé, faisons ici la même chose; par l'expression de quantités infiniment petites du premier odre, entendons, non ces quantités même, mais les rapports de ces quantités à la base infinitésimale, c'est-à-dire, leurs moments d'évanouissement (36). par quantités infiniment petites du second, troisième &c ordre, entendons les rapports de ces quantités à la seconde troisième &c puissance de la base infinitésimale; c'est-à-dire les moments d'évanouissements de leurs moments d'évanouissement &c; lesquels feront évidement toujours des quantités finies.

Cela posé, il est facile en suivant strictement la méthode des limites, non seulement, de la rendre aussi simple que l'analyse infinitésimale ordinaire, mais même de faire ensorte, que ces deux méthodes soient absolument identiques.

90. en effet, la difficulté de la méthode des limites consiste (85) en ce qu'on ne peut y séparer les quantités differentielles. or les remarques précédentes, fournissent le moyen de faire aisément cette séparation. car soit par exemple $L\frac{dy}{dz}$, pour séparer dy de dz, je commence par diviser l'une et l'autre par la base infinitésimale c'est-à-dire, que je mets la quantité précédente sous cette forme $L\frac{\left(\frac{dy}{dx}\right)}{\left(\frac{dz}{dx}\right)}$ in prennant dx pour base. or cette expression est évidement

{ 243 }

équivalante à la première, et d'un autre côté (88)
elle équivaut a $\dfrac{\mathcal{L}\left(\frac{\partial y}{\partial x}\right)}{\mathcal{L}\left(\frac{\partial z}{\partial x}\right)}$ expression dans laquelle le

numérateur et le dénominateur sont des quantités
finies, et où dy et dz se trouvent séparées.

91. on peut remarquer, qu $\mathcal{L}\left(\frac{\partial y}{\partial x}\right)$ et $\mathcal{L}\left(\frac{\partial z}{\partial x}\right)$ sont ce que
nous avons nommé (36) moments différentiels de y
et de z ou moments d'évanouissement de dy et de dz;
ainsi nous avons $\mathcal{L}\left(\frac{\partial y}{\partial x}\right) = Dy$ et $\mathcal{L}\left(\frac{\partial z}{\partial x}\right) = Dz$. Donc la
quantité proposée $\mathcal{L}\left(\frac{\partial y}{\partial z}\right)$ prend cette forme $\dfrac{Dy}{Dz}$. c'est-à-dire,
la même qu'elle auroit, si l'on avoit suivi la marche
ordinaire du calcul infinitésimal, à cela près, que la
caractéristique est D, au lieu d'être d ce qui ne fait
rien à la chose, puisque les caractéristiques sont
arbitraires et de convention. ainsi par ce moyen,
non seulement la méthode des limites devient aussi
simple que le calcul infinitésimal sans lui rien ôter
de sa vigueur; mais même, on rend les procédés de
ces deux méthodes absolument identiques; puisqu'il n'y
a pour cela, qu'à substituer aux quantités infiniment
petites, leurs moments d'évanouissement, qui sont des
quantités finies dont les rapports sont les mêmes, que
ceux des quantités évanouissantes qui leur répondent.

92. si l'on veut saisir encore mieux, les nuances qui
distinguent les diverses méthodes dont nous avons parlé
jusqu'ici quoique toutes les mêmes au fond, il n'y a qu'à

marquer d'un signe particulier, les différentielles sensible,
et d'un autre, les différentielles évanouissantes. Soit par
exemple, la différentielle dy, rien ne dénote dans cette
expression, s'il est question de l'infini absolu ou de l'infini
sensible: mettons donc pour le premier cas un point
à côté de la caractéristique d. en cette manière d˙y, et
pour le second cas, mettons un accent en cette manière d'y.
alors.

d˙y. sera la différentielle absolue ou évanouissante de y.

d'y. sera sa différentielle vraie, sensible ou assignable.

dy. signifiera indifféremment l'une ou l'autre, suivant qu'on
le jugera à propos.

Dy. sera enfin le moment différentiel de la même quantité,
ou son moment d'évanouissement.

or dans le calcul des évanouissantes ou de l'infini
absolu, on emploie des quantités de la nature de d˙y,
dans le calcul des erreurs compensées, on emploie des
quantités de la nature de d'y, dans le calcul de l'infini
en général, ce sont des quantités de la nature de dy, et
enfin dans le calcul des limites ou des première et
dernières raisons, ce sont des quantités de la nature de Dy.
Mais comme toutes ces expressions sont de même
forme à une caractéristique près qui est arbitraire,
il est aisé de voir que les méthodes dont on vient de
parler, ne diffèrent que dans la manière d'envisager la
question, et doivent toutes conduire aux mêmes procédés
et aux mêmes résultats.

l'analyse infinité-
simale, n'est autre
chose qu'une appli-
cation ou si l'on veut
une extension de la
méthode des indé-
terminées.

93. au reste il est aisé de sentir, que tous ces calculs étant fondés sur la comparaison de deux systèmes dont l'un n'est composé que de quantités désignées, et l'autre, de quantités dont les valeurs sont arbitraires il est aisé dis-je de sentir que tous ces calculs, ne doivent être qu'une application ou si l'on veut, une extension de la méthode des indéterminées, méthode lumineuse et qui peut en effet, les suppléer sans augmenter sensiblement les difficultés.

pour le prouver il suffira d'appliquer le calcul des indéterminées aux exemples déja traités par les méthodes dont on vient d'exposer les principes.

reprennons donc d'abord, celui où il s'agissoit de mener une tangente au point m du cercle dcmb.

nous avons trouvé (50) les deux équations imparfaites $\frac{tp}{mp} = \frac{mz}{rz}$ et $\frac{mz}{rz} = \frac{y}{a-x}$. pour les rendre exactes, je les mets sous cette forme $\frac{mz}{rz} = \frac{tp}{mp} + u$, et $\frac{mz}{rz} = \frac{y}{a-x} + u'$, u et u' étant des quantités telles qu'il les faut, pour que ces équations ayent lieu exactement égalant donc les deux valeurs de $\frac{mz}{rz}$ j'ai l'équation éxacte $\frac{tp}{mp} + u = \frac{y}{a-x} + u'$ ou $\left(\frac{tp}{mp} - \frac{y}{a-x}\right) + (u-u') = 0$.

cela posé, mz et rz pouvant être supposées aussi petites qu'on veut, il est clair, que u et u' pourront aussi être supposées aussi petites qu'on voudra, tandis que de toutes les quantités tp, mp, a, x, y, qui composent le premier terme, il n'y en a aucune, qui ne soit déterminée par la position prescrite du point m. donc l'équation précédente devant avoir lieu, quelles que soient les valeurs qu'on voudra attribuer à u et u', il est clair que suivant la méthode des indéterminées, chacun de ces termes doit être égal à zéro donc on a $\frac{tp}{mp} = \frac{y}{a-x}$ comme par les méthodes expliquées précédement. 75

on voit par là, que u et u', quoique quantités effectives, se détruisent d'elles-memes l'une par l'autre dans le résultat. on auroit donc pu, les négliger dans le cours du calcul, sans que ce résultat eût été affecté des erreurs commises. or c'est ce qu'on fait dans le calcul des erreurs compensées, la seule différence, est que dans celle des indéterminées qu'on vient d'expliquer, on attend, que l'équation finale soit trouvée, pour y supprimer u et u'; aulieu qu'en suivant l'autre, on traite ces quantités comme nulles, dans les équations partielles d'où l'équation finale est tirée.

94. Soit proposé maintenant, de trouver la surface du cercle dont il a été question (41)

la surface de tout polygone régulier, étant égale au produit de son périmètre par la moitié de l'apotême, si l'on nomme r le rayon du cercle c sa circonférence et s sa surface, il est clair que si l'équation $s = \frac{cr}{2}$ n'est pas exacte, en supposant que $s = \frac{cr}{2} + u$ ou $(s - \frac{cr}{2}) - u = 0$ le soit, u pourra être supposée aussi petite qu'on le voudra, tandis que de toutes les quantités s, c, r, qui composent le premier terme, il n'y en a aucune qui ne soit déterminée. Donc cette équation devant avoir lieu, quelle que soit la valeur qu'on voudra attribuer à u chacun de ses termes doit être égal à zéro. donc $S - \frac{cr}{2} = o$ est une équation exacte.

95. Soit enfin proposé de trouver le maximum de la fonction $2ax - xx$.

j'ai trouvé (69) l'équation exacte $(2a - 2z) - (u - u') = o$ laquelle doit avoir lieu quelques valeurs que je veuille attribuer à u et u'; donc suivant la méthode des indéterminées chacun des termes de cette équation est égal à zéro. Donc $2a - 2z = o$ ou $z = a$ Comme par les autres méthodes.

97. Ces exemples suffisent, pour faire voir que la méthode des erreurs compensées ne diffère de celle des indéterminées, qu'en ce qu'on y néglige comme nous l'avons déjà dit, dans le cours du calcul, des quantités qui se détruiroient toujours d'elles-mêmes dans le résultat, si on les laissoit subsister; au lieu que dans celle des indéterminées, on rend chaque équation exacte en lui ajoutant la quantité nécessaire pour cela, et l'on parvient ainsi, à une équation finale de deux termes, dont l'un est tout composé de quantités désignées, et dont l'autre ~~peut être supposée aussi petite~~ au contraire n'est composé que de quantités dont chacune peut être supposée aussi petite qu'on veut d'où il suit évidemment suivant la doctrine

des indéterminées*, que chacun de ces termes pris séparément est égal à zéro. Ce qui donne précisément le même résultat, que si dans le cours du calcul on eût opéré sur les équations imparfaites comme si elles eussent été exactes, sans se mettre en peine, de les rendre telles, par l'addition d'une nouvelle quantité.

On peut remarquer encore, que dans cette équation finale composée de deux termes, chacun égal à zéro, à laquelle on est conduit par la méthode des indéterminées, c'est la même chose, de traiter comme nul en général, le second terme, c'est-à-dire, la somme de tous les termes particuliers dont il est composé, ou de traiter comme nulle en particulier, chacune de ces dernières quantités. Par exemple, dans le cas dont il s'agit (93) c'est la même chose, de supposer dans l'équation finale $\left(\dfrac{tp}{mp} - \dfrac{y}{a-x}\right) + (u - u') = 0$, $u - u' = 0$, ou en particulier $u = 0$, $u' = 0$, puisque le résultat $\dfrac{tp}{mp} - \dfrac{y}{a-x} = 0$ est le même dans les deux cas; d'où l'on voit, que la méthode des indéterminées n'a pas moins d'analogie avec la méthode des évanouissantes, qui consiste à traiter, non pas seulement dans l'équation finale, mais dans tout le cours du calcul les quantités telles que u et u' comme nulles, chacune en particulier, qu'avec celle des erreurs compensées qui consiste à les négliger, non pas comme nulles chacune en

---

* en général, il est évident que si l'on a une équation exacte de cette forme $A + B + C + D + \&c = 0$ telle que $\dfrac{B}{A}$, $\dfrac{C}{B}$, $\dfrac{D}{C}$, &c ayent 0 pour limite c'est-à-dire puissent être rendues aussi petites qu'on voudra, on aura $A = 0$ $B = 0$ $C = 0$ &c —

particulier mais en somme, c'est-à-dire, comme produisant des erreurs qui doivent se détruire d'elles-mêmes par une suite nécessaire et infaillible des opérations du calcul.

il est possible, de comparer encore plus directement la méthode des indéterminées, avec celle du calcul infinitésimal, et de faire voir clairement que celui-ci n'est véritablement qu'une abréviation de l'autre.

par exemple, nous avons trouvé (59) l'équation exacte $\dfrac{tp+t't}{y} = \dfrac{2y+rz}{2a-2x-mz}$, laquelle contenant des quantités désignées et des quantités arbitraires, doit (57) pouvoir se décomposer en deux, l'une entre quantités désignées, l'autre contenant des quantités arbitraires; et en effet, il est aisé de voir, qu'on peut lui donner cette forme

$$\frac{tp}{y} + \frac{t't}{y} = \frac{y}{a-x} + \frac{y\cdot mz + a\cdot rz - x\cdot rz}{(a-x)(2a-2x-mz)}$$

qui se décompose visiblement en ces deux équations $\dfrac{tp}{y} = \dfrac{y}{a-x}$, et $\dfrac{t't}{y} = \dfrac{y\cdot mz + a\cdot rz - x\cdot rz}{(a-x)(2a-2x-mz)}$. Car si l'on fait passer tous les termes de la première dans le même membre, on aura

$$\left(\frac{tp}{y} - \frac{y}{a-x}\right) + \left(\frac{t't}{y} - \frac{y\cdot mz + a\cdot rz - x\cdot rz}{(a-x)(2a-2x-mz)}\right) = 0,$$ équation qui suivant la méthode des indéterminées ne peut avoir lieu généralement, c'est-à-dire, quelques valeurs qu'on attribue[*] à mz, rz, t't, sans que chaque terme ne soit égal à zéro; donc cette méthode employée de la manière la plus directe,

[*] ceci montre d'une manière bien évidente, ce que nous avons déjà dit tant de fois, savoir que les quantités t't, mz, rz, qu'on nomme infiniment petites, ne sont pas plus petites que les autres, puisque l'équation doit avoir lieu, quelles que soient les valeurs de ces quantités. ainsi le caractère de ces quantités réside, non dans la petitesse de leurs valeurs actuelles, mais dans la faculté qu'on a de les rendre moindres qu'aucune grandeur donnée.

conduit précisément au même résultat $\frac{tp}{y} = \frac{y}{a-x}$ entre
les quantités proposées, que le calcul infinitésimal, et
n'en diffère qu'en ce qu'on attend la fin du calcul, pour
négliger les quantités $t't$, $m z$, $r z$, que l'on néglige dans
le cours de ce même calcul en employant l'analyse
infinitésimale; soit comme absolument nulles, chacune
en particulier lorsqu'on suppose qu'il s'agit de l'infini
absolu; ou en somme, lorsqu'on suppose l'infini sensible
et produisant des erreurs qui ne tombent que sur l'équation
entre arbitraires, lesquelles doivent par conséquent se
détruire d'elles-mêmes, et ne point affecter l'équation
qui existe entre les quantités désignées. le calcul
infinitésimal, sous quelque point de vue qu'on le
considère, n'est donc qu'une abréviation de la méthode
des indéterminées; et l'on conçoit aisément par ces
remarques et les exemples rapportés ci-dessus que
celle-ci peut suppléer à l'analyse infinitésimale, sans
augmenter sensiblement la difficulté du calcul.

 on peut même rendre ces deux méthodes
absolument semblables quant aux procédés, il n'y a
pour cela, qu'à sous entendre à chaque équation imparfaite
la quantité nécessaire pour compléter l'égalité des deux
membres; c'est-à-dire, qu'il n'y a qu'à supposer cette
quantité ajoutée en effet à chaque équation imparfaite
mais n'on écrite. car alors il n'y aura évidemment plus
aucune différence, entre cette nouvelle méthode des

ceci peut suppléer
l'autre, sans aug-
menter sensiblement
difficulté du calcul.

indéterminées; ou plutôt, entre la méthode ordinaire des indéterminées ainsi modifiée; et l'analyse infinitésimale proprement dite.

Ce seroit peut-être ici le lieu de montrer par des exemples nombreux, l'application des principes contenus dans cet écrit; mais nous croyons les avoir expliqués assez clairement, pour que les commençants puissent les appliquer d'eux mêmes, aux divers exemples qu'ils trouveront dans les livres élémentaires, comparer les principes, connoître leur accord et saisir enfin l'esprit et la vraie métaphisique de ce calcul sublime.

Conclusion.

98. Si l'on demande maintenant laquelle de toutes les méthodes qu'on vient d'expliquer mérite la préférence je réponds qu'étant toutes bonnes exactes et lumineuses il convient je pense, de ne donner l'exclusion à aucune et qu'on doit tout simplement, employer en chaque cas particulier, celle que les circonstances rendent la plus simple. car il ne peut être qu'avantageux, d'envisager une question sous ses différents points de vue, de connoître les divers chemins qui conduisent au même but d'appercevoir enfin les liens et l'analogie par lesquels peuvent s'éclairer mutuellement et s'entre-aider des méthodes assez différentes au premier coup d'œil. est-il possible d'ailleurs de faire autrement? est-il possible de combiner des évanouissantes dont les rapports sont déterminés par la loi de continuité

sans comparer en même tems, au moins par la pensée, les quantités indéfiniment petites qui leur répondent? et réciproquement, peut-on calculer celles-ci, dans le dessein d'en découvrir les dernières raisons, et perdre en même tems de vue les dernières valeurs de ces quantités? ne sont-ce pas toujours en un mot, les mêmes idées qu'on a à rendre, les mêmes relations à exprimer, et toutes les méthodes diffèrent-elles autrement que dans la manière de rendre les idées et d'exprimer les relations? il faut regarder le calcul, comme une langue dans laquelle on peut s'exprimer avec d'autant plus de facilité, de clarté et de précision qu'il y a plus de mots ou de signes pour exprimer, non pas précisément les mêmes choses, mais leurs propriétés différentes et les nuances qui les caractérisent. si ces mots manquent, on se trouve forcé d'employer des expressions impropres et qui ne rendent pas exactement ce qu'on voudroit dire; ou bien il faut alonger et embarasser le discours par des periphrases. les langues, et celle des mathématiques comme toutes les autres, se forment peu à peu. bornée dans l'origine à un petit nombre d'expressions, il est nécessaire que les mêmes combinaisons de mots reviennent souvent; alors, on sent la nécessité d'en inventer de nouveaux pour remplacer les combinaisons par ces expressions plus courtes. ces nouveaux mots combinés avec les anciens donnent encore lieu à de nouvelles tournures de phrase qui se répètent elles-mêmes plus ou moins fréquemment, et déterminent de nouveau

2

l'invention de quelques expressions plus simples. et il ne faut pas regarder la perfection d'une langue propre à une science, comme une chose indifférente aux progrès de cette science. il y a un avantage considérable à pouvoir réunir beaucoup d'idées en peu de mots parcequ'on en apperçoit bien plus facilement, la liaison et les conséquences. eh! qu'est-ce autre chose au fond que le calcul, sinon l'art de s'exprimer d'une manière abrégée simple uniforme; de réunir presque en un point, des idées qui exprimées dans le langage ordinaire, se trouveroient étendues et comme délayées dans un long discours? lorsqu'aulieu de l'analyse infinitésimale, on employoit la méthode d'exhaustion on ne s'étoit pas encore apperçu de tous les usages auxquels on pouvoit l'appliquer; les expressions qu'elle employoit, revenoient trop rarement, pour qu'on sentit la nécessité de leur en substituer de plus simples; et quoique les principes fondamentaux du calcul infinitésimal, fussent dès lors connus et usités dans la pratique de la méthode d'exhaustion, c'est avec grande raison qu'on attribue l'invention de ce calcul à celui qui le premier, a eu le génie d'appercevoir la nécessité d'un langage nouveau, pour exprimer avec facilité, les idées qui se présentoient en foule à son esprit, et l'appliquer à tous les usages dont il prévoyoit qu'il

seroit susceptible, en quoi l'experience, a glorieusement
confirmé ses conjectures.

99. on peut remarquer ici, combien est admirable
la marche de l'esprit humain dans la recherche de
la vérité, on le voit s'élever par gradation, des problemes
les plus simples et indiqués par les besoins physiques,
à la connoissance des propriétés les plus générales
dont la quantité soit susceptible. d'abord, il a bien
fallu se borner à quelques combinaisons faites
sur des nombres déterminés et concrets; bientôt
après, on a remarqué que ces operations avoient
quelquechose de commun, delà sont venus les nombres,
abstraits: on s'est apperçu ensuite, qu'une opération
faite sur tel ou tel de ces nombres étoit souvent
détruite après, par une opération contraire: que par
exemple, après avoir été conduit par un raisonnement
à réunir deux nombres par addition, on étoit
conduit par la suite du même raisonnement à les
séparer l'un de l'autre par soustraction; on a conclu
de là, qu'il seroit avantageux de se borner d'abord,
à indiquer les opérations, à l'aide de quelques signes
de convention afin d'exécuter seulement, celles qui
dans la suite, seroient reconnues indispensables;
delà l'origine des signes algébriques. par la même raison
on a conclu, que jusqu'au moment où il seroit nécessaire

d'exécuter ces opérations, on pouvoit substituer aux nombres en question, d'autres simboles arbitraires, pour en tenir la place, de là, l'usage des caractères, alphabétiques dans le calcul. ces simboles étant arbitraires et pouvant être employés à représenter tel ou tel nombre, aussi bien que tel ou tel autre, les raisonnements et opérations qui doivent se faire, jusqu'au moment où il faut remettre les nombres à la place de ces caractères, indéterminés, appartiennent donc à une classe de grandeurs plus étendue que ces nombres; or on a fait de ces raisonnements et opérations, une science qu'on nomme arithmétique universelle ou algèbre. après cela, on a remarqué que la détermination des nombres représentés par ces caractères indéterminés, pouvoit être plus ou moins retardée, que par conséquent, ceux qu'on n'étoit forcé à déterminer que les derniers étoient ceux qui avoient le plus de propriétés applicables à toute espèce de nombres; de là, la distinction des constantes et des variables; les premières étant fixées par les conditions même du problème et les autres, ne l'étant que par des hypothèses subséquentes. de plus parmi ces variables, on s'est apperçu que quelques-unes ne se déterminoient point du tout, c'est-à-dire n'entroient ni implicitement ni explicitement, dans le résultat de ce genre, sont les quantités qu'on a nommées infinitésimales. mais de ces quantités, les unes s'éliminent plutôt que les autres et par conséquent, sont plus indéterminées; c'est-à-dire, ont des relations moins intimes ou moins de propriétés communes avec celles dont on cherche les rapports: de là, les différentielles de différents ordres. ce n'est pas tout encore, les quantités infinitésimales de différents ordres, sont liées entre elles par des relations quelconque, et ces relations sont assujetties à des loix plus ou — moins arbitraires; de là le calcul des variations. et c'est ainsi, que l'esprit ne cesse de s'élever continuellement à de nouvelles généralités, jusqu'à ce qu'enfin, ayant dépouillé pour ainsi dire successivement les quantités, de toute connexion, il ne lui reste plus, qu'un système d'êtres isolés, entre lesquelles il n'existe aucune relation, et qui par conséquent, cessent d'être l'objet des mathématiques.

100. la source de cette application continuelle des géomètres à généraliser, vient de ce qu'ils sentent, que, presque toujours généraliser c'est simplifier. généraliser en effet, c'est augmenter le nombre des choses comparées; c'est étendre des propositions particulières, à des classes de grandeurs plus étendues, or cela est aussi simplifier; parcequ'en augmentant le nombre des choses comparées ou des rapports

à considérer, on diminue la difficulté de chaque comparaison particulière. en effet, les propriétés communes à toutes les grandeurs sont les plus faciles à reconnaître parce qu'elles affectent tous les objets qui frappent nos sens: par la même raison, plus la classe de grandeurs à laquelle appartiennent ces propriétés est étendue, plus ces propriétés sont évidentes. c'est donc véritablement simplifier la recherche des propriétés d'une grandeur, que de la comparer d'abord, à la plus grande classe de quantités possible, pour voir ce qu'elle-a-été commun avec toutes, puis successivement, à des classes moins étendues, jusqu'à ce qu'en soit descendu par degrés, de ces propriétés générales aux propriétés particulières et caractéristiques.

c'est ainsi par exemple, que l'arithmétique des nombres abstraits, est plus générale, que celle des nombres concrets; parce que chacun y est comparé à une classe de nombres moins déterminée et par conséquent plus étendue, et qu'elle est en même tems plus simple, parce que les combinaisons ne s'y trouvent pas embarassée par la considération des propriétés particulières propres aux nombres concrets.

en algèbre, chaque grandeur se trouve comparée à une classe encore moins déterminée et par conséquent plus étendue; aussi les principes en sont-ils plus simples que ceux de l'arithmétique.

par les mêmes raisons, le calcul des variables est en même tems plus général et plus simple que celui des constantes; celui des quantités infinitésimales plus général et plus simple que celui des variables ordinaires*; celui des variations, plus général et plus simple

---

* ceci est peut-être un paradoxe, mais pour peu qu'on y fasse attention, on verra aisément, que rien n'est plus vrai. on n'employeroit pas l'analyse infinitésimale pour résoudre des questions proposées, si l'on pouvoit en venir à bout plus facilement sans son secours; et si elle paroit plus difficile que l'algèbre ordinaire, c'est uniquement, parceque les questions auxquelles on l'applique sont plus compliquées: mais toutes choses égales d'ailleurs, les calculs sont d'autant plus simples qu'ils sont plus généraux; et ils doivent l'être évidement, puisque ce qui les rend plus généraux, c'est qu'on y emploie un plus grand nombre de quantités auxiliaires pour découvrir les rapports du système désigné; or à quoi bon, auroit-on recours à ces quantités auxiliaires, si cela ne devoit servir qu'à augmenter les difficultés: il faut donc bien qu'on sente que cela doit simplifier. et en effet, plus une quantité est déterminée ou caractérisée par l'expression de ses propriétés, moins la classe de grandeurs auxquelles on la compare est étendue; donc généraliser, ou comparer à une classe de grandeurs plus étendue, c'est considérer moins de propriétés à la fois dans un même sujet, et par conséquent en simplifier l'examen.

il est encore évident, que ce qui distingue principalement les différentes branches du calcul, ce sont les divers degrés d'indétermination des quantités qui en font l'objet. d'après cela il est à présumer, que la perfection de la méthode des indéterminées, pourroit conduire peut-être, à quelque nouvelle branche intéressante de calcul. ce seroit une erreur puérile, que de croire la chose impossible, fondé sur ce que l'analyse infinitésimale décomposant les quantités jusque dans leurs éléments, on ne peut pénétrer plus qu'elle ne le fait dans les secrets de la nature. ces prétendus éléments ne sont rien moins que des êtres

que celui des différentielles; et qu'en un mot, plus on augmente le nombre des choses comparées, moins chacune d'elles, influe sur les rapports cherchés; plus il y a de combinaisons à faire, moins il y a de difficulté à faire chacune d'elles en particulier.

aussi lorsqu'on réfléchit, sur la nature des sciences, mathématiques, on s'apperçoit bientôt, que pour remplir leur but unique qui est de simplifier toujours, elles n'emploient pour ainsi dire, qu'un seul procédé, qui est de généraliser toujours. pour simplifier une question, elles la décomposent en plusieurs autres questions plus simples. c'est-à-dire, qu'elles la généralisent en augmentant le nombre des rapport à considérer, pour en diminuer la complication. c'est ce qu'on fait en arithmétique, lorsqu'on décompose les nombres, en unités, dixaines centaines &c afin de pouvoir exécuter en plusieurs opérations partielles et simples, une seule opération difficile: c'est ce qu'on fait en algèbre, lorsqu'au lieu d'opérer tout d'un coup sur un assemblage de quantités réunies par différents signes

d'une nature inconcevable, des espèces de monades qui n'offrent point de choses fixes et sensibles à l'esprit; je me suis efforcé de prouver le contraire et de montrer que ce sont des quantités tout comme les autres, et qui n'en diffèrent que par leur limite qui est zéro: c'est-à-dire que ces quantités ne sont autre chose que les parties du système auxiliaire, qui ont 0 pour parties correspondantes dans le système désigné on opère successivement sur chacune d'elles: c'est ce qu'on fait en géometrie, lorsque pour trouver une distance sur un terrein on la compare par une suite de triangles, à diverses lignes intermediaires dont la connoissance conduit par degres à celle qu'on a en vue: c'est ce qu'on fait dans l'analyse infinitésimale, lorsque pour trouver les relations qui existent entre les diverses quantités qui composent un système désigné, on emploie pour y parvenir plus aisément l'entremise de plusieurs quantités auxiliaires: c'est ce qu'on fait enfin, à toute occasion et en toutes circonstances: à cela seul, se réduit l'esprit des mathématiques; le grand art du calculateur, est de savoir choisir les termes de comparaison, auxquels il doit rapporter les quantités inconnues, pour en trouver la valeur: s'il n'est aucune quantité connue, à laquelle on puisse comparer directement l'inconnue, l'adresse consiste à découvrir quelques quantités auxiliaire dont il soit facile de trouver d'une part la relation avec la quantité inconnue cherchée; et d'une autre part, la relation qu'elle a avec quelque quantité donnée, ce qui fournit deux rapports particuliers, d'ou résulte celui que l'on cherche. Si cela est encore trop difficile, on a recours à de nouvelles quantités intermédiaires, on augmente le nombre des relations, pour diminuer la difficulté de chacune d'elles: on généralise en un mot, pour obtenir une plus grande simplification: partout enfin, on voit que l'objet des mathématiques est de faire passer par gradation des rapports simples et qu'on apperçoit pour ainsi dire, intuitivement, à ceux qui exigent une suite de combinaisons délicates et de raisonnements serrés; de ramener peu à peu par des analogies, aux questions déja résolues, celles qui ne le sont pas encore; de substituer à plusieurs propositions particulières, ou qui ne peuvent s'appliquer qu'a des classes bornées de quantité des proposition générales. C'est-à-dire, qui embrassent tout le systême de ces quantités. or tout cela n'est évidement, qu'augmenter le nombre des choses comparées, pour en faciliter les combinaisons et résoudre par parties, des questions trop compliquées, pour être résolues tout d'un coup et directement. en un mot, c'est tout à la fois, simplifier et généraliser. mais c'est parler trop longtems métaphisique dans un ouvrage dont le but est de ramener à des principes clairs et lumineux, des notions qui peuvent paroître d'abord, vagues et peu précises.

fin.

Notes
à la "Dissertation" de Carnot

A. P. YOUSCHKEVITCH

La "Dissertation sur la théorie de l'infini mathématique" a été envoyée par L. Carnot au Secrétaire Perpétuel de l'Académie de Berlin J.H.S. Formey ci-joint à la lettre suivante:

"à Monsieur

Monsieur Formey

à Arras le 8 7bre 1785.

Monsieur,

L'académie ayant proposé pour sujet du prix de mathématique qu'elle doit adjuger en 1786, la théorie de l'infini; je vous serai infiniment obligé si vous voulez bien présenter au concours la pièce ci-jointe.

J'ai laissé au bureau des diligences 12 livres argent de France pour affranchir le paquet; cette somme étant à peu près ce que coûteroit ici un pareil envoi."

Le manuscrit et la lettre sont conservés au Zentrales Archiv der deutschen Akademie der Wissenschaften zu Berlin (Sign. AAW, 1261-1262).

Préambule. La partie de la "Dissertation" qui précède le "Sommaire" est presque entièrement reproduite dans le n°I des deux éditions des *Réflexions sur la métaphysique du calcul infinitésimal,* celle de 1797 et de 1813 (citées plus loin comme *Réflexions,* I, et *Réflexions,* II).

Note[x]. Cette note est assez réduite dans les *Réflexions,* I, mais de par son contenu reste en substance la même: l'idée d'une quantité infinitésimale y est ramenée à celle de la limite. Dans les *Réflexions,* II, la note est largement modifiée. En particulier, on y lit: "Qu'est-ce, en effet, qu'une quantité dite infiniment petite en Mathématiques? Rien d'autre chose qu'*une quantité que l'on peut rendre aussi petite qu'on le veut . . .*"; le lien entre les notions de limite et d'infiniment petit n'est établi dans les *Réflexions,* II, que dans les § 131-132.

Pour les expressions "première raison" et "dernière raison" voir le § 6.

Enfin, l'expression "indéfiniment petite" recommandée à la fin de la note remonte à B. Pascal (1658); cf. les n°n° 115-116 des *Réflexions,* II, où un morceau correspondant de Pascal est cité et commenté.

Sommaire. Ce texte n'est pas reproduit dans les *Réflexions.*

§ 2-3. Dans les *Réflexions,* I, la notion de quantité désignée apparait dans le n°14 et celle de quantité non-désignée ou auxiliaire dans le n°15; on les retrouve dans le n°17 des *Réflexions,* II.

§ 4-8. Les expressions "première raison" et "dernière raison" remontent à Newton qui distinguait la limite du rapport des quantités naissantes (prima ratio quantitatum nascentium) d'avec la limite du rapport des quantités évanouissantes (ultima ratio quantitatum evanescentium); d'ailleurs ces deux limites ont toujours pour Newton la même valeur. Cf. ses *Principia mathematica philosophiae naturalis*, 1687, lib.I, sec.I, lemma XI, scholium et *Tractatus de quadratura curvarum*, 1704, Introductio.

Dans le n°17 des *Réflexions*, I, Carnot définit la notion de limite comme il suit: "une *limite* n'est autre chose qu'une quantité désignée de laquelle une quantité auxiliaire est supposée s'approcher perpétuellement, de manière qu'elle puisse en différer aussi peu qu'on voudra, et que leur dernière raison soit une raison d'égalité." Il est à noter que Carnot ne présuppose pas que la quantité auxiliaire dont il s'agit est toujours croissante ou toujours décroissante,—restriction sous-entendue dans les définitions de limite données par d'Alembert (*Encyclopédie*, t.I, 1765, art. "Limite") et par S. L'Huilier (*Exposition élémentaire des principes des calculs supérieurs*, Berlin, 1786, p. 7). Plus tard L'Huilier a levé cette restriction (*Principiorum calculi differentialis et integralis expositio elementaris*, Tübingen, 1795, p. 17).

La variation continue, Carnot la décrit comme celle qui se produit "par degrés insensibles suivant une loi quelconque" (§ I). La définition moderne d'une fonction univoque continue remonte à B. Bolzano (1817) et A. Cauchy (1821). L'expression "loi de continuité" (lex continuationis) appartient à Leibniz.

§ 10.   Nous donnons ici la note en bas du § 10, très difficile à lire dans le manuscrit:

* par exemple si l'on se propose de trouver les rapports qui existent entre a, b, c, etc x, y, z, etc et que pour y parvenir, on ait recours a un systeme auxiliaire composé des quantités a, b, c, etc $(x + dx)$, $(y + dy)$, etc celles a, b, c, qu'on suppose les mêmes dans le systeme auxiliaire que dans le systeme désigné seront les *constantes*, celles $(x + dx)$, $(y + dy)$ etc qui changent par le rapprochement des deux systèmes seront les *variables auxiliaires*, et leurs limites x, y, z, etc seront les variables désignées.

§ 12.   Presque les mêmes définitions et explications sont reproduites dans les n°n° 21-23 des *Réflexions*, I.

Tant dans la "Dissertation" que dans les *Réflexions*, I, Carnot a pour la première fois ramené d'une manière explicite la notion d'une quantité infiniment petite à celle de limite,—ce qu'il n'a pas répété dans les *Réflexions*, II, (voir plus haut, p. 163). L'Huilier dans son *Exposition* ... (1786), a défini une quantité infiniment petite directement, comme une variable qui peut devenir plus petite qu'une quantité donnée arbitraire (*ibid.*, p. 22); d'ailleurs il évitait ici l'emploi des termes "infiniment petit" (ou "grand") qu'il réservait à l'infini actuel. Plus loin il donne aux quantités infinitésimales potentielles les noms "infiniblement petites" et "infinibles" (*ibid.*, p. 147).

§ 15.   On retrouve ce texte, avec quelques modifications, dans le n°41 des *Réflexions*, I.

"Le grand géomètre" mentionné par Carnot, c'est Newton (voir la célèbre scolie citée plus haut, dans ma note au § 4-8).

§ 19.   La définition d'une équation imparfaite dans les *Réflexions*, I (n°31), reste presque la même; dans les *Réflexions*, II (n°33), Carnot se passe de la notion de limite.

§ 28.   Newton appelait fluxion (fluxio) d'une quantité fluente sa vitesse instantanée, ce qui correspond à notre dérivée. A la notion de différentielle de Leibniz correspondait chez Newton la notion de moment (momentum) ou de l'accroissement instantané d'une fluente. Néanmoins, au XVIIIe siècle on employait parfois les mots différentielle et fluxion comme synonymes. Dans les *Réflexions*, II, la méthode des fluxions est décrite correctement (n°n°136-144).

§ 28-33.   Les éléments du calcul infinitésimal sont plus amplement exposés et appliqués à quelques exemples dans les n°n°49-67 des *Réflexions*, I, et d'une manière encore plus détaillée dans le 2e chapitre des *Réflexions*, II (n°n°46-94).

§ 34-35.   Ce texte est supprimé dans les *Réflexions*, I. Dans les *Réflexions*, II, Carnot a consacré au calcul des variations les n°n°95-105 du 2e chapitre.

§ 36.   Par la suite Carnot n'a retenu de ce texte que le signe de limite employé tantôt sous la même forme $\mathcal{L}$ (*Réflexions*, I, n°38 et *Réflexions*, II, n°135) tantôt sous la forme *lim* (*Réflexions*, I, n°56, 62-65).

Ce dernier symbole, proposé et publié pour la première fois par S. L'Huilier (*Exposition . . .*, 1786, pp. 24, 31), fut adopté par de nombreux savants: A. Cauchy (1821), K. Weierstrass, qui écrivait comme à present $\lim_{n=\infty}$ (1854), et d'autres; la flèche ne fut introduite—à ce qu'il paraît—que dans les premières années de notre siècle ($\lim_{n\to\infty}$ ou $\lim_{x\to a}$, etc.).

Ainsi, ce sont Carnot et L'Huilier qui ont inventé le symbole de passage à la limite.

Carnot exprime son "moment différentiel" par $\mathcal{L}\frac{dy}{dx}$ parce qu'il définit la différentielle comme la différence de deux valeurs infiniment proches d'une variable (§ 27). La notation de L'Huilier était plus moderne: il écrivait $\lim.\frac{\Delta P}{\Delta x}$ (*Exposition . . .*, 1786, pp. 31-32), en employant le signe Δ de l'accroissement fini introduit par Euler (1755); le symbole $\frac{dP}{dx}$ n'est utilisé, selon L'Huilier, que pour abréger et faciliter les calculs.

Enfin on trouve occasionnellement la caractéristique Dy chez Jean I Bernoulli, et L.F.A. Arbogast s'en servit plus tard constamment dans son *Calcul des Dérivations*, Strasbourg, 1800.

§ 37-38.   Ce texte est reproduit presque intégralement dans le n°2 des *Réflexions*, I et II, mais la note y est supprimée. Cf. remarques de Carnot sur la priorité de l'invention du calcul infinitésimal faites plus loin, § 68 et dans les *Réflexions*, II, n°n° 144 et 165.

§ 39-49.   Le texte de ces paragraphes est reproduit sans grands changements dans les n°n°3-11 des *Réflexions*, I et II, après quoi Carnot passe à la théorie des équations imparfaites, en commençant par des définitions de la quantité constante, variable, désignée, etc.

§ 51-56.   Aux théorèmes I-III de la "Dissertation" correspondent les théorèmes I-III des *Réflexions*, I, formulés et argumentés pareillement (n°n°32-35). La suite des raisonnements dans les *Réflexions*, II, est tout à fait différente, la notion de limite n'y figure pas explicitement (nᵒnᵒ24-33) et en fin de compte Carnot déclare que toute la théorie de l'infini peut être regardée comme renfermée dans le théorème suivant (nᵒ34):

"Pour être certain qu'une équation est nécessairement et rigoureusement exacte, il suffit de s'assurer: Iᵒ Qu'elle a été déduite d'équations vraies ou du moins imparfaites, par des transformations qui ne leur ont point ôté le caractère d'équations au moins imparfaites. 2ᵒ. Qu'elle ne renferme plus aucune quantité infinitésimale, c'est-à-dire, aucune quantité autre que celles dont on s'est proposé de trouver la relation." L'argumentation est semblable à celle du nᵒ55 de la "Dissertation."

§ 57.   Le texte de la scolie reapparaît presque intact au commencement du nᵒ36 des *Réflexions*, I.

§ 58.   A partir d'ici, Carnot passe au calcul des quantités évanouissantes développé pour la première fois par Euler dans ses *Institutiones calculi differentialis*, 1755. Aux § 58-83 de la "Dissertation" correspondent les nᵒnᵒ41-47 des *Réflexions*, I, où toutefois l'exposé est de beaucoup plus sommaire (les nᵒnᵒ36-39 des *Réflexions*, I, sont consacrés à la méthode des indeterminées et à la méthode des limites; voir plus loin).

Le calcul des quantités évanouissantes est analysé aussi dans les nᵒnᵒ145-153 des *Réflexions*, II, et ici Carnot s'en réfère à l'ouvrage mentionné d'Euler.

§ 65.   Le théorème IV manque dans les *Réflexions*, I et II.

§ 67.   L'égalité 2.0 = 2.0 +3.0.0, qui ne sert à rien, Carnot l'a remplacée dans le nᵒ46 des *Réflexions*, I, par une autre 6.0 = 2.0 + 4.0 aussi inutile que la première (de plus il y a des fautes d'impression, notamment: 60 = 20 + 40).

§ 72.   Dans les *Réflexions*, II, Carnot a écrit (nᵒ153): "Il suit de ce que nous venons de dire qu'on peut à volonté considérer les quantités infiniment petites ou comme absolument nulles, ou comme de véritables quantités; un motif cependant me ferait préférer cette dernière manière d'envisager l'analyse infinitésimale, c'est que ceux qui la considèrent ainsi me semblent traiter la question d'une manière plus générale que les autres. Car ceux-ci, en attribuant aux quantités infiniment petites la valeur o, font une opération inutile . . ." (parce que l'élimination se produit indépendamment de cette supposition).

§ 77.   Le théorème V manque dans les *Réflexions*, I et II.

§ 83.   La méthode des indivisibles n'est pas mentionnée dans les *Réflexions*, I, mais dans les *Réflexions*, II, Carnot en donne l'analyse succincte (nᵒnᵒ113-118).

§ 84.   Dès le § 84 Carnot entreprend l'examen de la méthode des limites qu'il tâche d'adapter aux besoins du calcul infinitésimal (§ 85-91). Evidemment non satisfait de cette tentative, il ne la reprend pas dans les *Réflexions*, I et II, où la méthode des limites est analysée, respectivement, dans les nᵒnᵒ38-39 et nᵒnᵒ129-135.

§ 86.   Le théorème VI manque dans les *Réflexions*, I et II. Cependant dans le nᵒ17 des *Réflexions*, I, Carnot explique qu'on peut parler dans un certain sens de la limite d'une quantité désignée qui est elle-même sa propre limite, "ce qu'on ne peut refuser d'accorder, puisque la dernière valeur d'une quantité déterminée ne peut être que cette quantité elle-même.''

Carnot a formulé le premier la proposition sur la limite d'une constante.

§ 87.   Le théorème VII, connu depuis longtemps et formulé dans les termes de la théorie des limites par d'Alembert (1765), ne figure plus dans les *Réflexions*, I, et n'est qu'effleuré dans le n°133 des *Réflexions*, II.

§ 88.   Le théorème VIII sur la limite d'un rapport appartient à Carnot et à L'Huilier, qui l'a énoncé dans son *Exposition* . . . , 1786 (p. 24). Cette proposition, dont Carnot n'a plus besoin dans les *Réflexions*, I et II, n'y figure pas.

D'autre part, dans le n°38 des *Réflexions*, I, et le n°135 des *Réflexions*, II, Carnot reprend l'exemple avec la tangente au point donné de la circonférence $y^2 = 2ax - x^2$ et calcule la valeur numérique de la sous-tangente moyennant la méthode des limites, ce qu'on ne trouve pas dans la "Dissertation."

§ 93-97.   La méthode des coefficients indeterminés est exposée dans les n°n°36-38 des *Réflexions*, I, et dans les n°n°119-128 des *Réflexions*, II, où le nom de son inventeur R. Descartes (1637) est mentionné.

Les deux exemples traités dans les § 94 et 95 de la "Dissertation" manquent dans les *Réflexions*, I, et sont examinés d'une manière détaillée dans les n°n° 124, et 125 des *Réflexions*, II.

§ 98-100.   Les considérations générales sur "la marche de l'esprit humain" et, en particulier, sur le développement des sciences mathématiques ont été omises dans les *Réflexions*, I et II. Dans la brève "Conclusion," constituée par le dernier n°68 des *Réflexions*, I, Carnot résume derechef les avantages de sa théorie par rapport à la méthode des limites et il y emploie certaines expressions tirées du § 98 de la "Dissertation."

Dans la "Conclusion générale" (n°n°159-174) des *Réflexions*, II, il reprend le sujet de ce même § 98 de la "Dissertation," compare de nouveau les diverses méthodes infinitésimales qui ne sont à proprement parler que la méthode d'exhaustion des anciens présentée sous divers points de vue, et proclame enfin la supériorité de la méthode de Leibniz dont la justesse est basée, selon Carnot, sur l'élimination nécessaire des erreurs commises dans le cours du calcul.

Appendix B

The memoir that Carnot submitted to the Academy in 1778 in competition for its prize of 1779 consists of 85 sections in some 64 folio sheets. It is organized into two parts. Part I occupies § 1 through 26, and consists of experiments on friction. Part II has three further subdivisions. The first discusses the principles of "machines en général" in § 27 through 50. The second, in § 51 through 79, discusses the application of the foregoing considerations to the seven classes of simple machines in equilibrium. The third, in § 80 through 85, discusses simple machines in a state of motion.

By kind permission of the Permanent Secretaries of the *Académie des sciences, Institut de France*, we reproduce here the title page and preamble followed by the theoretical discussion that Carnot gave in Part II, § 27 through 50. It seems worthwhile to include also § 51 through 60 from the second portion of Part II since it was there that Carnot first demonstrated his method of representing the projection of the magnitude of directed quantity upon another direction as the product of the magnitude by the cosine of the angle between them. As will be seen, the demonstration he gives applies to the resultant of a number of intersecting vector quantities. Unfortunately, all figures to which he makes reference are missing from this manuscript, but it is a simple matter to reconstruct these from the context, and when that is done at the end of this appendix it turns out, interestingly enough, that they are identical with those that he reproduced almost twenty-five years later in Plate I, figures 2 and 3 of the *Principes fondamentaux de l'équilibre et du mouvement* of 1803.

Despite the large format in which these facsimiles are reproduced, the legibility of several footnotes continues to be less than ideal. These have been transcribed and they appear in type following Appendix C.

MÉMOIRE

Sur la théorie des machines

pour concourir au prix de 1779 proposé

par l'académie Royale des sçiences de paris

par Carnot (Lazare Nicolas Marguerite)

videndum

qua ratione fiant et qua vi quaeque gerantur

lucret

MÉMOIRE
sur la théorie des machines.

l'academie Royale des sciences de paris propose pour sujet du prix qu'elle doit adjuger en 1779 la theorie des machines simples en ayant égard au frottement et a la roideur des cordages, mais elle exige que les loix du frottement et l'examen de l'effet resultant de la roideur des cordages soient determinés d'apres des experiences nouvelles et faites en grand elle exige de plus que les experiences soient applicables aux machines usitées dans la marine telles que la poulie le cabestan le plan incliné

pour repondre aux vues de cette illustre societé j'ai fait plusieurs experiences dont je rendrai compte dans la premiere partie de ce memoire, et dans la seconde j'en ferai l'application aux machines.

* * * * * * *

27. imaginons un systeme quelconque de corps parfaitement durs dont les uns soient mobiles les autres fixes et dont plusieurs si l'on veut se tiennent par des verges inflexibles que l'on imprime a ce systeme un mouvement quelconque et que par l'action réciproque de ces corpuscules le mouvement soit changé en un autre qu'il s'agit de trouver je dis que si l'on appelle m l'un quelconque de ces corpuscules c'est-a-dire sa masse v la vitesse qu'il auroit prise s'il eut été libre c'est adire sans la reaction qu'il eprouve des autres parties du systeme, u la vitesse qu'il a réellement prise et y l'angle formé par les vitesses V, u, on aura $\int mu(V\cos y - u) = 0$

c'est-a-dire que la somme des produits de la quantité

{ 272 }

de mouvement de chacun des corpuscules par la vitesse quil a perdue estimée dans le sens de celle quil a prise est egale a zero

pour le demontrer nous observerons d'abort que lorsqun corps dur en pousse ou choque un autre 1° le corps choquant et le corps choqué vont apres le choc tous les deux de compagnie c'est adire que leurs vitesses estimées dans le sens perpendiculaire a leur surface commune au point de contact sont egales 2° que si ces corps se tiennent par une verge inflexible leurs vitesses estimées suivant une droite menée de l'une a l'autre sont egales 3° que tout cela est vrai soit que les deux corps soient ou non mobiles tous les deux 4° que si ces corpuscules sont liés par une verge inflexible leur action reciproque s'exercera suivant la direction de la ligne droite menée de l'un a l'autre de ces corpuscules et que s'ils sont detachés l'un de l'autre cette action s'exercera suivant une perpendiculaire elevée a leur surface commune au point de contact 5° que si ces corpuscules sont mobiles l'un et l'autre la reaction sera egale et directement opposée a l'action entre eux. il suit d'abort de tout cela comme il est aisé de le voir que dans tous les cas leurs vitesses estimées dans le sens de leur action reciproque sont egales cela posé

Soient m et m' deux molécules quelconques du systema V, V' les vitesses virtuelles de ces corpuscules c'est adire celles quils auroint prises s'ils avoint été libres u et u' les vitesses réeles de ces memes corpuscules y l'angle formé par V et u; y' celui qui est formé par V' et u'

si l'on decompose la vitesse V en deux dont l'une soit u l'autre est ce que j'entends par la vitesse perdue et cette vitesse estimée dans le sens de u est evidement $V \cos y - u$

{ 273 }

F la force imprimée a la molecule m par la molecule m'
F' la force imprimée a la molecule m' par la molecule m
z l'angle formé par F et u z' l'angle formé par F' et u'

il est d'abord evident qu'on aura pour chacun de
ces corpuscules $m(u - V\cos y) = \int F\cos z$ car $F\cos z$ est la force
que m' communique a m estimée dans le sens de u
et $\int F\cos z$ est la somme de celles que cette molecule m
reçoit au meme instant dans le meme sens de toutes
les parties du Systeme or cette Somme est visiblement
égale a $m(u - V\cos y)$ gagnée au meme instant
par cette molécule dans le meme sens donc en effet
$m(u - V\cos y) = \int F\cos z$ multipliant par u nous aurons
$mu(u - V\cos y) = \int F u\cos z$ donc pour tout le Systeme on aura
$\int mu(u - V\cos y) = \int\int F u\cos z$ maintenant je dis que le dernier
membre de cette equation est egal a zero en effet il est

clair que $\int\int F u\cos z$ et $\int\int F'u'\cos z'$ sont deux quantités
identiques donc il s'agit de prouver que $\int\int F u\cos z + \int\int F'u'\cos z' = 0$
pour cela considerons deux a deux toutes les molecules
du Systeme: par ce qui a été dit au commencement
de la démonstration il est evident quelque soient
les molecules m et m' 1° que si elles agissent l'une sur
l'autre on aura $u\cos z = -u'\cos z'$ car $u\cos z$ est la vitesse de m
estimée dans le Sens de F et $-u'\cos z'$ la vitesse de m'
estimée dans le sens de F' or ou ces molecules m m' Sont
toutes les deux fixes ou l'une est fixe et l'autre est
mobile oubien elles sont toutes les deux mobiles
Si elles sont fixes toutes deux on a $u = 0, u' = 0$, et partant
$F u\cos z + F'u'\cos z' = 0$. Si l'une est fixe et l'autre mobile on a
ou $u = 0$ ou $u' = 0$ donc $u\cos z = 0$ et $u'\cos z' = 0$ puisque $u\cos z = -u'\cos z'$
donc $F u\cos z + F'u'\cos z' = 0$ Si toutes deux sont mobiles on a $F = F'$

$u \cos z = -u' \cos z'$ Donc $F u \cos z + F' u' \cos z' = 0$ Donc 1° Si les deux molecules agissent l'une sur l'autre on a toujours $F u \cos z + F' u' \cos z' = 0$ 2° Si les deux molecules n'agissent point l'une sur l'autre on aura $F = 0$, $F' = 0$, donc on aura encore $F u \cos z + F' u' \cos z' = 0$ donc pour toutes les molecules du systeme prises deux à deux on a $F u \cos z + F' u' \cos z' = 0$ donc pour tout le systeme on a $\iint (F u \cos z + F' u' \cos z') = 0$ et partant $\iint F u \cos z = 0$ donc enfin $\int m u (V \cos y - u) = 0$ ce qu'il falloit prouver.

Corollaire

18. la meme chose a lieu quelque soit le nombre des corpuscules donc elle a lieu pour un systeme quelconque de corps durs. Si les corps agissent les uns sur les autres par des fils ou par des leviers la meme chose aura encore lieu Car ces fils et ces leviers peuvent etre regardés comme des corps d'une masse infiniment petite faisants partie du systeme et puisque cette proposition est encore vraie quelque soit le nombre des corpuscules fixement arretés il est clair qu'elle le sera aussi lorsque les corps agiront les uns sur les autres par l'entremise d'une machine Car cette machine n'est jamais qu'un assemblage de points fixes de fils de leviers ou de verges inflexibles j'en excepte le cas ou il y entreroit des ressorts parceque j'ai supposé des corps parfaitement durs. de tout cela nous pouvons donc conclure la proposition suivante sur laquelle nous etablirons toute notre theorie ⸺

theoreme fondemental

29. si dans un systeme quelconque de corps durs agissants les uns sur les autres d'une maniere quelconque soit immediatement soit par

l'entremise d'une machine on appelle m une molécule quelconque du système V la vitesse qu'auroit eu cette molécule si elle eut été libre a un instant donné u celle quelle prend réellement a cause de l'action réciproque des différentes parties du système y l'angle compris entre les directions des vitesses V et u je dis qu'on aura $\int mu(V\cos y - u) = 0$

Corollaire 1er

30 la vitesse V est évidement la résultante de la force motrice du corpuscule m et de la vitesse qu'il avoit l'instant précédent et par laquelle il est arrivé au point m donc si l'on appelle cette vitesse u' z l'angle quelle fait avec u et l'élément du tems on aura (53) $V\cos y = u'\cos z + pdt\cos x$ donc l'équation deviendra $\int mu(u'\cos z - u) - \int mupdt.\cos x = 0$

Coroll. 2

31 Si il y a choc entre les corps du système a l'instant proposé il est clair que pour cet instant pdt pourra être censé nul en comparaison de u donc dans ce cas on a pour l'instant $\int.mu(u'\cos z - u) = 0.$

un pourroit de la tirer facilement toutes les loix du choc des corps mais ce seroit trop nous écarter de notre objet principal

Coroll. 3

32 Si l'on suppose y infiniment petit il est clair qu'on aura $\cos y = 1$ $\cos z = 1$ donc l'équation du corollaire premier devient $\int mu(u'-u) + \int mupdt\cos x = 0$

Si l'on suppose de plus que u diffère infiniment peu de u' c'est a dire que le mouvement de chaque point varie infiniment peu d'un instant a l'autre soit en grandeur soit en direction on aura $u'-u = -du$ donc en supposant $ds = udt$ on aura $\int.mpds\cos x - \int mudu = 0$ ou $\int.mupdt\cos x - \int mudu = 0$ ————

Coroll. 4.

33. en supposant comme dans le corollaire precedent que le systeme change par degres insensibles s'il passoit par une position ou les forces p se detruisissent mutuellement c'est a dire ou il y auroit équilibre sans le mouvement aquis il est clair que pendant un instant le mouvement seroit le meme que si les forces p etoint nulles donc dans le cas on aura pour l'instant $\int mu \, du = 0$ et $\int mp \, ds \cos x = 0$

Coroll. 5.

34. Si le systeme proposé est en équilibre c'est a dire que nonseulement les forces p se detruisent mutuellement mais encore qu'il n'y ait aucun mouvement aquis l'équation $\int mp \, ds \cos x = 0$ trouvée pour le cas dans le corollaire precedent paroit devenir inutile parceque le systeme etant en repos on a $u = 0$ et par consequent l'équation se reduit a $0 = 0$ qui n'apprend rien cependant on peut tirer grand parti de cette équation meme dans le cas ù

pour cela concevons qu'on imprime a toutes les parties du systeme en equilibre des mouvements tels qu'il n'en resulte aucun changement dans l'action reciproque des differentes parties du systeme ensorte que les mouvements soient reçus sans alteration par les corps auxquels ils sont imprimés les forces p continueront de se faire mutuellement équilibre et rien alors n'empechera de regarder le mouvement comme aquis depuis un certain tems donc on retombe dans le cas du corollaire precedent et l'équation $\int mp \, ds \cos x = 0$ ne se reduira plus a l'équation identique $0 = 0$

Coroll. 6.

35. je suppose par exemple que le systeme proposé en équilibre soit libre c'est a dire ne soit gené par aucun point fixe ni autre obstacle quelconque imprimons a toutes les parties

de ce Systeme un mouvement Commun c'est a dire des vitesses toutes egales et paralleles il est clair que les vitesses ne tendant ni a raprocher les differentes parties du Systeme ni a les eloigner l'une de l'autre les Corps recevront les vitesses imprimées Sans alteration et les forces p continueront de se faire mutuellement equilibre on aura donc $\int mp\,ds\cos x = 0$ et parceque ds est Constante par hypotese c'est a dire la meme pour toutes les molecules on aura $\int mp\cos x = 0$ or mp est la force motrice absolue de la molecule m $mp\cos x$ est Cette force estimée dans le sens de la vitesse u de Cette molecule. donc dans le Systeme en équilibre la somme des forces estimées dans un sens quelconque est egale a zero; proposition tres Connue.

Coroll 7

36 imaginons encore un Systeme de Corps en equilibre qui ne Soit gené par aucun obstacle Comme dans le Corollaire precedent ayant imaginé a volonté un axe fixe imprimons au Systeme un mouvement tel que Sans les forces p il eut tourné autour de Cet axe Sans que les differentes parties de ce Systeme eussent tendu a Se raprocher ou a S'eloigner l'une de l'autre il est donc evident que Ces forces n'alterent en rien l'action réciproque des differentes parties du Systeme en vertu des forces p donc Ces forces Continueront de se faire mutuellement équilibre et le Systeme tournera autour de l'axe de maniere que chaque point Se mouvra dans un plan perpendiculaire a Cet axe avec une vitesse proportionelle a la distance de Cet axe nous retomberons donc encore dans le Cas du Corollaire Cinquieme et l'équation $\int mp\,ds\cos x = 0$ aura lieu or si l'on appelle R la distance de la molecule m a l'axe il est clair que $ds = aR$ a etant une Constante donc l'équation deviendra $\int mp\,R\cos x = 0$ c'est a dire que la somme des moments des forces gagnées par quelquesuns des Corps du Systeme Sera perdue par les autres.

{ 278 }

Si le systeme etoit obligé de tourner autour d'un point ou d'un axe fixe il est aisé de voir ce qui lui arriveroit j'en parlerai a l'article du levier.

Corolla: 8.

37. l'équation $\int mpds\cos x = 0$ ou $\int mp\dot{u}\cos x = 0$ ayant lieu dans le cas d'équilibre entre les forces p quelque soit le systeme proposé supposons qu'il n'y ait que deux corpuscules m, m'; que leurs vitesses soient u et u' leurs forces motrices p et p' et les angles que forment les forces avec leurs vitesses x, x' l'equation sera $mpu\cos x + m'p'u'\cos x' = 0$ donc $mp : m'p' :: u'\cos x' : -u\cos x$ c'est adire (abstraction faite des signes) que les forces $mp, m'p'$ qui se font équilibre sont en raison réciproques des vitesses estimées dans le sens de ces forces, ce qui est le fameux principe de *descartes* pour l'équilibre des machines

Coroll: 9.

38 Dans un systeme de corps en equilibre il arrive souvent que le systeme des forces appliquées a ce systeme de corps peut se decomposer en plusieurs autres dont chacune en particulier soit tel que s'il etoit seul il y auroit encore équilibre; dans ce cas on a donc pour chacun de ces systemes consideres comme seul $\int \cdot mpds\cos x = 0$ et de cette maniere cette seule équation fournit dans chaque cas particulier toutes celles qui sont necessaires pour la solution du probleme

Coroll: 10.

39 Supposons maintenant le systeme en mouvement d'une maniere quelconque pourvu toujours qu'il ne passe pas brusquement d'un etat a l'autre nous aurons par le corollaire 3 $\int mpds\cos x - \int mudu = 0$. Si l'on prend l'integrale de cette équation pour avoir le mouvement aubout d'un tems indeterminé on aura

$\int\int mpds\cos x - \int mu^2 + C = 0$ le signe $\int$ etant relatif a la durée du mouvement et le signe $\int$ a la figure du systeme pour determiner la constante C nommons K la vitesse initiale de la molecule m cela posé au commencement du mouvement on a $\int\int mpds\cos.x = 0$ donc lequation se reduit pour cet instant a $\int mK^2 = C$ donc $2\int\int mpds\cos.x = \int mu^2 - \int mK^2$ equation qui exprime visiblement le principe de la conservation des forces vives

40 tout ce que nous avons dit d'un systeme de corps durs doit aussi s'entendre d'un systeme de corps elastiques en faisant entrer la force de l'elasticité dans la valeur de p car on peut regarder un corps elastique comme l'assemblage d'une infinité de corpuscules durs separés par des ressorts

41 on peut aussi regarder un fluide comme un amas de corpuscules durs ou elastiques donc tout ce que nous avons dit s'applique egalement aux fluides soit incompressibles soit elastiques

42 a propos des fluides je prie le lecteur de vouloir bien me permettre ici une observation etrangere a mon objet mais que je n'aurai peutetre point occasion de placer ailleurs on peut la passer si l'on veut

on regarde ordinairement le principe d'égalité de pression dans les fluides en équilibre comme une verité experimentale je crois cependant qu'on peut le démontrer a priori et c'est l'objet de cette remarque.

soit (Fig p) ABQ une masse fluide sans pesanteur ni force motrice quelconque enfermée de tous cotes dans un vase et supposons qu'ayant fait en A une ouverture infiniment petite on y applique un piston Amn lequel soit poussé perpendiculairement a la surface du fluide par une puissance quelconque A je considère pour plus de facilité le fluide comme étant dans un seul plan ABQ mais ce que je dirai s'applique sans difficulté a une masse fluide quelconque imaginons par le milieu K de mn une courbe continue KpsB laquelle soit perpendiculaire a la paroi du vase aux points K et B ensuite ayant décomposé la longeur de cette courbe en une infinité d'elements Kp ps imaginons par chaque point de division une perpendiculaire a sa courbure enfin par les points m n ayant imaginé les perpendiculaires ml no sur mn concevons par les points o et l ou les perpendiculaires sont rencontrées par la ligne opl perpendiculaire a la courbe concevons dis je par ces points les perpendiculaires lq or a ol et soit cette construction continuée jusqu'aux points D et F de la paroi du vase cela posé je dis que DF diffère infiniment peu de mn c'est adire que DF-mn est infiniment petite du second ordre ce qui n'a ce me semble pas besoin d'etre prouvé

maintenant j'observe que le fluide mlno ne peut agir sur celui qui l'environne que suivant des directions perpendiculaires a leur surface commune par consequent les forces avec lesquelles le petit prisme mlno pressera le fluide adjacent auront des directions perpendiculaires aux lignes ml no lo or ml et no sont parallèles ainsi la force resultante des pressions qui s'exercent sur les deux lignes est perpendiculaire a la direction de la puissance A quant a la pression sur lo puisqu'elle est perpendiculaire a lo elle est infiniment peu inclinée a la puissance A donc la puissance A se décompose en deux dont l'une agit perpendiculairement sur lo et fait avec elle un angle infiniment petit l'autre agit sur ml et no et lui est perpendiculaire mais une telle décomposition ne peut se faire évidemment si la première de ces deux forces composantes n'est infiniment grande relativement a l'autre d'ou il suit visiblement que la seconde de ces deux forces ne diffère de la force A que d'une quantité infiniment petite du second ordre ce qui est clair a cause de l'angle droit que fait la puissance A avec les forces qui agissent sur ml et no donc enfin la force qui agit sur lo est a une quantité infiniment petite du second ordre pres egale a la puissance A on prouvera par le meme raisonnement que la force qui agit sur gr ne diffère de celle qui agit sur lo que d'une quantité infiniment petite du second ordre donc la pression se transmet de proche en proche a la base DF en diminuant seulement d'une quantité infiniment petite du second ordre a chaque division de la courbe donc la pression sur mn ne peut différer de celle sur DF que d'une quantité infiniment petite du premier ordre et comme on peut prendre DF en quel endroit on veut de la surface du vase et meme de l'interieur du fluide il s'en suit que la pression A se repend egalement en tout sens Desorte que chaque molecule est egalement en tous ses points suivant une perpendiculaire a sa surface

Coroll. 13.

43 Comme la pluspart des machines ont pour objet d'enlever des poids je m'arreterai quelques moments sur le cas ou dans l'equation $\int mu\,du - \int mp\,ds\cos x = 0$ trouvée Coroll. 3 la force p est la gravité soient donc M la masse totale du systeme H la hauteur dont est descendu le centre de gravité pendant le tems t V la vitesse due a la hauteur H h la hauteur dont la molecule m est

est descendüe dans le meme tems; il est donc clair qu'on aura
$ds \cos x = dh$ et $\int mp ds \cos x = \int mp dh = Mp dH$ donc $2\int\int mp ds \cos x = 2\int Mp dH = 2MpH$.
or $2MpH = MV^2$ donc $\int mu^2 = \int mK^2 + MV^2$

Coroll 4

44 si le systeme des poids passe par une position dans laquelle
toutes les forces se detruisent c'est-a-dire ou tous les poids
fussent deumeures en équilibre Sans le mouvement aquis
les corps se mouvront pendant un instant. Comme si l'on avoit
$p = o$ donc l'équation $Mp dH - \int mu du = o$ donnera les deux autres
$\int mu du = o$ $Mp dH = o$ donc $dH = o$ donc le centre de gravité
Sera fixe ou Stationaire ou decrira une petite ligne horisontale
Ce qu'on exprime ordinairement en disant que le centre de
gravité est au point le plus bas possible

Coroll. 15

45 puisque $\int mu^2 = \int mK^2 + MV^2$ il s'en Suit que si le centre de gravité
est fixe ou se meut horisontalement on aura $\int mu^2 = \int mK^2$
Donc le Systeme se mouvra comme s'il n'etoit pas pesant

Si l'on a en outre $K = o$ on aura $\int mu^2 = o$ Ce qui ne peut etre
amoins qu'on n'ait $u = o$. Donc pour prouver que plusieurs
poids appliques a une machine doivent se faire mutuellement
equilibre il Suffit de prouver que le centre de gravité ne
descendra pas

par exemple dans un treuil Si le poids applique au Cilindre
est au poids appliqué a la roue Comme le rayon de la roue est
a Celui du Cilindre il doit y avoir equilibre parceque le Centre
de gravité ne peut pas descendre puisqu'il resteroit fixe quand
meme il y auroit du mouvement dans la machine

on tire de la tres facilement la loi d'equilibre dans
chaque Machine meme lorsque les puissances ne sont pas
des poids parceque il est aisé de substituer un poids a une force
quelconque par une poulie de renvoi

46 Supposons qu'outre les poids appliqués a la machine il y ait encore d'autres puissances pour les mouvoir comme des hommes des animaux et voyons ce que deviendra l'équation générale $\int m u^2 = \int m K^2 + 2 \int\int m p \, ds \cos x$ le terme $2 \int\int m p \, ds \cos x$ se décompose en deux parties dont l'une est MV^2 et l'autre est le double de la somme des produits de chacune des forces mouvantes multiplié par l'élement du chemin qu'elle décrit dans le sens de cette force partant si F est l'une de ces forces et u sa vitesse on aura $\int m u^2 = \int m K^2 + MV^2 + 2 \int\int F u \, dt$

47 Supposons $u = o$ $K = o$ en sorte que tous les corps soient en repos au commencement et a la fin du mouvement notre équation devient $2 \int\int F u \, dt + MV^2 = o$ et nommant H la hauteur a laquelle est monté le centre de gravité du Systeme M on aura $\int\int F u \, dt = M p H$ donc pour elever le centre de gravité d'un Systeme de corps a une hauteur donnée il faut quelque chemin qu'on lui fasse prendre quelque Machine qu'on employe soit qu'on l'eleve par parties ou tout a la fois soit enfin qu'on employe une force unique ou plusieurs ensemble il faut disje que la quantité $\int\int F u \, dt$ soit toujours la même

Si les forces F sont constantes pendant tout le mouvement et se meuvent uniformement la quantité precedente deviendra $\int F u t$ et l'équation sera $\int F u t = M p H$ enfin s'il n'y a qu'une force employée on aura $F u t = M p H$

L'équation precedente suppose qu'il n'y ait ni frottement ni autre résistance. Si cela n'etoit pas il est clair que $F u t$ devroit etre plus grande que $M p H$ donc toutes choses d'ailleurs egales la force F devroit augmenter donc il seroit avantageux toutes choses d'ailleurs egales que la puissance fut appliquée immédiatement au poids.

faisant abstraction des frottements si aulieu de la puissance F je voulois en employer une autre f qui n'en fut que la moitié par exemple je pourrois y parvenir en employant une autre machine ou en appliquant la nouvelle force a un autre point de la premiere tel qu'en nommant t' le tems qu'il faudroit alors pour elever Mp a la hauteur H et u' la vitesse de f on eut Fut $= fu't'$ et puisqu'on suppose f sousdouble de F il faudroit que $u't'$ fut double de ut et si l'on vouloit que le tems fut le meme il faudroit qu'on eut $u' = 2u$ l'avantage que donnent les machines est donc seulement de pouvoir varier les facteurs F,u,t, mais le produit doit toujours etre egal a MpH lorsqu'il s'agit d'un poids ou a une quantité analogue s'il s'agit d'une autre resistence

avés vous reconnu par exemple que pour elever cent muids d'eau a une certaine hauteur il vous falloit 100 bras capables chacun d'une force de 10^{tb} et d'une vitesse de 10 pieds par minute et employés pendant 10 jours en supposant les forces appliquées a tirer immédiatement cette eau vous pourrés a l'aide d'une machine faire ensorte qu'il n'y ait que 50 bras employés avec la meme vitesse mais il vous faudra 20 jours pour faire ce que vous auries fait en 10 voulés vous avec 50 bras n'employer que 10 jours il faudra que ces bras soient capables d'une vitesse de 20 pieds par minute ne peuvent ils prendre qu'une vitesse de 10 pieds par minute il faudra que vous en ayes davantage ou que vous les employés plus longtems toujours dans le meme rapport

le simple bon sens indique tout cela et cependant c'est faute d'y faire attention qu'on se trompe tous les jours sur l'effet des machines et qu'on en attend souvent des effets si extraordinaires; la grande source des erreurs que commettent a cet egard les mecaniciens peu instruits qui promettent tant de merveilles vient de ce quils se bornent a considerer les machines dans leur etat dequilibre; par exemple ils scavent en gros qu'a l'aide d'un levier une certaine puissance bornée peut faire équilibre a un tres grand poids et croient par cette raison qu'avec un levier on peut rendre une force capable de quel effet on veut dans un tems donne mais un peu plus de connoissance les rammeneroit bientot de cette folle pretention en leur apprenant qu'on perd en vitesse tout ce qu'on gagne en force.

49 avec ce principe il ne faut ordinairement qu'un coup d'oeil pour juger a peu pres de l'effet d'une machine et comme on s'en tient la pour l'ordinaire a moins qu'on ait besoin d'une grande precision il sera bon d'ajouter ici quelques reflections propres a guider ceux qui ne veulent pas entrer dans un calcul détaillé

1° quoi qu'on puisse a volonté faire varier les facteurs de la quantité Fu, lorsqu'on employe des hommes ou des animaux il y a un certain rapport a mettre entre les facteurs si l'on veut epargner autant qu'il est possible les forces du moteur et profiter de la maniere la plus avantageuse de celles qu'on a a sa disposition

par exemple : l'on a reconnu qu'un homme applique pendant 8 heures par jour a une manivelle de 14 pouces de rayon peut faire continuellement un effort de 25^{lb} en faisant 30 tours par minute mais si l'on vouloit forcer cet homme a faire beaucoup plus de 30 tours par minute croyant par la avancer beaucoup la besogne on la retarderoit parceque l'homme ne seroit pas en etat alors de faire un effort de 25^{lb} ni meme moindre que 25^{lb} dans la meme raison que la vitesse a augmenté c'est adire en un mot que Fu diminueroit; si au contraire vous diminues la vitesse

{ 286 }

la force augmente Mais dans un moindre rapport
de sorte que Fu diminue encore ainsi pour faire que Fu
soit un Maximum il faut lui conserver la vitesse
de 30 tours par minute

2° l'on observera que par F j'entends la force mouvante
estimée dans le sens de sa vitesse u donc si l'on appelle F'
la véritable force et x l'angle compris entre cette force
et sa vitesse on aura $F = F'\cos x$ on voit donc qu'on produiroit
le même effet en substituant au lieu de F' une force
égale a F pourvu qu'on la dirigeat dans le sens même
de la vitesse u or comme F est moindre que F' puisqu'on a
$F = F'\cos x$ on voit combien il est avantageux que la force
mouvante soit dirigée dans le sens même de la vitesse
Si par exemple c'est un homme appliqué a une manivelle
il faut pour ne point consommer inutilement sa force
qu'il agisse perpendiculairement au rayon de sa manivelle
et l'on doit se dispenser autant qu'il est possible de
l'appliquer a une béquille qui agissant toujours a peupres
dans le même sens n'est presque jamais perpendiculaire
au rayon de la manivelle.

3° lorsqu'on n'a d'autre but que d'elever un corps a une
certaine hauteur il faut pour menager ses forces faire
en sorte que le mouvement du corps s'eteigne en arrivant
a cette hauteur ou que sa vitesse soit nulle au moment
de cette arrivée on en sera convaincu en jetant les yeux
sur l'équation $\int mu^2 - \int mK^2 + MV^2 = 2\int\int F u dt$ trouvée (46)
car si l'on y suppose K = o elle se réduira a $2\int\int F u dt = \int mu^2 + 2MpH$
donc la quantité $2\int\int F u dt$ employée dans le cas pour elever Mp
a la hauteur H est plus grande de toute la quantité $\int mu^2$
(laquelle est évidemment positive) que si chaque point du système
eut eu une vitesse nulle en arrivant a cette hauteur et plus

« Sera petite moins la quantité $2\int\int \cdot F u \, dt$ sera grande plus en un mot toutes choses égales d'ailleurs la force F sera petite et partant moins il faudra d'effort pour élever le poids à la hauteur H

ainsi dans les pompes par exemple il faut autant qu'il est possible que l'eau en arrivant dans le reservoir superieur ait très peu de vitesse toute celle qu'elle a ayant inutilement consommé l'effort de la puissance Motrice 50 tout ce que j'ai dit des forces en general doit s'entendre egalement du frottement et des autres resistences les resistences pouvant etre regardées comme des forces actives et j'observerai a ce propos en finissant ce qui regarde les machines en general que toutes choses egales d'ailleurs plus on peut faire grande la vitesse respective des surfaces qui se frottent plus on donne d'avantage a la force mouvante car nous avons vu que la vitesse contribue a diminuer le frottement.

Section Seconde
des machines simples en équilibre.

51 les cordes et le levier sont les seules machines vraiment simples encore peut on ramener assés facilement l'une a l'autre mais on donne ordinairement le nom de simples aux sept machines suivantes; les cordes le levier la poulie le treuil le plan incliné la vis et le coin.

je vais traiter successivement de ces machines dans leur etat d'equilibre en ayant egard autant que je le pourrai au frottement et a la roideur des cordes mais je n'en dirai que ce qui me paroitra le plus essentiel les details n'etant pas fait pour un ouvrage tel que celui ci ou je dois supposer le lecteur instruit des exelents ouvrages que nous avons

sur cette matiere ainsi je ne m'etendrai gueres au dela de
ce qui est absolument necessaire pour indiquer les changements
que me paroissent exiger les resultats de mes experiences
   je commencerai par la machine funiculaire sur laquelle
j'entrerai dans un certain detail quoique je n'aye rien
d'essentiellement nouveau a endire mais j'en userai ainsi 1°
parcequ'on peut ramener toutes les autres machines a cellela
comme ont fait quelques auteurs 2° affin de montrer comment
je reduis a un meme principe la theorie de toutes les
machines le principe n'est autre chose que celui de descartes
auquel je donne plus d'extension et que j'ai demontré et
expliqué asses au long dans la section précédente il se
reduit a l'equation $\int mp\, ds \cos x = 0$ trouvée ( ) entre un nombre
quelconque de puissances mp qui se font mutuelement
équilibre soit immediatement soit par l'entremise d'une
machine le meme principe nous servira également
dans la section suivante ou il s'agira du mouvement
en réduisant par le principe de Mr d'alembert la theorie
du mouvement a celle de l'équilibre

de la machine funiculaire.

Lemme

5. pour trouver la resultante AB (Fig. 4) de plusieurs
forces AD, AC, AF, AG soit construit un poligone AF g, d B
dont un coté AF soit egal et parallele a AF un autre
g d egal et parallele a AD ainsi de suite jusqu'a la
derniere des forces proposées alors si l'on mene la
droite AB du point A au dernier point B qu'on
obtient par cette construction je dis que cette droite AB
representera tant pour sa grandeur que pour sa
direction la resultante de toutes les forces proposées
c'est ce qu'on verra aisement pour peu qu'on fasse

attention a la marche quon suit ordinairement pour trouver la resultante de plusieurs forces données.

on tire de la comme il est évident une methode bien simple pour décomposer une force donnée en tant d'autres qu'on voudra et réciproquement

Coroll. 1er

53 il suit dela que la resultante estimée dans un sens quelconque est egale a la somme des forces composantes estimées dans le meme sens car soit (Fig 5) un poligone ABCDF dont un cote AB represente la resultante et les autres cotés les forces composantes estimons toutes ces forces dans le sens d'une ligne quelconque AK. pour cela soient menées de tous les angles du poligone des perpendiculaires a cette ligne alors il est clair que Ab est la force AB estimée dans le sens AK cb est la force CB estimée dans le meme sens &c or on a Ab = bc + cd + df &c donc &c

il est clair qu'on a Ab = AB cos BAK Af = AF cos FAf &c donc la resultante multipliée par le cosinus de l'angle qu'elle fait avec une ligne quelconque est egale a la somme des forces composantes multipliées chacune par le cosinus de l'angle qu'elle fait avec cette meme ligne j'ai fait usage de ce corollaire dans plusieurs endroits de la premiere section

Coroll. 2ond.

54 Soit AB (Fig 6) la resultante des forces AB, AF, AC supposons que le point E represente un axe quelconque perpendiculaire au plan de ces forces AE une droite menée par le point A et le point E et AB' une perpendiculaire a cette droite soient aussi du point E menées les perpendiculaires Eb' Ec' Ed' Ef' sur les directions de ces forces et les

perpendiculaires $Bb, Cc, \&c$ sur AE. Cela posé a cause des triangles semblables $Ab'E \; ABb$ on aura

$AB \cdot Eb' = AE \cdot Bb$ on aura dememe les equations suivantes

$$o = -AC \cdot Ec' + AE \cdot Cc$$
$$o = AD \cdot Ed' - AE \cdot Dd$$
$$o = AF \cdot Ef' - AE \cdot Ff$$

ajoutant ensemble toutes les equations et observant que la somme des quantités qui multiplient AE est nulle par le corollaire précédent il vient $AB \cdot Eb' = AD \cdot Ed' + AE \cdot Ef' - AC \cdot Ec'$ c'est a dire que le moment de la resultante relativement au point E est egal a la somme des moments des forces composantes qui tendent a faire tourner dans le meme sens quelle moins la somme des moments des forces composantes qui tendent a faire tourner en sens contraire et la meme demonstration a lieu quelque soit le nombre des forces composantes

Si les forces n'etoint pas dans un meme plan alors en imaginant un axe quelconque et faisant la projection du systeme des forces sur un plan perpendiculaire a cet axe on trouveroit dememe en considerant les moments relativement a l'axe en question que le moment de la resultante projeteé est egal a la somme des moments des forces composantes aussi projetées sur le meme plan de celles disje qui tendent a faire tourner dans le meme sens que la resultante moins la somme des moments de celles qui tendent a faire tourner en sens contraire

probleme

55 Determiner les conditions generales de l'equilibre dans une machine funiculaire en supposant les cordons sans pesanteur et parfaitement flexibles

Sol. Concevons que la machine proposée soit mise en

mouvement de telle maniere que tous ses points parcourent des chemins egaux et paralleles dans le meme sens ces puissances n'en agissent pas moins les unes sur les autres comme auparavant donc elles continuent a se faire mutuellement équilibre donc l'equation $\int mp\, ds \cos x = 0$ trouvée (35) exprimera les conditions de cet equilibre mp exprimant chacune de ces puissances ds le chemin quelle parcourt pendant un instant x l'angle formé par la direction de cette puissance et le chemin quelle parcourt or par hypotese ds est la meme pour tous les points du Systeme donc l'equation se reduira a $\int mp \cos x = 0$ donc la somme des forces appliquées a la machine estimees dans un sens quelconque est egal a zero de plus chaque cordon doit etre egalement tendu dans toute sa longueur dou il suit quil y a equilibre entre les forces appliquées a chaque noeud donc la somme des forces appliquées a chaque noeud estimées dans un sens quelconque est egale a zero. telle est la condition generale de l'equilibre

Coroll. 1

56 Si l'on represente les forces du Systeme par des lignes proportionelles prises sur leurs directions et qu'on fasse la projection de la machine sur un plan quelconque je dis que si l'on regarde cette projection comme une machine funiculaire et les projections des puissances données comme les forces appliquées a cette machine il y aura equilibre entre toutes les puissances il est clair en effet que dans cette nouvelle machine la resultante qui repond a chaque noeud est nulle aussi bien que dans la machine proposée

Coroll. 2

57 Si l'on construit un poligone dont le nombre des

cotés soit egal au nombre des forces et que les cotés soient égaux et parallèles chacun a chacune des lignes qui representent les forces je dis que le poligone sera fermé cest adire quele premier point du premier coté et le dernier point du dernier se confondent et la meme chose aura lieu soit quil sagisse de toute la machine ou seulement dun ou plusieurs noeuds (52)

Coroll. 3.

58 Si lon considere les moments de toutes les forces appliquées a la machine relativement a un axe quelconque on trouvera que la somme des moments de celles de les forces qui tendent a faire tourner dans un sens autour de cet axe est egale a la somme des moments de celles qui tendent a faire tourner en sens contraire (54)

Coroll 4

59 Donc si quelque noeud rassemble seulement trois cordons les puissances appliquées a les cordons seront dirigées dans un meme plan et deux quelconque dentreelles seronten raison réciproques des perpendiculaires menees dun point quelconque de la troisieme sur leurs directions car en rapportant le moment des forces a cepoint celui de la puissance dont la direction y passe est nul et partant ceux des deux autres doivent etre egaux et conspirer en sens contraire

Donc les trois forces peuvent etre representées en meme tems chacune par le sinus de langle compris entre les directions des deux autres car en prenant pour rayon la distance du noeud au point dont on a parlé dans le corollaire precedent cest adire dont les perpendiculaires sont abbaissées il est clair que ces perpendiculaires sont proportionelles aux sinus en question

par moment dune force relativement a un axe jentends la projection de cette force sur un plan perpendiculaire a cet axe multipliée par la distance de sa direction a cet axe.

coroll. 5

60 Supposons maintenant qu'on veuille avoir égard à la pesanteur des cordes, il est d'abord evident que cette pesanteur, fera prendre une certaine courbure aux cordons si cette courbure etoit connue il est visible qu'on auroit aussi la direction et la quantité de tension de chacun de ces cordons a ses extrémités Donc on auroit toutes les forces appliquées a chaque noeud et puisqu'on suppose qu'il y a équilibre les puissances auront entre elles les rapports trouvés dans le probleme donc toute la difficulté se reduit a connoître la courbure d'un cordon attaché par ses extrémités a deux points fixes cette courbe qui s'appelle chainette est trop connue pour qu'il soit nécessaire d'en parler encore ici

il seroit a propos maintenant de resoudre la meme question en ayant egard au deffaut de flexibilité des cordes mais la loi que suit cette roideur est trop peu connue pour qu'on puisse se flater de resoudre un tel probleme d'une maniere non hypotetique il faudroit d'abord pour cela sçavoir trouver la courbure que doit prendre une telle corde dont les bouts sont attachés a deux points fixes; or 1° cette question est indeterminée car si vous attachés une corde neuve un peu roide a deux points fixes et que vous lui fassies prendre une courbure quelconque et meme la forme d'une courbe a double courbure vous verres que votre corde restera a peu pres dans l'état ou vous l'aures mise 2° lorsque les cordes sont un peu vieilles elles prenent presque la meme courbure qu'elles auroint si elles etoint parfaitement flexibles or comme c'est dans cet etat que se trouvent les cordes apres quelque temps de service il s'en suit qu'il est plus que suffisant dans la pratique d'avoir egard ala pesanteur des cordes.

au reste si l'on connoissoit la ~~force de la~~ roideur ou si l'on vouloit se contenter d'une solution hypotetique il seroit aisé d'etre satisfait car les geometres savent trouver l'équation d'une courbe parfaitement flexible et tirée en chaque point par deux forces dont la loi est donnée et dont l'une est parallele a une ligne donnée de position et l'autre perpendiculaire a la courbe.

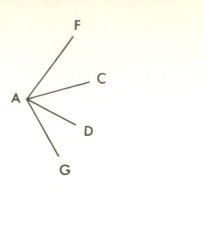

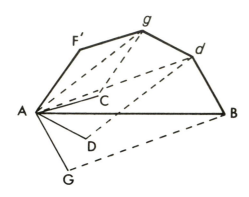

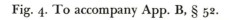

Fig. 5. To accompany App. B, § 53, Cor. 1.

Fig. 4. To accompany App. B, § 52.

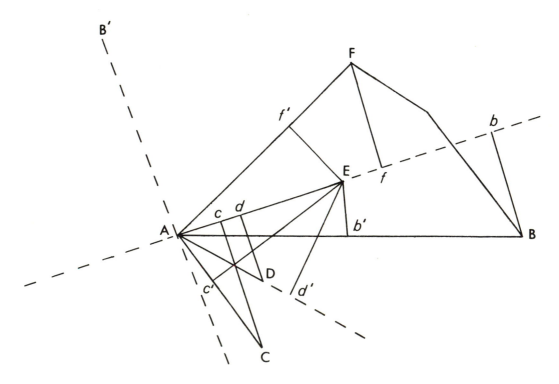

Fig. 6. To accompany App. B, § 54, Cor. 2.

Appendix C

The second memoir that Carnot submitted to the Academy, this one for the prize to be adjudged in 1781, consists of 191 sections in 107 folio pages. It is divided into two parts. Part I consists of experimental work on friction and occupies § 1 through 100. Part II is set off more decisively than in the 1778 memoir, and also consists of two parts. The first, on "Machines en général," § 101 through 160, is that which we reproduce here. The second, "Application des expériences aux machines simples," completes the memoir in § 161 through 191. As in the case of the earlier memoir, these pages are reproduced by the kind permission of the Permanent Secretaries of the *Académie des sciences, Institut de France,* in whose archives the originals are conserved. Microfilm copies of the complete manuscripts are also deposited in the Firestone Library of Princeton University.

Despite the large format in which these facsimiles are reproduced, the legibility of several footnotes continues to be less than ideal. These have been transcribed and they appear in type following Appendix C. There being minimal punctuation, the places where sense requires a new sentence have been indicated by a space and a double slash.

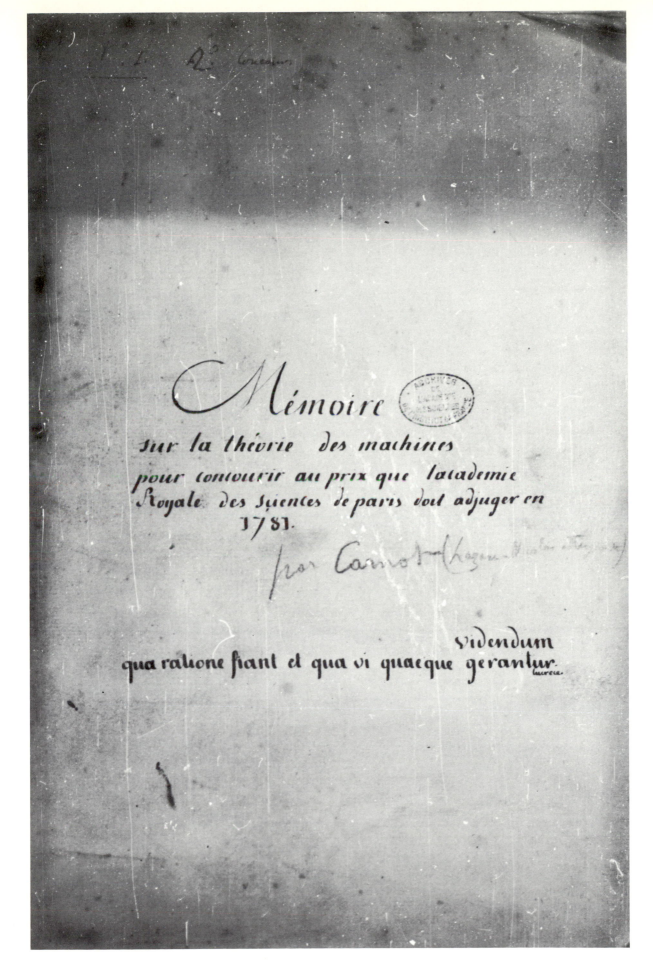

Mémoire
sur la théorie des machines

Seconde partie.
application des experiences a la pratique

101. cette seconde partie sera divisée en deux sections
dans la premiere on traitera des machines en général et dans la
seconde on traitera en particulier de chacune des machines simples

Section premiere
des machines en général.

102. chaque machine peut avoir des proprietés particulieres
il ne s'agit ici que de celles qui sont communes a toutes, en
examinant les machines sous ce point de vue général nous
eviterons les répétitions qu'il auroit fallu faire pour rappeller a
l'occasion de chacune desquelles nous avons a traiter les propriétés
qui lui sont communes avec toutes les autres il est d'ailleurs avantageux
de reduire les principes au plus petit nombre possible et d'appercevoir
la liaison qui reunit des verites asses disparates au premier coup
d'oeil or malgré la difference des machines toutes leurs proprietés
sont en quelque sorte comprises dans une meme loi fort simple, et
de laquelle on déduit avec une grande facilité tout ce qui les concerne
chacune en particulier. cette loi sera expliquée et démontrée
rigoureusement apres que nous aurons donné en peu de mots une
idée générale et simple de notre sujet.

103. imaginons donc d'abord une machine a laquelle il n'y ait
d'autres forces appliquées que des poids je la suppose d'ailleurs d'une
forme arbitraire mais qu'on ne lui ait imprimé aucun mouvement
cela posé il est clair quelque puisse etre la disposition des corps
du systeme que s'il y a équilibre la somme des résistances des
points fixes ou obstacles quelconques estimées dans le sens vertical
contraire a la pesanteur sera égale au poids total du systeme
mais s'il y a du mouvement une partie de la pesanteur sera

employée a le produire et ce n'est qu'avec le surplus que
les points fixes pourront se trouver chargés donc dans ce cas
la somme des résistances verticales des points fixes sera
moindre au premier instant que le poids total du systeme
donc de ces deux forces combinées il en résultera une seule force
qui poussera le systeme de haut en bas donc le centre de gravité
descendra nécessairement avec une vitesse egale a cette force divisée
par la masse totale du systeme ; donc si le centre de gravité
du systeme ne descend pas il y aura nécessairement équilibre
donc en géneral

     pour s'assurer que plusieurs poids appliqués a une machine
quelconque doivent se faire mutuellement équilibre il suffit de
prouver que si l'on abandonne cette machine a ellememe le
centre de gravité du systeme ne descendra pas au premier instant
proposition par laquelle on trouvera tres facilement la loi
d'équilibre dans une machine donnée a laquelle il n'y a d'autres
forces appliquées que des poids.

104. par exemple je dis qu'il y aura nécessairement équilibre
entre deux poids A et B appliqués a une machine quelconque
telle cependant que l'un de ces poids ne puisse descendre sans
que l'autre soit forcé de monter il y aura disje équilibre si ces
deux poids sont en raison reciproque des vitesses verticales qu'ils
auroint s'il y avoit du mouvement. en effet si cela n'etoit pas
supposons que A descendit alors au premier instant avec la vitesse V
tandis que la vitesse de B aussi estimée dans le sens vertical
seroit u on aura donc par hypotese $A \cdot B = u \cdot V$ ou $AV = Bu$
donc $\dfrac{AV - Bu}{A + B} = 0$ or le premier membre de cette équation est
visiblement la vitesse verticale du centre de gravité donc ce centre
de gravité ne descendra pas au premier instant donc par la
proposition précédente il doit y avoir équilibre.

105. par exemple encore je dis qu'il y aura nécessairement
équilibre entre plusieurs poids appliqués a une machine
quelconque si le centre de gravité du systeme est au point le
plus bas possible car par notre proposition il suffit de prouver
que ce centre ne descendra pas; or comment descendroit il puisque
par hypotese il est au point le plus bas possible ?

106. s'il entroit dans le systeme d'autres forces que des poids
on rameneroit aisément ce cas au premier en substituant
un poids a chacune des forces a l'aide d'une poulie de renvoi

on pourroit donc regarder la proposition précédente comme le principe général de l'équilibre dans les machines, et quant au mouvement on en ramene aisément les loix à celles de l'équilibre, en decomposant le mouvement virtuel du Système en deux autres dont le premier Soit le mouvement reel; le second Sera parconséquent tel que S'il etoit Seul il y auroit équilibre: or on connoit les loix de cet équilibre par la proposition precedente on parviendroit donc a connoitre aussi celles du mouvement.

107. la loi d'équilibre entre deux forces appliquées a une machine est comme on Sçait que ces forces Soient en raison réciproque de leurs vitesses estimées dans le sens de ces forces; cette loi se déduit tres facilement de ce que nous venons de dire au Sujet de deux poids en equilibre, et notre proposition S'étend même aux cas ou il y auroit plus de deux forces mais je n'ai voulu donner jusqu'ici qu'une idée generale de la théorie que je vais essayer de developer dans les articles Suivants ou je me propose d'examiner les machines sous un point de vue plus général encore et plus direct.

108. lorsqu'un corps agit Sur un autre c'est toujours immédiatement ou par l'entremise de quelque corps intermédiaire ce corps intermédiaire est en général ce qu'on appelle une machine le mouvement que perd a chaque instant chacun des corps appliqués a cette machine est en partie absorbé par la machine même et en partie reçu par les autres corps du Systeme mais comme il peut arriver que l'objet de la question Soit uniquement de trouver l'action réciproque des corps appliqués au corps intermédiaire Sans qu'on ait besoin d'en connoitre l'effet Sur le corps intermédiaire même on a imaginé pour Simplifier la question de faire abstraction de la masse de ce corps en lui conservant d'ailleurs toutes les autres proprieté de la matiere, des lors la Science des machines est devenue en quelque sorte une branche isolée de mechanique dans laquelle il s'agit de considerer l'action reciproque des differentes parties d'un Systeme de corps parmis lesquelles il s'en trouve qui privés de l'inertie commune a toutes les parties de la matiere telle quelle existe dans la nature ont retenu le nom de machines.

Cette abstraction pouvoit simplifier dans certains cas particuliers ou les circonstances indiquoint ceux des corps dont il convenoit de négliger la masse pour arriver plus facilement au but mais on conçoit aisément que la théorie des machines en général est devenue réellement plus compliquée qu'auparavant car alors cette théorie étoit renfermée dans celle du mouvement des corps tels que la nature nous les offre mais a present il faut considerer a la fois deux Sortes de corps les uns tels qu'ils existent réellement les autres dépouillés en partie de leurs proprietés naturelles or il est clair que le premier de ces problemes n'est qu'un cas particulier de celui-ci donc ce dernier est plus compliqué que l'autre aussi quoiqu'on parvienne aisément par de pareilles hipoteses a trouver les loix de l'équilibre et du mouvement dans chaque machine particuliere telle que le levier le treuil la vis il en resulte un assemblage de connoissances dont la liaison s'apperçoit difficilement et seulement par une espece d'analogie; ce qui doit necessairement arriver tant qu'on aura recours a la figure particuliere de chaque machine pour démontrer une proprieté qui lui est commune avec toutes les autres ces proprietés etant celles que nous avons principalement en vue dans cette premiere Section il est clair que nous ne parviendrons a les trouver qu'en faisant abstraction des formes particulieres commençons donc par simplifier l'etat de la question en cessant de considerer dans un meme Systeme des corps de differentes natures, rendons enfin aux machines leur force d'inertie il nous sera facile apres cela d'en négliger la masse dans le resultat nous serons maitres d'y avoir égard ou non et partant la Solution du probleme sera aussi generale en meme tems qu'elle sera plus simple.

109 la science des machines en général ainsi que toute la mechanique se réduit donc a la question suivante.

connoissant le mouvement virtuel d'un systeme
de corps c'est-a-dire celui que prendroit chacun des
corps S'il etoit libre trouver le mouvement reel
qu'il aura l'instant suivant a cause de l'action réciproque
des corps en supposant que chacun d'eux soit doué de l'inertie
commune a toutes les parties de la matiere.

et puisque ce probleme est plus simple que si parmis
les corps il s'en trouvoit quelquns qui fussent privés
de cette inertie il est clair qu'on ne peut avoir une
théorie générale des machines sans avoir resolu ce
probleme dans toute son etendue c'est ce que nous allons
essayer.

110. imaginons donc un Systeme quelconque de corps
dont le mouvement virtuel donné soit changé par
leur action réciproque en un autre qu'il s'agit de
trouver. Si les corps Sont elastiques on peut evidement
considerer le Systeme comme composé d'une infinité
de corpuscules durs separés par de petits ressorts au moyen
des quels le mouvement se transmet de proche en proche
dans toute l'etendue du Systeme; la force de ces ressorts
depend du degré d'elasticité des corps. Si ces corps etoint
parfaitement durs les petits ressorts deviendroint des
verges absolument incompressibles et partant les
corpuscules appliqués aux extremités de chacune d'elles
auroint necessairement apres l'action réciproque la
meme vitesse estimée dans le sens de cette verge
comme il est évident ou ce qui revient au meme la
vitesse relative de ces corpuscules apres leur action
reciproque Seroit nulle. si les corps etoint parfaitement
elastiques les petits ressorts se restitueroint avec une
force egale a celle qui les a comprimés ainsi la force
perdue dans le choc par chacun des corpuscules du Systeme
Seroit apres la restitution des ressorts double de ce qu'elle
auroit eté si les corps avoint eté durs et en general quelque

soit le Degré d'elasticité on pourra toujours le representer par le rapport de la force avec laquelle se restituent les petits ressorts a celle qui les a comprimés ainsi la nature des corps etant donnée on pourra toujours ramener les loix de leur mouvement a celles du mouvement des corps durs: la difficulté se réduit donc a sçavoir ce qui doit arriver a un Systeme de corpuscules durs séparés par de petites verges parfaitement incompressibles ou telles qu'en comparant deux a deux ceux qui se trouvent aux extremités de chacune de ces verges les vitesses relatives ou estimées dans le meme sens suivant la longeur de cette verge soient egales apres l'action réciproque, mais pour embrasser la question dans toute sa généralité il faut supposer que le mouvement puisse changer subitement ou varier par degrés insensibles; enfin comme il peut se rencontrer des points fixes ou autres obstacles quelconques il faudra les considerer comme des corps faisant euxmemes partie du Systeme proposé mais fixement arretés dans le lieu de l'espace ou ils sont placés.

III. le seul principe qui puisse nous conduire a la solution de ce probleme est celuici.

la réaction est toujours égale et contraire a l'action voila cette loi simple et incontestable d'ou nous partirons loi universelle qui assujetit également tous les corps soit dans le choc la pression l'attraction meme et tous les phenomenes connus de la nature mais il ne s'agit ici que de connoitre son effet dans ceux du choc et de la pression.

112. pour y parvenir considerons a part deux quelconques des corpuscules du Systeme separés par une de ces petites verges incompressibles par lesquelles comme nous l'avons dit le mouvement se communique de proche en proche dans toute l'etendue

du Systeme Soient donc

m' et m" _ les masses de ces corpuscules.

V' et V" _ les vitesses qu'ils doivent avoir l'instant suivant.

F' _ _ _ _ l'action de m" Sur m' c'est a dire la force ou quantité de mouvement que le premier imprime a l'autre.

F" _ la réaction de m' Sur m".

q' et q" _ les angles formés par les directions de V' et F' et par celles de V" et F".

Cela posé la vitesse reele de m' etant V' cette vitesse estimée dans le sens de F' Sera visiblement $V' \cos q'$, dememe la vitesse de m" estimée dans le sens de F" Sera $V'' \cos q''$, donc la vitesse relative de m' et m" Sera $V' \cos q' + V'' \cos q''$ donc puisque les corps doivent aller de compagnie on aura $V' \cos q' + V'' \cos q'' = 0$, (A) Donc par le principe de la réaction égale et contraire a l'action on aura aussi $F' V' \cos q' + F'' V'' \cos q'' = 0$, (B) car Si m' et m" Sont mobiles tous les deux il est clair par le principe qu'on a $F' = F''$ donc a cause de l'équation (A) on aura aussi l'équation (B) et Si l'un des deux m' par exemple est fixe ou fait partie d'un obstacle on aura $V' \cos q' = 0$ donc a cause de l'équation (A) on aura aussi $V'' \cos q'' = 0$ donc l'équation (B) aura encore lieu donc cette équation (B) est vraie pour tous les corpuscules du Systeme pris deux a deux imaginant donc une pareille équation pour tous les corps pris en effet deux a deux ou ce qui revient au meme intégrant l'équation (B) on aura pour tout le Systeme $\int F' V' \cos q' + \int F'' V'' \cos q'' = 0$ c'est a dire que la Somme des produits des quantités de mouvement que S'impriment réciproquement les corpuscules séparés par chacune des petites verges incompressibles de ces quantités disje multipliées chacune par la vitesse du corpuscule auquel elle est imprimée estimée dans le sens de cette force est égale a zero

Cela posé abandonnant les denominations precedentes nommons.

la masse de chacun des corpuscules du Systeme ___ m

Sa vitesse virtuelle ___ ___ ___ ___ ___ ___ W

Sa vitesse réele ___ ___ ___ ___ ___ V

la vitesse qu'il perd desorte que W soit la
resultante de V et de cette vitesse ___ ___ ___ U .

la force qu'imprime a m chacun des corpuscules
adjacents et par lesquels il reçoit évidement tout
le mouvement qui lui est transmis des differentes
parties du Systeme ___ ___ ___ ___ ___ F

l'angle compris entre les directions de W et V ___ ___ X

l'angle compris entre les directions de W et U ___ Y

l'angle compris entre les directions de V et U ___ Z

l'angle compris entre les directions de V et F ___ q

on aura donc pour tout le Systeme $\int F.V.\cos q = 0$, ou $\int V.F.\cos q = 0$, (C)
a present il faut observer que la vitesse de m avant l'action
réciproque étant W cette vitesse estimée dans le sens de V
sera $W\cos X$ donc $V - W\cos X$ est la vitesse gagnée par m
dans le sens de V donc $m(V - W\cos X)$ est la somme des
forces F qui agissent sur m estimées chacune dans le sens
de V donc $mV(V - W\cos X)$ est la meme somme multipliée
par V or a chaque molecule repond une pareille somme
et deplus la somme totale de toutes ces sommes particulieres
est visiblement pour tout le Systeme $\int V.F.\cos q$ donc
$\int mV(V - W\cos X) = \int V.F.\cos q$ ajoutant a cette équation
l'équation (C) il vient $\int mV(V - W\cos X) = 0$ (D). mais W
étant la resultante de V et U il est clair qu'on aura
$W\cos X = V + U\cos Z$ substituant donc cette valeur de $W\cos X$
dans l'équation (D) elle se réduira a $\int m.V.U.\cos Z = 0$, (E), équation
fondamentale.

113. imaginons maintenant qu'au moment ou le choc va
Se faire le mouvement virtuel du systeme Soit tout a coup
detruit et qu'on lui fasse prendre a la place Successivement
deux autres mouvement arbitraires mais égaux et directement
opposés c'est adire qu'on le fasse partir successivement de sa
position actuelle avec deux mouvements tels qu'en vertu du second

chaque point du Systeme ait au premier instant une vitesse égale et directement opposée a celle quil auroit eue au premier instant en vertu du premier cela posé il est clair 1° que la figure du Systeme etant donnée cela peut se faire d'une infinité de manieres differentes et par des operations purement géometriques; c'est pourquoi j'appellerai ces mouvements mouvements géometriques c'est a dire que si un systeme de corps part d'une position donnée avec un mouvement arbitraire mais tel quil eut été possible aussi de lui en faire prendre un autre tout a fait égal et directement opposé chacun de ces mouvements sera nommé mouvement géometrique.ᵃ 2° il est clair qu'en vertu de l'un quelconque de ces mouvements géometriques les

ᵃ pour distinguer par un exemple tres simple les mouvements que j'appelle géometriques de ceux qui ne le sont pas imaginons deux globes parfaitement egaux agissants l'un sur l'autre par un contact immediat et du reste libres et degagés de tout obstacle imprimons a ces globes des vitesses egales et dirigées dans le meme sens suivant la ligne des centres ce mouvement est géometrique parceque les corps pourroient demeure etre mus en sens contraire avec la meme vitesse comme il est evident mais supposons qu'on imprime a ces corps des mouvements egaux si l'on veut et meme dirigés dans la ligne des centres mais qui au lieu d'etre comme precedemment dirigés dans le meme sens tendent au contraire a les eloigner l'un de l'autre ces mouvement quoique possibles ne sont pas ce que j'entends par mouvements géometriques parceque si l'on vouloit faire prendre a chacun de ces mobiles une vitesse egale et contraire a celle qu'il reçoit dans ce premier mouvement on en seroit empeché par l'impénétrabilité des corps.

de meme si deux corps sont attachés aux extremités d'un fil inextensible et qu'on fasse prendre au systeme un mouvement arbitraire mais tel que la distance des deux corps soit constement egale a la longueur du fil ce mouvement sera géometrique parceque les corps peuvent prendre un pareil mouvement dans un sens tout contraire mais si ces mobiles se rapprochent l'un de l'autre le mouvement n'est point géometrique parceque'ils ne pourroint prendre un mouvement egal et contraire sans s'eloigner l'un de l'autre ce qui est impossible a cause de l'inextensibilité du fil.

en général quelque soit la figure du systeme et le nombre des corps, s'il est possible de lui faire prendre un mouvement tel qu'il n'en resulte aucun changement dans la position respective des corps ce mouvement sera géometrique, comme il est evident après tout ce qui vient d'etre dit; il est clair en effet qu'il ne s'exerce aucune pression ni action quelconque entre les differentes partie du systeme et que par conséquent la connoissance des loix de la communication des mouvements est inutile pour les determiner; la determination de ces mouvements est donc une affaire de pure géometrie et c'est pour cette raison que je les appelle géometriques.

il ne s'ensuit pourtant pas de cela que le mouvement d'un systeme de corps ne puisse etre géometrique sans que toutes ses parties conservent entre elles la meme position respective; c'est ce qui sera facile a comprendre par quelques exemples imaginons un treuil a la vone et au cilindre duquel soient attachés des poids suspendus par des cordes si l'on fait tourner la machine de maniere que le poids attaché a la voue descende d'une hauteur egale a sa circonference tandis que celui du cilindre montera d'une quantité egale a la sienne ce mouvement sera géometrique parceque'il est egalement possible de faire descendre le poids attaché au cilindre d'une hauteur egale a sa circonference tandis que le poids de la voue montera d'une hauteur egale a la sienne.

de meme si plusieurs corps sont attachés aux extremités de differents fils reunis par les autres extremités a un meme noeud et qu'on fasse prendre au systeme un mouvement tel que chacun des corps varie constement eloigné du noeud d'une meme quantité egale a la longueur du fil auquel il est attaché le mouvement sera géometrique quand meme ces differents corps se rapprocheroint ou s'eloigneroint les uns des autres.

si deux corps sont attachés aux extremités d'un fil dans lequel soit enfilé un grain mobile il suffira pour que le mouvement soit géometrique que la somme des distances du grain mobile a chacun des deux autres corps soit constement egale a la longueur du fil desorte que si ces autres corps etoint fixes le grain ne sortiroit pas d'une courbe elliptique

si un corps se meut sur une surface courbe par exemple sur la convexité d'une calote spherique le mouvement sera géometrique tant que le corps sera mu tangentiellement a la surface mais s'il s'en ecarte le mouvement n'est plus géometrique parceque le mouvement directement opposé est visiblement impossible

si deux corps viennent a se choquer le mouvement du systeme après le choc sera toujours géometrique et partant si l'on decompose le mouvement virtuel en deux dont l'un soit tel que s'il etoit seul il y auroit equilibre et dont l'autre soit géometrique celui-ci sera le vrai mouvement du systeme après le choc mais il ne suffiroit pas pour l'assurer de prouver que ce mouvement est possible c'est a dire compatible avec l'impénétrabilité ou que les corps pussent le prendre sans se gener mutuellement il faut de plus que le mouvement egal et contraire soit possible c'est a dire que le mouvement soit etre géometrique.

il resulte visiblement de tout cela que si a la place du mouvement réel d'un systeme on en substitue un autre tel qu'il n'en resulte aucun changement dans l'action réciproque des corps ce mouvement sera ce que j'entends par géometrique mais j'ai preferé une définition suivant laquelle il fut clair par l'enoncé meme que ces mouvements sont reellement géometriques c'est a dire absolument indépendans des regles de la dinamique.

corpuscules placés aux extrémités de chacune des petites verges inflexibles ne tendent ni a se rapprocher ni a s'éloigner l'un de l'autre car nommant alors u la vitesse de chaque corpuscule m 1° Sils tendoint a se rapprocher ils se pousseroint necessairement par la petite verge qui les sépare donc le mouvement du Systeme seroit alteré par cette action réciproque donc u ne pourroit pas etre la vraie vitesse de m ce qui est contre l'hypotese 2° Sils tendoint a s'éloigner il s'ensuivroit visiblement qu'en vertu du mouvement egal et directement opposé ou de la vitesse −u ils tendroint a se rapprocher donc −u ne pourroit pas etre la vraie vitesse de m ce qui est encore contraire a l'hypotese donc ils ne tendent en vertu de u ni a se rapprocher ni a s'éloigner; donc si chaque corpuscule etoit animé en meme tems de la vitesse u et de la vitesse V ces corps ne tendroint pas a se rapprocher ou a s'éloigner davantage l'un de l'autre que Sils etoint animés des seules vitesses V donc l'action réciproque exercée entre les parties du Systeme sera la meme Soit que chaque molécule soit animée de la seule vitesse V ou des deux vitesses V et u mais si elles etoint animées des seules vitesses V il y auroit équilibre donc si chaque molecule est animée des deux vitesses V et u ou d'une vitesse unique qui en soit la résultante V sera encore la vitesse perdue par m et partant u sera la vitesse réele donc en nommant z l'angle compris entre les directions de V et u on aura $\int m \cdot u \cdot V \cos z = 0$ (F) par la meme raison qu'on a eu l'équation (E).

114. il est bien facile a present de resoudre le probleme que nous nous sommes proposés car l'équation (F) devant avoir lieu quelque soit la valeur de u et sa direction pourvu que le mouvement auquel elle se rapporte soit géometrique il est clair qu'en attribuant Successivement a cette arbitraire differentes valeurs et differentes directions on obtiendra tant d'équations qu'on voudra entre les quantités inconnues V d'ou depend la solution du probleme et des quantités ou connues ou arbitraires.

115. pour achever de mettre cette solution dans tout son jour il suffira d'en donner un exemple.

Supposons donc que tout le systeme se réduise a un assemblage de corps liés entreux par des verges inflexibles de sorte que toutes les parties du systeme soient forcées de conserver toujours

leurs memes positions respectives, mais qu'il n'y ait
aucun point fixe ou obstacle quelconque l'équation (F)
va nous donner la solution de ce probleme en attribuant
successivement a u differentes valeurs et differentes
directions arbitraires.

1° comme les vitesses u ne sont assujeties a aucune autre
condition sinon que le mouvement du systeme en vertu
duquel les corpuscules m ont ces vitesses u soit géometrique
il est évident que nous pouvons dabort les supposer toutes
égales et paralleles a une meme ligne donnée alors u étant
constante ou la meme pour tous les points du systeme
l'équation (F) se réduira a $\int m.U.\cos z$ ce qui nous apprend
que la somme des forces perdues par l'action réciproque des
corps dans le sens arbitraire de u est nulle et que
parconséquent celle qui reste est la meme que si chaque
mobile eut été libre.

2° imaginons maintenant qu'on fasse tourner tout
le systeme autour d'un axe donné desorte que chacun
des points décrira une circonference autour de cet axe et
dans un plan qui lui sera perpendiculaire ce mouvement
est visiblement géometrique donc l'équation (F) a lieu
mais alors en nommant R la distance de m a l'axe il
est clair qu'on a u=AR, A étant la meme pour tous les
points donc l'équation (F) se réduit a $\int m.R.U.\cos z = 0$, c'est a dire
que la somme des momments des forces perdues par l'action
réciproque relativement a un axe quelconque est nulle.

3° on pourra achever si l'on veut la solution du probleme
en attribuant encore aux arbitraires u differentes autres valeurs
et directions mais on peut aussi s'en tenir aux précédentes
car on sçait quelles suffisent pour resoudre la question ou
dumoins pour la réduire a une affaire de pure géometrie.

il sera bon d'ajouter quelques reflections utiles pour simplifier
l'application de ce probleme aux differents cas particuliers qui peuvent
se rencontrer, et surtout aux machines: je vais dabort récapituler en
deux mots la solution pour montrer d'un coup d'oeil la suite
des opérations qu'on vient d'indiquer.

probleme.

116 Connoissant le mouvement virtuel d'un systeme quelconque donné de corps durs (c'est a dire celui qu'il prendroit si chacun des corps etoit libre) trouver le mouvement reel qu'il doit avoir l'instant suivant

Solution. nommons

la masse de chaque molécule du systeme	m
Sa vitesse virtuelle donnée	W
Sa vitesse réele cherchée	V
la vitesse qu'elle perd dans le choc desorte que W soit la résultante de V et de cette vitesse	U
imaginons maintenant qu'on fasse prendre au Systeme un mouvement géometrique arbitraire et soit la vitesse qui aura alors m	u
l'angle formé par les directions de W et V	X
l'angle formé par les directions de W et U	Y
l'angle formé par les directions de V et U	Z
l'angle formé par les directions de W et u	x
l'angle formé par les directions de V et u	y
l'angle formé par les directions de U et u	z

Cela posé on aura pour tout le systeme (113) $\int m \cdot u \cdot U \cos z = 0 \quad (\Lambda\Lambda)$ équation qui aura lieu pour toutes les valeurs de u et qui fera trouver dans chaque cas particulier toutes celles dont on aura besoin pour la solution du probleme en attribuant successivement aux indeterminées u differents rapports[b] ou differentes directions arbitraires.

117. S'il y avoit dans le systeme quelques parties qui restassent fixes c'est a dire pour lesquelles la vitesse u qui répond a chacun de leurs points fut nulle il est clair que cette portion du systeme n'entreroit pour rien dans l'équation $(\Lambda\Lambda)$ ainsi quoique nous ayons regardé les points fixes et autres obstacles comme des corps faisant eux memes partie du systeme (ce qui dans le fait est entierement conforme a la nature) cette hypotese ne peut faire naitre aucun embaras puisque ces corps n'entrent pour rien dans notre équation la quelle est par conséquent la meme que si l'on eut fait abstraction de la masse de ces corps.

de meme S'il y avoit quelque fil verge levier ou autre portion

---

[b] je dis qu'il faut attribuer aux indeterminées u differents rapports ou differentes directions parceque si l'on se contentoit de leur attribuer differentes valeurs sans que les rapports ni les directions fussent changés on auroit des équations qui seroient toutes justes a la vérité mais qui deviendroient identiques en les multipliant par differentes constantes comme on le verra aisément en jettant les yeux sur l'équation $(\Lambda\Lambda)$.

quelconque du systeme : dont la masse put être supposée infiniment petite alors on pourroit supposer m=o c'est a dire faire abstraction de la masse de cette portion et l'équation (ΛΛ) deviendroit visiblement independante de cette meme portion il sera donc inutile d'y avoir égard et par la on parvient aisement a trouver la théorie des machines avec les abstractions qu'on a imaginées pour simplifier cette théorie dans chaque cas particulier c'est principalement a cette partie du probleme que nous devons nous arreter et c'est ce que nous ferons apres avoir mis la formule générale sous les differentes formes qui pourront nous être utiles.

118. il faut commencer par distinguer en général deux cas, le premier est celui ou le systeme passe subitement d'un etat a l'autre le second est celui ou le mouvement change par degrés insensibles.

1er cas. parmis les differentes autres formes qu'on peut donner dans ce cas a l'équation (ΛΛ) je me bornerai a trois, voici la premiere.

puisque W est la resultante de V et U on a $W\cos x = V\cos y + U\cos z$ ou $U\cos z = W\cos x - V\cos y$. substituant cette valeur de $U\cos z$ dans l'équation (ΛΛ) on aura $\int mu.W\cos x = \int mu.V\cos y$, (BB) équation dans laquelle les vitesses detruites U n'entrent plus mais les vitesses virtuelles données W et les vitesses réelles cherchées y entrent a la place.

cette équation nous met a meme d'exprimer aisement la loi d'équilibre car alors on a V=o donc l'équation se réduit a $\int mu.W\cos x = o$ (CC) qui donnera toutes les conditions de l'équilibre en attribuant aux vitesses u successivement differents rapports ou directions arbitraires.

119. voici les autres formes que nous nous sommes proposés de donner a l'équation (ΛΛ). imaginons a volonté trois axes perpendiculaires entre eux ayant ensuite decomposé les vitesses W, V, U, u. chacune en trois autres paralleles a ces axes je suppose que

celles qui répondent a W soient	W' W'' W'''	
celles qui répondent a V	V' V'' V'''	
celles qui répondent a U	U' U'' U'''	
celles qui répondent a u	u' u'' u'''	

cela posé il est aisé de voir qu'on a $U.u\cos z = U'u' + U''u'' + U'''u'''$, on verra donc aisément que l'équation (ΛΛ) peut se mettre sous cette forme $\int mu'U' + \int mu''U'' + \int mu'''U''' = o$ (DD) et que l'équation (BB) pourra s'exprimer ainsi

$\int mu'W' + \int mu''W'' + \int mu'''W''' = \int mu'V' + \int mu''V'' + \int mu'''V''' = o$ (EE)

120. enfin par les memes raisons l'équation (CC) qui exprime les loix de l'équilibre pourra se mettre sous cette forme.

$\int mu'W' + \int mu''W'' + \int mu'''W''' = o$ (FF).

ces différentes équations serviront donc a trouver les loix du choc des corps dans tous les cas particuliers soit que ce choc soit immédiat ou qu'il se fasse par l'entremise d'une machine quelconque en attribuant successivement différents rapports arbitraires aux vitesses u', u'', u''' qui sont connues de direction

121. je n'abandonnerai pas le premier cas dont il s'agit sans faire mention d'une belle propriété du mouvement qui s'y rapporte

dans l'équation (AA) on peut évidement supposer $u = V$ ce qui donne $\int m V \cdot U \cos Z = 0$ qui n'est autre chose que l'équation fondamentale (E) trouvée (112) maintenant puisque W est la résultante de V et U il est clair que $W, V,$ et U sont proportionnelles aux trois cotés d'un certain triangle donc $W^2 = V^2 + U^2 + 2 V \cdot U \cos Z$ donc $\int m \cdot W^2 = \int m V^2 + \int m U^2 + 2 \int m V U \cos Z$ or nous venons de trouver $\int m V \cdot U \cos Z = 0$ donc $\int m \cdot W^2 = \int m V^2 + \int m U^2$ c'est a dire que

dans le choc des corps durs soit que ce choc soit immédiat ou qu'il se fasse par l'entremise d'une machine quelconque sans ressort soit enfin (117) qu'on veuille ou non avoir égard a la masse même de la machine la somme des forces vives avant le choc est égale a la somme des forces vives après le choc plus la somme des forces vives qui auroit lieu si la vitesse de chaque mobile étoit égale a celle qu'il a perdue par le choc.

122. cette proposition est analogue au principe de la conservation des forces vives dans le choc des corps élastiques lequel se déduit tout simplement du précédent en supposant U qui est la vitesse perdue par le choc double de ce quelle est lorsque les corps sont durs.

123. cette même proposition prouve aussi la conservation des forces vives dans un système de corps durs dont le mouvement change par degrés insensibles car alors U^2 est infiniment petite du second ordre et partant l'équation se réduit a $\int m W^2 = \int m V^2$ mais ceci se rapporte au second cas qui nous reste a examiner.

124. 2nd cas je suppose maintenant que le mouvement change par degrés insensibles. nommons

la masse de chaque corpuscule _____ m

sa vitesse _____ V

sa force motrice _____ p

imaginons de plus qu'on fasse prendre au système un mouvement quelconque géométrique et nommons

la vitesse qu'aura alors m _____ u

l'angle formé par les directions de V et p _____ x

l'angle formé par les directions de V et u _____ y

l'angle formé par les directions de u et p _____ z

l'élement du tems _____ ∂t

cela posé si m étoit libre il auroit évidement l'instant d'après la vitesse V qu'il a actuellement plus la vitesse qu'il recevroit par l'action de la force p c'est-à-dire que sa vitesse absolue seroit la resultante de V et pdt, cette resultante estimée dans le sens de u est visiblement V.cos y + pdt.cos z or la vitesse réele de m aussi estimée dans le sens de u étant V.cos y deviendra au bout d'un tems infiniment court V.cos y + d(V.cos y) donc la vitesse perdue par m dans le sens de u et pendant dt est pdt.cos z − d(V.cos y) donc cette quantité est celle qu'il faut substituer pour V.cos Z dans l'équation (AA) qui devient par ce moyen

$$\int m.u.pdt.\cos z = \int m.u.d(V.\cos y), \ (GG)$$ ainsi pour trouver l'état du systeme après un tems infiniment court il n'y a qu'à attribuer successivement aux arbitraires u differents rapports et differentes directions jusqu'à ce qu'on ait toutes les équations necessaires a la solution du probleme.

125. en supposant comme on le peut u = V l'équation devient

$$\int m.V.pdt.\cos x = \int m.V.dV \ (HH)$$ intégrant cette équation pour avoir l'état du systeme après un tems indéfini et nommant K la vitesse initiale de m on aura $\int m.V^2 = \int m.K^2 + 2\int\int m.V.pdt.\cos x$ le signe d'intégration $\int$ étant relatif a la figure du systeme et $\int$ relatif a la durée du mouvement: et cette équation exprime visiblement le principe de la conservation des forces vives.

126. pour avoir le cas d'équilibre il n'y a qu'à supposer V = 0 dV = 0 ce qui fait évanouir le second membre de l'équation (GG) laquelle se réduit donc a $\int m.u.pdt.\cos z = 0$ ou $\int m.u.p.\cos z = 0$ (JJ) d'où l'on tirera toutes les loix de l'équilibre en attribuant successivement a u differentes valeurs arbitraires.

127. enfin si l'on imagine a volonté trois axes perpendiculaires entre eux. et qu'ayant décomposé V, u, et p chacune en trois autres paralleles a ces axes on appelle V', V", V"' les vitesses composantes de V, u', u", u"' les composantes de u, p', p", p"' les forces composantes de p: il est aisé de voir que l'équation (GG) peut etre mise sous cette forme

$$\int m.u'.dV' + \int m.u".dV" + \int m.u"'.dV"' = \int m.u'.p'dt + \int m.u".p"dt + \int m.u"'.p"'dt \ (KK)$$ et que celle (JJ) qui exprime les conditions de l'équilibre deviendra

$$\int m.u'.p' + \int m.u".p" + \int m.u"'.p"' = 0 \ (LL)$$ je passe a ce qui regarde plus particulierement les machines proprement dites

128. lors qu'il y a des forces vives appliquées aux machines leur théorie se rapporte au choc des corps et nous en avons dit assés là-dessus pour résoudre les questions de ce genre mais ce qu'on entend ordinairement par la théorie des machines ne concerne que le cas ou les forces employées pour les mouvoir sont des forces mortes telles que les poids la force des animaux le vent et les autres agents de cette nature c'est aussi a ce dernier cas que je me bornerai dans le reste de cette section et pour me conformer a quelques expressions qui semblent affectées a cette partie de la mechanique je commencerai par fixer exactement dans les Définitions suivantes le sens que j'y attache.

Définitions

129. parmis les forces appliquées a une machine en mouvement les unes sont telles que leurs directions font chacune un angle aigu avec la direction de la vitesse du point ou elle est appliquée tandis que les autres forment des angles obtus avec les leurs cela posé j'appellerai les premieres forces *sollicitantes* et les autres forces *résistantes*. par exemple si une force telle que celle d'un homme fait monter un poids par le moyen d'un levier d'une poulie, d'un treuil, &c. il est clair que la direction de la pesanteur et celle de la vitesse du poids forment necessairement par leur concours un angle obtus autrement il est visible que le poids descendroit aulieu de monter mais la direction de la puissance et celle de sa vitesse forment un angle aigu ainsi suivant notre définition le poids sera la force *résistante* et la force de l'homme sera la force *sollicitante* il est visible en effet que celleci tend a favoriser le mouvement actuel de la machine tandis que l'autre s'y oppose.

on observera que les forces sollicitantes peuvent etre dirigées dans le sens meme de leurs vitesses puisqu'alors l'angle formé par leurs directions est nul et par conséquent aigu et que les forces résistantes peuvent agir dans le sens directement opposé a celui de leurs vitesses puisqu'alors l'angle formé par leurs directions est de 180° et par conséquent obtus.

130. Si une force motrice absolue p se meut avec la vitesse u la quantité $P.u.dt$ sera nommée quantité d'action consommée pendant dt par cette force c'est a dire que la quantité d'action consommée dans un tems infiniment court par une force motrice absolue est le produit de cette force par le chemin que parcourt dans ce tems infiniment court le point ou elle est appliquée.

j'appellerai quantité d'action consommée dans un tems donné par une force la somme des quantités d'action consommées par elle a chaque instant desorte que nommant ds le chemin parcouru par cette force pendant dt on aura $\int p.ds$ pour la quantité d'action consommée par p aubout d'un tems indéterminé et partant si cette force est constante comme seroit le poids d'un corps donné cette quantité sera ps au bout d'un tems donné c'est a dire qu'elle sera le produit de la force par le chemin qu'elle a parcouru.

131. Si une force motrice absolue p se meut avec la vitesse u et que l'angle formé par le concours des directions de u et p soit z la quantité $p.u.dt.\cos z$ sera nommée quantité d'action produite par la force p pendant dt c'est a dire que la quantité d'action produite par la force p dans un tems infiniment court est le produit de cette force estimée dans le sens de sa vitesse par le chemin qu'elle parcourt pendant cet instant

j'appellerai quantité d'action produite par cette force dans un tems donné la somme des quantités d'action produites par elle a chaque instant desorte que $\int p.\cos z.u dt$ est cette quantité d'action produite apres un tems indéterminé

par exemple si p est un poids donné la quantité d'action produite dans un tems indéterminé sera $P \int u dt . \cos z$. Supposons donc qu'au bout du tems t le poids P soit descendu de la quantité H on aura évidement $dH = u dt . \cos z$ donc la quantité d'action produite pendant t sera PH tandis que la quantité d'action consommée dans le meme tems est PS, S étant le chemin effectivement parcouru par le poids P desorte que si le corps etoit descendu verticalement la quantité d'action consommée seroit égale à la quantité d'action produite; et en général toutes les fois que la force sera dirigée dans le sens meme de sa vitesse z sera nul on aura $\cos z = 1$ et partant la quantité d'action produite sera égale à la quantité d'action consommée. Sinon elle sera toujours moindre.

132. lorsqu'il s'agira d'un systeme de forces appliquées à une machine j'appellerai quantité d'action consommée ou produite la somme des quantités d'action consommées ou produites dans le tems par chacune des forces du systeme ainsi la quantité d'action consommée ou produite par les forces sollicitantes sera la somme des quantités d'action consommées ou produites par chacune de ces forces et la quantité d'action consommée ou produite par les forces résistantes sera la somme des quantités d'action consommées ou produites par chacune d'elles; et comme chaque force résistante fait un angle obtus avec la direction de sa vitesse le cosinus de cet angle est négatif la quantité d'action produite par les forces résistantes est donc aussi une quantité négative et partant la quantité d'action produite par toutes les forces du systeme est la meme chose que la différence entre la quantité d'action produite par les forces sollicitantes et la quantité d'action produite en meme tems par les forces résistantes.

Ces définitions nous seront utiles pour donner une forme plus simple aux propositions que nous allons donner sur la théorie des machines; mais on observera de ne point confondre ce que nous venons d'appeller quantités d'action produites et consommées avec ce que les géometres connoissent depuis M^r de maupertuis sous le nom de quantité d'action du systeme ou quantité d'action exercée dans le systeme et dont je ne ferai point usage dans ce memoire.

theoreme 1er.

principe général de l'équilibre dans les machines.

133. lorsqu'une machine est en équilibre si on vient à lui faire prendre un mouvement géometrique arbitraire sans rien changer d'ailleurs aux forces qui lui sont appliquées la quantité d'action produite alors au premier instant par les forces sollicitantes sera égale à la quantité d'action produite dans le meme tems infiniment court par les forces résistantes.

dem. il est clair que la question se réduit à prouver que la quantité d'action produite par les forces sollicitantes moins la quantité d'action produite par les forces résistantes est nulle ou ce qui revient au meme (132) que la quantité d'action produite produite par toutes les forces du systeme est nulle or cela est clair par l'équation (jj) (126) car mp y représente chacune des forces du systeme et partant mp cos z.udt est la quantité d'action produite par cette force pendant dt donc la quantité d'action produite au meme instant par toutes les forces du systeme est ∫mp.cos z.udt ou ∫mupdtcosz mais par l'équation (jj) cette quantité est nulle donc &c.

remarque.

134. il est clair que la proposition précédente auroit pu s'énoncer ainsi

lorsqu'une machine est en équilibre si on vient a lui faire prendre un mouvement géométrique arbitraire sans rien changer d'ailleurs aux forces qui lui sont appliquées la quantité d'action produite alors dans un tems infiniment court par toutes les forces du systeme est égale a zero.

mais alors il faudra faire attention a l'espece d'angle formé par le concours des directions de chaque force avec sa vitesse et prendre négativement les cosinus de ceux qui se trouveront obtus au lieu que suivant le premier énoncé après avoir distingué les forces sollicitantes des forces resistantes on prend positivement les cosinus de tous les angles formés par chacune des forces et la direction de sa vitesse.

il est encore évident que cette meme proposition peut aussi etre énoncée très simplement comme il suit.

lorsqu'une machine est en équilibre si on vient a lui faire prendre un mouvement quelconque géometrique sans rien changer d'ailleurs aux forces qui lui sont appliquées la somme des produits de chacune des forces du systeme multipliée par sa vitesse estimée dans le sens de cette force est nulle

cela est clair en divisant par dt l'équation (jj) (126)

de quelque maniere qu'elle soit énoncée cette proposition fera trouver toutes les conditions de l'équilibre en faisant prendre successivement a la machine differents mouvements géométriques arbitraires dont chacun donnera une équation entre les forces appliquées a cette machine.

on pourra aussi employer cette proposition pour trouver les loix du mouvement en décomposant le mouvement virtuel en deux autres dont l'un soit le mouvement réel car l'autre sera détruit et par conséquent assujetti aux conditions de l'équilibre énoncées dans le théoreme précédent.

Corollaire 1er.

principe général de l'équilibre entre deux puissances.

135. lorsque deux puissances appliquées a une machine quelconque sont en équilibre si on lui fait prendre un mouvement géometrique arbitraire 1° elles seront en raison réciproque de leurs vitesses estimées dans le sens de ces forces 2° leurs directions feront l'une un angle aigu avec la direction de sa vitesse et l'autre un angle obtus avec la sienne. cela est clair par le theoreme précédent.

coroll. 11.

principe général de l'équilibre dans les machines a poids.

136. lorsqu'une machine a laquelle il n'y a d'autres forces appliquées que des poids est en équilibre si on lui fait prendre un mouvement quelconque géométr: que la vitesse verticale du centre de gravité sera nulle au premier instant.

c'est a dire qu'au premier instant du mouvement la vitesse du centre de gravité estimée dans le sens vertical sera nulle quelques soient celles des differentes parties du systeme: pour le prouver nommons M la masse totale du systeme H la hauteur que parcourera le centre de gravité dans un tems donné t, h la hauteur parcourue dans le meme tems par le corpuscule m et g la gravité il s'agit donc de prouver que dH est nulle au premier instant or par le théoreme on a $\int m.g.dh = 0$ ou $\int m dh = 0$ mais par les proprietés du centre de gravité on a $M dH = \int m dh$ donc $M dH = 0$ ou $dH = 0$

pour avoir toutes les conditions de l'équilibre il n'y aura donc qu'à faire prendre successivement a la machine differents mouvements géometriques arbitraires et egaler dans chacun de ces cas la vitesse verticale du centre de gravite a zero.

remarque.

137. on prend ordinairement pour principe de l'équilibre dans les machines a poids qu'alors le centre de gravité est au point le plus bas possible mais on scait que ce principe n'est pas generalement vrai car outre que ce point pourroit dans certains cas etre au point le plus haut il y en a une infinité d'autres ou il n'est ni au point le plus haut ni au point le plus bas par exemple si tout le systeme se reduit a un corpuscule pesant et que ce mobile soit placé sur une courbe qui ait un point d'inflexion dont la tangente soit horisontale le corpuscule sera en équilibre si on le place a ce point qui n'est cependant ni le plus haut ni le plus bas possible.

on peut observer encore que pour demontrer ou expliquer la loi presente il ne suffiroit pas de prouver que dans une machine a poids infiniment proche de la position ou il y auroit équilibre le centre de gravité doit descendre infiniment moins au premier instant qu'il ne le feroit dans toute autre position. cette assertion est vraie mais ne conduit ce me semble a rien car ce n'est pas précisément la vitesse verticale du centre de gravité qu'il faut prouver etre nulle c'est le rapport de cette vitesse a la vitesse absolue du meme point ou des autres points du systeme or lors qu'on abandonne a elle meme une machine a poids dans une position infiniment proche de l'équilibre il est asses clair que le centre de gravité doit descendre infiniment moins que dans toute autre position mais il est clair aussi que sa vitesse horisontale et les vitesses absolues des

des autres points du systeme doivent egalement etre infiniment moindres dans ce cas que dans tout autre cela ne prouve donc point que le rapport de l'une a l'autre soit infiniment petit et c'est cependant ce qu'il faut demontrer pour assurer que le centre de gravité est au point le plus bas ou le plus haut ou sur un point d'inflection pour assurer enfin qu'il y a équilibre

138. le principe général de l'équilibre dans une machine a poids peut encore s'exprimer d'une maniere tres simple de la maniere suivante

pour s'assurer que plusieurs poids appliqués a une machine quelconque doivent se faire mutuellement équilibre il suffit de prouver que si l'on abandonne cette machine a ellememe le centre de gravité du systeme ne descendra pas.

nous avons deja fait mention de cette proposition (103), en voici la demonstration rigoureuse: je conserve les dénominations du corollaire second et je suppose que si ayant abandonné la machine a ellememe elle ne demeuroit point en équilibre comme je prétends qu'elle doit le faire la vitesse de m apres le tems t fut V on auroit (125) $\int m.V.dV$ dont l'integrale est $\int m.g.h = \frac{1}{2}\int m.V^2$ a laquelle il n'y a point de constante a ajouter puisqu'on suppose au premier instant V=0, mais $\int m.g.h = MgH$ donc $MgH = \frac{1}{2}\int m.V^2$ or par supposition on a H=0 donc $\int m.V^2=0$ de plus V^2 est necessairement positive soit que ce carré vienne de $(+V)^2$ ou de $(-V)^2$ donc l'équation $\int m.V^2=0$ ne peut avoir lieu sans qu'on ait V=0 c'est a dire sans qu'il y ait équilibre.

on conçoit mieux par la le principe ordinaire de l'équilibre dans les machines a poids qui est comme nous l'avons dit que le centre de gravité est au point le plus bas possible en effet si il etoit au point le plus bas il ne pourroit descendre davantage donc par la proposition précédente il doit y avoir équilibre mais il est vrai comme on l'a prouvé que cet équilibre peut encore avoir lieu dans d'autres cas ainsi l'on peut dire en général que si le centre de gravité est au point le plus bas possible il y aura necessairement équilibre mais il n'est pas toujours vrai que dans le cas d'équilibre le centre de gravité soit au point le plus bas possible

Corollaire III.
principe général de l'équilibre entre deux poids
139. lorsque deux poids se font mutuellement équilibre si l'on fait prendre a la machine un mouvement quelconque géometrique
1° les vitesses verticales de ces corps seront en raison réciproque de leurs poids
2° l'un de ces corps montera necessairement tandis que l'autre descendra
cette proposition est clair par le corollaire premier.

theoreme II
principe général du mouvement dans les machines.
140. Si l'on vient a changer subitement le mouvement actuel d'une

machine en un autre mouvement quelconque géometrique et qu'on abandonne aussitôt la machine à ellememe la conservation des forces vives aura lieu ensuite à chaque instant du mouvement qu'elle prendra quelque changement qu'il arrive d'ailleurs aux forces motrices.

dem. il est prouvé (125) que la conservation des forces vives a lieu dans tout Systeme dont le mouvement change par degres insensibles. or si l'on n'avoit rien changé au mouvement de la machine elle auroit selon l'hypotese (128) changé par degres insensibles mais (113) lorsqu'on Substitue au mouvement réel un mouvement géometrique, l'action réciproque des corps ne change pas au moment de cette substitution donc le mouvement doit varier encore par degres insensibles au moins pendant quelque tems après le changement du mouvement réel en un autre mouvement géometrique et cela quelques forces motrices qu'on substitue à celles qui animent réellement les differentes parties du Systeme.

remarque.

141. cette proposition Suffira dans tous les cas pour avoir le mouvement cherché mais on peut l'employer pour cela de plusieurs manieres, je me bornerai à celle qui me paroit en general la plus simple ; Soient donc

la masse de chacun des corpuscules	m
Sa vitesse réele	V
Sa force motrice	p
Sa vitesse apres le changement du mouvement réel en autre mouvement quelconque géometrique	u
l'angle formé par V et p	x
l'angle formé par V et u	y
l'angle formé par u et p	z
la force motrice apres avoir Subi un changement arbitraire	p'
l'angle formé par p' et u	z'

on aura donc par la presente proposition $\int mudu = \int mudt.p.\cos z$ équation qui aura lieu pour chaque valeur de u quelques soient d'ailleurs p' et z' imaginons que les forces p' soient décomposées chacune en deux autres p'' et p''' dont les premieres p'' soient telles que si elles étoient Seules il y auroit équilibre. nommons z'' l'angle formé par p'' et u et z''' l'angle formé par p''' et u on aura donc $p'.\cos z' = p''.\cos z'' + p'''.\cos z'''$ donc l'équation précédente devient $\int mudu = \int mp''.udt.\cos z'' + \int mp'''.udt.\cos z'''$. or il est clair que pendant un instant le mouvement est le meme que si l'on avoit $p'' = 0$ donc pour ce premier instant l'équation se réduit à $\int mudu = \int mp'''.udt.\cos z'''$; otant cette équation de la précédente et réduisant il reste $\int mp''.udt.\cos z'' = 0$ équation qui a lieu dans le premier instant du mouvement quelques puissent etre u et p' pourvu que le mouvement substitué au mouvement réel soit géométrique.

maintenant on observera que la vitesse virtuelle de m estimée dans le sens de u étant comme il est clair $V \cos y + p\,dt \cos z$ et la vitesse réelle estimée dans le même sens $V \cos y + d(V \cos y)$ il s'ensuit que la vitesse perdue par m pendant dt estimée dans le sens de u est $p\,dt \cos z - d(V \cos y)$ or il est clair que si les vitesses perdues par les corpuscules avoint été les seules forces du système il y auroit eu équilibre. donc ces vitesses peuvent être prises pour p'' donc ces vitesses estimées dans le sens de u peuvent être substituées a $p'' \cos z''$ dans l'équation $\int m p'' u\,dt \cos z'' = 0$ ce qui donne $\int m u p\,dt \cos z - \int m u\,d(V \cos y) = 0$ équation qui ne renferme plus que les inconnues d'ou dépend la solution du probleme et des quantités ou données ou arbitraires ainsi en attribuant successivement a ces arbitraires différentes valeurs on aura toutes les équations nécessaires a la solution du probleme. ce qui s'accorde avec ce que nous avons déja trouvé (124 et 127).

il est clair qu'a la place de mp on peut substituer une autre force quelconque qui lui soit égale donc si outre les corps que nous avons supposés appliqués a la machine il y en avoit encore d'autres telles que celle d'un homme ou d'un animal ou même des forces passives telle que le frottement nommant alors F chacune de ces forces U sa vitesse et Z l'angle formé par F et U l'équation deviendroit $\int F U\,dt . \cos Z + \int m p . u\,dt \cos z = \int m u\,d(V \cos y)$. on imprimeroit donc successivement au système différents mouvements géométriques arbitraires ce qui donneroit différentes valeurs et directions aux indéterminées U et u et par ce moyen on obtiendroit toutes les équations nécessaires a la solution.

Corollaire 1ᵉʳ.

du mouvement dans les machines a poids.

142. si l'on vient a changer subitement le mouvement actuel d'une machine a poids en un autre mouvement quelconque géométrique et qu'on abandonne aussitôt le système a ses propres forces, la somme des forces vives qui aura lieu ensuite a chaque instant du mouvement qu'il prendra est égale a la somme des forces vives initiales (c'est a dire immédiatement après le changement) plus la somme des forces vives qui auroit lieu si chaque point du système avoit une vitesse égale a celle qui est due a la hauteur dont est descendu le centre de gravité depuis le changement. dem. soient M la masse totale du système, H la hauteur dont est descendu le centre de gravité depuis le changement W la vitesse due a la hauteur H, h la hauteur dont est descendu chaque corpuscule m du système depuis le même instant K sa vitesse initiale V sa vitesse au bout du tems, z l'angle formé par V et par la direction de la pesanteur g la gravité.

je dis donc qu'on aura $\int m.V^2 = \int m.K^2 + M.W^2$ équation qui sera évidente si sa differentielle $\int m.V.\partial V = M.W.\partial W$ est une equation exacte mais on a $W^2 = 2gH$ donc il s'agit de prouver que $\int m.V.\partial V = Mg\partial H$ ou $\int m.V.\partial V = \int mg\partial h$ ou $\int m.V.\partial V = \int mg.V.\partial t.\cos z$ or cette équation est celle que donne la conservation des forces vives qui a été démontrée cy devant.

Corollaire II.

des machines qui se meuvent uniformement.

143. lorsqu'une machine se meut uniformement (c'est a dire que chaque point du systeme a une vitesse constante) la quantité d'action produite dans un tems donné par les forces sollicitantes est egale a la quantité d'action produite en meme tems par les forces resistantes.

dem. il est clair (132) qu'il me suffit de prouver que la quantité d'action produite a chaque instant par toutes les forces du systeme est nulle or en conservant les denominations de l'article (124) on a (125) $\int mp.V.\partial t.\cos x = \int m.V.\partial V$ et puisque ∂V est nulle par hypotese on aura $\int mp.V.\partial t.\cos x = 0$ ce qu'il falloit prouver.

Coroll. III.

des machines a poids qui se meuvent uniformement.

144. lorsqu'une machine a laquelle il n'y a d'autres forces appliquées que des poids se meut uniformement le centre de gravité du systeme reste constament a la meme hauteur sans monter ni descendre.

dem. en conservant les denominations de l'article (142) on aura a chaque instant $V = K$ donc par le meme article $M.W^2 = 0$ donc $W = 0$

Coroll. IV.

des machines assujeties a des retours periodiques.

145. dans une machine assujetie a des retours periodiques la quantité d'action produite pendant chaque periode par les forces sollicitantes est egale a la quantité d'action produite en meme tems par les forces resistantes

Cela est clair par l'article (125) puisque par hypotese on a $V = K$ au commencement de chaque periode.

Coroll. V.

des machines sujetes au frottement.

146. tout ce que nous avons dit des machines en général doit s'entendre aussi de celles qui sont sujetes au frottement ou autres résistances quelconques en regardant ces résistances comme des forces actives appliquées a la machine mais quelque soit le

mouvement il est clair que ces forces doivent toujours etre complées parmis celles que nous avons appellées résistantes c'est à dire parmis celles qui sont exercées dans un sens contraire aux vitesses réeles des points où elles sont appliquées ainsi par exemple si les corps n'etoint animés d'aucune force motrice il est clair que les vitesses iront toujours en diminuant et que par conséquent on chercheroit en vain une machine sujete au frottement et qui put conserver perpetuellement son mouvement primitif sans alteration: il y a meme plus car nous avons trouvé par experience que le frottement augmente lorsque la vitesse relative des corps diminue; et partant les degrés de vitesse perdus à chaque instant seroint de plus en plus grands desorte que le mouvement ne peut manquer non seulement de s'affoiblir petit à petit mais meme de s'eteindre totalement au lieu que si le frottement étoit par exemple proportionnel à la vitesse comme l'ont cru quelques phisiciens célebres le mouvement s'affoibliroit à la vérité de plus en plus mais pourroit néanmoins etre perpétuel.

le meme raisonnement fait sentir que dans les machines à poids il arrivera la meme chose car chaque fois que le centre de gravité montera pour descendre ensuite ce sera toujours avec une moindre vitesse et par conséquent il montera toujours de moins en moins et finira par ne point monter du tout.

scholie.

147. nous venons de voir (143) que si le mouvement d'une machine est uniforme la quantité d'action produite dans un tems donné par les forces sollicitantes est égale à la quantité d'action produite en meme tems par les forces résistantes: cette proposition est la plus utile de toutes dans la théorie des machines parceque la plupart d'entr' elles telles que la poulie la vis et le cabestan parviennent ordinairement bien vite à l'uniformité de mouvement; en voici la raison: les agents destinés à mouvoir ces machines comme les hommes les animaux les courrants &c se trouvant d'abord un peu au dessus des forces résistantes la machine prend un mouvement qui s'accelere pendant un certain tems par l'action continuée de la force sollicitante; mais soit que par une suite nécessaire de ce mouvement la force sollicitante diminue soit que la résistance augmente soit enfin qu'il survienne quelque variation dans leurs directions il arrive presque toujours que le rapport de ces deux forces s'approche de plus en plus de celui en vertu du quel elles pourroint se faire équilibre; alors ces deux forces se detruisent à chaque instant l'une par l'autre et la machine ne se

ment plus qu'en vertu du mouvement acquis lequel reste ordinairement uniforme acause de l'inertie de la matière.

pour comprendre encore mieux comment cela doit arriver il n'y a qu'à faire attention au mouvement que prend un navire qui a le vent en poupe c'est une espece de machine animée par deux forces contraires qui sont l'impulsion du vent et la force de l'eau si la premiere de ces deux forces qu'on peut regarder comme sollicitante est la plus grande le mouvement du navire s'accelere. mais cette acceleration a necessairement des bornes pour deux raisons car plus le mouvement du navire s'accelere plus il est soustrait a l'impulsion du vent 2° plus au contraire la resistance de l'eau augmente; par consequent ces deux forces tendent a l'egalité, lorsqu'elles y seront parvenues elles se detruiront mutuellement et partant le navire sera mu comme un corps libre, c'est a dire que sa vitesse sera constante. si le vent venoit a baisser la resistance de l'eau surpasseroit la force sollicitante, le mouvement du navire se ralentiroit mais par une suite necessaire de ce ralentissement le vent agiroit plus efficassement sur les voiles et la resistance de l'eau diminueroit en meme tems ces deux forces tendroint donc encore a l'egalité et la machine arriveroit dememe a l'uniformité de mouvement.

la meme chose arrive et par des raisons peu differentes lorsque les forces mouvantes sont des hommes ou des animaux. dans les premiers instants le moteur est un peu au dessus de la resistance ce qui fait naitre un petit mouvement qui s'accelere ensuite peu a peu par les coups repetés de la force mouvante mais l'agent luimeme est obligé de se mouvoir d'un mouvement acceleré affin de rester attaché au corps auquel il imprime le mouvement ce mouvement qu'il se procure a luimeme consomme une partie de sa force desorte que celui de la machine s'accelere de moins en moins et devient bientot sensiblement uniforme un homme par exemple qui pouvroit faire un certain effort dans le cas d'equilibre en fera un beaucoup moindre si le corps auquel il est appliqué lui cede et qu'il soit obligé de le suivre pour agir sur lui s'il tourne une manivelle de 14 pouces de rayon avec une vitesse de 30 tours par minute il fera suivant l'experience un effort d'apeupres 25# a chaque instant aulieu que si la machine etoit en repos il pourroit je crois en faire un de plus de 100; ce n'est pas que l'effort absolu de cet homme c'est a dire son travail soit moindre dans le premier cas que dans le second mais c'est que cet effort est partagé en deux dont l'un est employé a mettre l'homme luimeme en mouvement et l'autre transmis a la machine : or c'est de ce dernier seul que l'effet se manifeste dans l'objet qu'on s'est proposé.

C'est pour les raisons qu'on vient de voir que dans la théorie des machines simples contenue dans la section suivante j'ai presque toujours supposé ces machines parvenues a l'uniformité de mouvement non qu'elles ne puissent absolument avoir un mouvement varié mais la considération de ces mouvements est presque un objet de pure curiosité tant a cause du petit nombre de cas ou cela se rencontre dans la pratique qu'a cause des calculs qui deviennent presque intraitables. Je placerai cependant encore dans ce scholie quelques réflexions sur les machines considerées sous un point de vue plus général

148. Imaginons donc a volonté une machine a laquelle soient appliqués differents corps et forces quelconques je suppose seulement comme cy devant que le mouvement change par degres insensibles et que par conséquent la conservation des forces vives aura lieu a chaque instant de ce mouvement ou meme (140) de quel autre mouvement géometrique on voudra lui substituer or il suit visiblement de ce principe (125) que la quantité d'action produite a chaque instant par toutes les forces du systeme est égale a la moitié de la quantité dont augmente dans le meme tems infiniment court la somme des forces vives donc au bout d'un tems donné la quantité d'action produite pendant ce tems par toutes les forces du systeme est egale a la moitié de celle dont la somme des forces vives aura augmenté en meme tems c'est a dire (132) que la quantité d'action produite pendant un tems donné par les forces sollicitantes est égale a la quantité d'action produite en meme tems par les forces resistantes plus la moitié de l'augmentation que reçoit aussi dans le meme tems la somme des forces vives du systeme.

D'ou il suit que si la somme des forces vives est la meme au commencement et a la fin de ce tems comme cela arrive dans les machines dont le mouvement est uniforme et dans celles qui sont assujeties a des retours périodiques alors la quantité d'action produite dans ce tems par les forces sollicitantes sera égale a la quantité d'action produite en meme tems par les forces resistantes.

De plus il est bien remarquable que la moitié de l'augmentation de la somme des forces vives n'est autre chose que la quantité d'action produite par la force d'inertie des corps du systeme ; en effet cette inertie est une vraie force qui s'oppose au changement d'état de ces corps or si l'on veut mettre cette force en ligne de compte on trouvera que dans tous les cas le mouvement etant varié ou non la quantité d'action produite par les forces sollicitantes dans un tems donné sera justement égale a la quantité d'action

produite en même tems par les forces résistantes en effet lorsque
le corpuscule m animé de la vitesse V change de mouvement
par degrés insensibles il est visible que sa vitesse estimée dans le sens de V
devenant V+dV la quantité m.dV est sa force d'inertie estimée dans
ce même sens c'est à dire celle qu'il oppose dans le sens de V à son
changement d'etat et que partant m.V.dV est la quantité d'action
produite (131) par cette force dans un tems infiniment court
donc la quantité d'action produite par cette même force dans
un tems donné est $\frac{1}{2}(mV^2 - mK^2)$ en nommant K la vitesse
initiale de m c'est à dire que la moitié de l'augmentation de la
force vive de m est la même chose que la quantité d'action
produite par sa force d'inertie donc on peut établir pour regle générale
qu'en ayant égard à toutes les forces soit actives soit passives
appliquées à une machine en mouvement et même à l'inertie
de la matière considérée comme une force réele qui résiste
au changement d'etat des corps; la quantité d'action produite
par les forces sollicitantes dans un tems donné, sera toujours
quelque soit le mouvement de la machine égale à la quantité
d'action produite en même tems par les forces résistantes
et comme la même chose a lieu soit qu'il s'agisse du mouvement
réel de la machine ou d'un autre mouvement quelconque géometrique
substitué au mouvement réel il s'ensuit qu'on peut réduire à
cette seule proposition toute la théorie des machines soit en
repos soit en mouvement.

149 cela posé désignons par Q la quantité d'action produite par
les forces sollicitantes dans un tems donné t et par q la quantité
d'action produite en même tems par les forces résistantes l'inertie
comprise nous aurons donc dans toutes les machines possibles et
quelque soit le mouvement q = Q

je suppose par éxemple qu'ayant désigné chacune des forces
sollicitantes par F et sa vitesse estimée dans le sens de cette force par V
ces quantités F et V soient constantes pendant tout le mouvement
on aura donc (132) $\int F.V.t = q$ et si toutes les forces sollicitantes se
réduisent à une seule on aura F.V.t = q

on peut regarder q comme l'effet produit: par exemple lorsqu'il
s'agit d'elever un poids p à une hauteur H il est tout simple de
regarder l'effet produit comme en raison directe du poids et de la hauteur
de sorte que p.H est ce qu'on entend alors naturellement par l'effet
produit or d'un autre coté p.H est précisément ce que nous avons appellé
quantité d'action produite (131) c'est à dire que dans ce cas q = p.H.

il est visible que dans tous les autres cas q est une quantité analogue qu'on peut de meme regarder comme l'effet produit et que nous appellerons effectivement ainsi. pour fixer les termes: nous entendrons donc par effet produit la quantité d'action produite par les forces résistantes, et la quantité d'action produite en meme tems par les forces sollicitantes sera nommée la cause.

150. on voit par l'équation $F.V.t = q$ que nous venons de trouver qu'il n'est pas nécessaire de connoitre la figure d'une machine pour scavoir de quel effet est capable une force donnée qui lui est appliquée suposons par exemple qu'un homme soit capable d'exercer continuellement 25# de forces avec une vitesse uniforme d'un pied par seconde; lorsqu'on l'appliquera a une machine pendant un tems t, la quantité FVt sera donc $25^{#} \cdot 1^{p.} \cdot t$ ainsi quelque soit la machine on aura toujours l'effet produit ou $q = 25^{#} \cdot 1^{p.} \cdot t$ donc la quantité de cet effet est indépendante de la figure de la machine et ne peut jamais surpasser la quantité determinée $25^{#} \cdot 1^{p.} \cdot t$ par exemple si cet homme avec son effort de 25# et sa vitesse d'un pied par seconde. est en etat avec une machine donnée ou sans machine d'elever dans un tems donné un poids p a une hauteur H on ne peut inventer aucune machine par laquelle il soit possible d'elever dans le meme tems le meme poids a une plus grande hauteur ou un poids plus grand a la meme hauteur ou enfin le meme poids a la meme hauteur dans un tems plus court

151. l'avantage que procurent les machines n'est donc pas de produire de grands effets avec de petits moyens mais de donner a choisir entre differents moyens egaux celui qui convient le mieux a la circonstance présente; pour forcer un poids donné a monter a une hauteur proposée ou telle autre resistance qu'on voudra imaginer a produire une certaine quantité d'action q il faut que la force F qui y est destinée produise elle meme une quantité d'action égale a la première aucune machine ne peut en dispenser mais comme cette quantité resulte de plusieurs termes ou facteurs on peut les faire varier a volonté en diminuant la force au dépend du tems ou la vitesse au dépend de la force, oubien en employant deux ou plusieurs forces au lieu d'une ce qui donne une infinité de ressources pour produire la quantité d'action nécessaire par des moyens differents mais quoi qu'on fasse il faut toujours que ces moyens soient egaux c'est a dire que la cause ou quantité d'action produite par les forces sollicitantes soit égale a l'effet ou quantité d'action produite par les forces résistantes et c'est dans ce sens qu'on peut admettre ce principe que *les causes sont proportionnelles a leurs effets.*

152. ces reflexions paroissent suffisantes pour desabuser ceux qui croient qu'avec des machines chargées de leviers arrangés misterieusement on pourroit parvenir a mettre un agent si foible qu'il soit en etat de produire les plus grands effets; l'erreur vient de ce qu'on se persuade.

pouvoir appliquer aux machines en mouvement ce qui n'est vrai que pour
le cas d'equilibre; la moindre force peut faire équilibre au plus grand obstacle
mais si le mouvement vient a naitre la puissance perd peu a peu son avantage
parceque pour continuer de résister a l'obstacle il faudroit qu'elle se procurat une
vitesse au dessus de ses facultés ou qui du moins lui feroit perdre une partie d'autant
plus grande de son effort qu'elle seroit obligée de se mouvoir plus vite par la raison
deja dite que cet effort est décomposé en deux dont le premier est employé a mouvoir
l'agent luimeme desorte que le second seul est employé a vaincre les résistances

en un mot quoiqu'on puisse regarder l'equilibre comme un mouvement évanouissant
ou plutot comme le dernier terme ou la limite du mouvement de la machine; ce
sont cependant deux etats tres differents et en quelque sorte heterogènes; dans le
premier cas il s'agit de detruire d'empecher le mouvement, dans le second l'objet est
de le faire naitre et de l'entretenir or il est clair que l'examen de ce dernier cas
exige une consideration de plus que le premier sçavoir la vitesse réele de chaque
point du systeme et c'est sans doute faute de faire attention a ces differents effets
d'une meme machine que des personnes auxquelles la saine théorie n'est point
entierement inconnue s'abandonnent quelque fois aux idées les plus chimeriques
tandis qu'on voit de simples ouvriers faire valoir par une espece d'instinct les
proprietes réeles des machines et juger tres bien de leurs effets.

mais si c'est un abus de croire les machines capables de tout parce que
archimede ne demandoit qu'n levier et un point fixe pour déplacer le globe de la
terre ce seroit une erreur plus insoutenable encore de ne les croire propres a rien
fondé sur ce qu'un homme par exemple qui sans machine aura elevé un poids
donné a une certaine hauteur n'en pourroit pas elever a la meme hauteur
une once de plus avec le meme travail c'est a dire en employant a chaque instant
la meme quantité de force et la meme vitesse que dans le premier cas, quand
meme il seroit muni de la plus belle machine du monde: archimede avoit raison
et ce que nous venons de dire de la force d'un homme n'est pas moins vrai
il n'y a point la de contradiction; contrebalancer un poids et l'elever sont
deux choses fort differentes: avec une machine un enfant va soulever un rocher
faut il le transporter en un lieu et dans un tems donné toutes les forces
humaines n'y suffiront pas.

mais dira t on comment se fait il qu'n levier puisse donner tant d'energie
a une petite force dans le cas d'équilibre tandis qu'elle devient si impuissante
desqu'il s'agit de produire le moindre mouvement: si comme vous l'aves dit
l'objet des machines n'est pas de produire de grands effets avec de petits moyens
comment archimede auroit il soulevé le monde avec son levier, et s'il est vrai
au contraire que les machines multiplient réelement les forces pourquoi
ne pourroit on tirer aucun parti de cette proprieté dans le cas du mouvement?

quoique nous ayons démontré en rigueur tout ce que nous avons avancé nous essayerons de rendre une raison sensible de ce paradoxe affin de dissiper le nuage qui malgré les démonstrations offusque beaucoup de gens auxquels il semble toujours quil y a dans les machines quelque chose de magique.

les points fixes ou obstacles sont des forces purement passives qui peuvent absorber un mouvement si grand quil soit. mais en produire un si petit quon voudra l'imaginer dans un corps en repos c'est chose impossible. or c'est improprement que dans le cas d'équilibre on dit qu'une petite force en détruit une grande. ce n'est pas par la petite force que la grande est détruite c'est par les points fixes. la petite force ne contrebalance véritablement qu'une petite partie de la grande et les obstacles font le reste. si archimede avoit eu ce quil demandoit ce n'est pas lui qui auroit soutenu le globe de la terre c'est son point fixe. tout son art auroit consisté non a redoubler d'effort pour luter contre l'univers mais a mettre en opposition les grandes forces l'une active l'autre passive quil auroit eu a sa disposition. quil eut été question de faire naitre un mouvement effectif archimede auroit été obligé de le tirer tout entier de son propre fond aussi n'auroit il pu etre que tres petit meme apres plusieurs années parce qu'un point fixe encore un coup peut bien absorber le mouvement d'un corps mais non pas lui en donner. la table d'un billard peut arreter dans sa chute un poids de mille livres mais elle ne soulevera jamais d'une ligne une petite boule d'ivoire placée sur le tapis. n'attribuons point aux forces actives ce qui n'est du qu'aux obstacles et l'effet ne nous paroitra pas plus disproportioné a la cause dans les machines en équilibre que dans les machines en mouvement.

153. quel est donc enfin le véritable objet des machines? nous l'avons deja dit c'est de pouvoir faire varier a volonté et suivant que les circonstances l'exigent les termes de la quantité d'action ∫FVl ou Q produite par les forces sollicitantes si le tems est precieux que l'action doive etre produite dans un tems tres court et qu'on n'ait cependant qu'une force capable de peu de vitesse mais d'un grand effort on pourra trouver une machine pour suppléer la vitesse nécessaire par la force s'il faut au contraire elever un poids tres considérable et qu'on n'ait qu'une foible puissance mais capable d'une grande vitesse on pourra imaginer une machine avec laquelle l'agent pourra compenser par sa vitesse la force qui lui manque enfin si la puissance n'est capable ni d'un grand effort ni d'une grande vitesse on pourra encore a l'aide d'une machine convenable lui faire surmonter la résistance en question mais alors on ne pourra se dispenser d'employer beaucoup de tems et c'est la ce qu'il faut entendre par ce principe si connu que dans les machines en mouvement on perd toujours en tems ou en vitesse ce qu'on gagne en force.

les machines sont donc tres utiles non en multipliant les forces mais en les modifiant c'est adire en les transmettant suivant des loix données; on ne parviendra jamais par elles il est vrai a diminuer la depense de quantité d'action necessaire pour produire un effet proposé mais elles pourront aider a faire de cette quantité une répartition convenable au dessein qu'on a en vue c'est par leur secours qu'on reussira sinon a determiner la quantité absolue d'action produite par chaque partie du systeme dumoins a établir entre ces quantités les rapports qui conviendront le mieux, c'est par elles, enfin qu'on pourra parvenir a donner aux forces mouvantes, les situations et directions les plus commodes les moins fatiguantes les plus propres a proportionner les facultés de l'agent de maniere a produire toute la quantité d'action dont elles sont Susceptibles car plus cette quantité d'action qui est la cause sera grande plus la quantité d'action produite par les forces résistantes qui est l'effet sera grand, mais il faut bien distinguer la quantité d'action *produite* de la quantité d'action *consommée* une force pourroit consommer beaucoup de quantité d'action et en produire tres peu, dans ce cas elle feroit peu d'effet parce qu'elle ne feroit pas employée de la maniere la plus avantageuse, le travail de la force mouvante° doit s'estimer par la quantité d'action quelle consomme et son effet par la quantité d'action quelle produit mais (131) la quantité d'action produite n'est égale a la quantité d'action consommée que dans le cas seul ou la force agit dans la direction meme de sa vitesse dans tous les autres cas elle est moindre, donc pour tirer d'un agent tout l'effet qu'on peut en attendre il faut le faire agir dans le sens meme de sa vitesse, Si par exemple l'agent est un homme qui fasse tourner une manivelle il faut pour ne point consommer inutilement sa force qu'il agisse perpendiculairement au rayon de la manivelle et l'on doit se dispenser toutes choses égales d'ailleurs (c'est a dire en supposant dans les deux cas une égale facilité d'agir et une situation également commode) de l'attacher a une bequille qui agissant toujours a peupres dans la meme direction n'est presque jamais perpendiculaire au rayon de la manivelle. dememe un agent capable de produire une quantité d'action F.V.t dans un tems donné pourroit avec une machine convenable obliger les forces résistantes a consommer une quantité d'action beaucoup plus grande et meme aussi grande qu'on le voudroit mais il ne les forcera jamais a produire une quantité d'action plus grande que celle qu'il produit luimeme.-

par exemple si la force F animée d'une vitesse V oblige le poids P a monter a la hauteur H la quantité d'action produite par le poids est PH quelque chemin qu'on lui fasse prendre mais si la longeur du chemin est S la quantité d'action consommée

° il faut prendre garde de confondre le travail absolu ou la peine de l'agent avec ce que j'appelle ici le travail de la force mouvante ou sollicitante car celle ci n'est qu'une partie de l'autre ce n'est dirje que la partie de l'effort absolu de l'agent qui est transmise aux forces résistantes par l'entremise de la machine. mais l'agent est en outre obligé comme on l'a deja dit de faire un autre effort pour se mouvoir luimeme et l'on doit avoir égard a cet effort aussibien qu'a celui qui est transmis a la machine et qui est proprement la force mouvante pour juger de son travail absolu. nous disons ici que la force mouvante doit produire autant d'action qu'elle en consomme mais ce n'est point la seule condition a remplir pour tirer d'un agent dont le travail absolu est determiné tout l'effet qu'on a droit d'en attendre il faut encore que la partie de ce travail employée a mouvoir l'agent luimeme soit la moindre possible afinque la partie transmise soit la plus grande et c'est a quoi l'on ne peut parvenir que par des experiences multipliées sur la force des animaux comme on verra dans l'article suivant.

sera ps et comme la quantité d'action produite par F ne peut surpasser PH il s'ensuit que cette force avec la quantité d'action limitée PH pourroit faire consommer a la résistance une quantité d'action aussi grande quelle le voudroit ce qui peut etre avantageux dans bien des circonstances.

ainsi par exemple il est visible qu'abstraction faite du frottement il n'en couteroit pas plus de travail a une puissance employée convenablement pour mener un corps donné A au point C que pour le conduire au point B qui répondroit verticalement au point A et horisontalement au point C.

155. ceci nous conduit naturellement a cette question interessante quelle est la maniere la plus avantageuse d'appliquer des puissances motrices données aux machines en mouvement pour leur faire produire le plus grand effet possible?

la condition générale est que Q soit un maximum ce qui fait naitre deux reflections la premiere concerne les directions qu'il convient de donner aux forces l'autre la maniere de proportionner convenablement les facultés dont elles sont capables le premier point est résolu par ce que nous avons dit précédement car nous y avons prouvé que les forces doivent etre dirigées dans le sens meme de leurs vitesses. la seconde reflection se presente aussi tres naturellement un agent est succeptible de deux facultés l'une est sa force l'autre sa vitesse or pour obtenir de cet agent tout l'effet dont il est capable il y a un certain rapport a mettre entre sa force et sa vitesse rapport qui ne peut etre connu que par experience par exemple on a reconnu je suppose qu'un homme attaché pendant 8^{heures} par jour a une manivelle peut faire continuellement un effort de 25^{tb} avec une vitesse d'un pied par seconde mais si l'on forçoit cet homme a aller beaucoup plus vite croyant par la avancer la besogne on la retarderoit parce que l'homme ne seroit pas alors en etat de faire un effort continuel de 25^{tb} pendant 8^{heures} par jour desorte que la quantité d'action FVt qu'il produit diminueroit si au contraire on diminuoit la vitesse la force augmenteroit mais dans un moindre rapport et FVt diminueroit encore ainsi suivant l'experience pour que FVt soit un maximum il faut proportionner la machine de maniere a lui conserver la vitesse d'a peupres un pied par seconde et ne le faire travailler qu'environ 8^{heures} par jour.

au reste l'art d'employer de la maniere la plus avantageuse la force des animaux qui est sans contredit la partie la plus utile de la mechanique pratique exige beaucoup d'experiences et quoique de grands phisiciens ayent travaillé avec succes sur cette matiere il reste encore beaucoup a faire pour la porter au degré desirable de perfection.

156. parmis plusieurs observations que nous pourrions faire encore sur les machines en général il y en a deux qui sont trop importantes a la pratique pour que nous puissions les ommetre. voici la premiere

Selon ce qui vient d'être dit deux hommes par exemple appliqués ensemble a une même machine pendant une heure agissants de concert et de la manière la plus avantageuse possible produiront exactement le même effet que s'ils étoint appliqués successivement chacun pendant une heure a cette machine en supposant qu'ils agissent toujours de la manière la plus avantageuse mais cela n'est point ainsi lorsque la machine est sujete au frottement. pour en faire sentir la raison qui est simple je suppose par exemple qu'il s'agit d'élever un poids par le moyen d'un treuil si les deux hommes agissent de concert ils auront a vaincre a chaque instant outre l'effort du poids celui du frottement, lequel est composé du frottement produit par la force des hommes du frottement produit par le poids a enlever et du frottement produit par le poids de la machine mais lorsqu'il n'y a qu'un homme il faudroit pour qu'il put élever un poids égal a la moitié de ce que faisoint les deux hommes réunis qu'il surmontât outre ce poids un frottement composé du frottement que produiroit ce poids de celui que produiroit la puissance et de celui que produiroit le poids de la machine or le frottement du poids ᵃ ᵉⁿˡᵉᵛᵉʳ et celui de la force seroint a peupres chacun sous double de ce qu'il étoit lorsque les forces agissoint de concert, mais le poids de la machine étant resté le même le frottement qu'il occasionne n'a pas changé sensiblement Donc le frottement total seroit plus grand que la moitié de celui qu'avoint a vaincre les deux hommes réunis donc si chaque homme agissant seul ne fait pas plus d'effort qu'il en feroit en agissant de concert avec l'autre il n'enlevera pas un poids égal a la moitié du premier donc il est avantageux qu'ils agissent de concert et c'est le cas de dire avec vérité *vis unita fortior* cet exemple suffit pour montrer ce qui doit arriver dans les cas analogues. 157. la seconde observation que nous avons a faire concerne les machines a poids mais elle s'applique aussi aux autres cas a quelques modifications pres la voici

lorsqu'on n'a pas d'autre but que d'élever un poids donné a une certaine hauteur peu importe le chemin qu'on lui fasse prendre, la quantité d'action a produire ne sera pas plus grande pour lui faire prendre une route tortueuse que pour le conduire par le chemin le plus court mais il faut s'il est possible que le mouvement de ce corps s'éteigne en arrivant au point ou l'on veut le faire parvenir car en nommant P le poids et H la hauteur a laquelle il doit être élevé il est aisé de voir par ce qui a été dit (148) sur le mouvement des machines en général que dans le cas en question la quantité d'action produite par les forces sollicitantes sera seulement PH si la vitesse du mobile s'éteint en arrivant a la hauteur H au lieu que si cela n'étoit pas elle seroit PH plus la demi somme des forces vives c'est adire qu'en nommant m chaque molecule et u sa vitesse en arrivant a cette hauteur la quantité d'action produite par les forces sollicitantes seroit $PH + \frac{1}{2}\int mu^2$ or u^2 est necessairement positive donc $\int mu^2$ est aussi toujours positive donc l'action produite par les forces mouvantes seroit plus grande dans le second cas que dans le premier

ainsi par exemple dans les pompes il faut autant qu'il est possible que l'eau en arrivant dans le réservoir superieur ait trespeu de vitesse toute celle qui lui reste ayant inutilement consommé l'effort de la puissance motrice.

aureste toutes ces remarques ont seulement pour objet les machines dont le merite est de favoriser autant qu'il est possible les forces mouvantes telles sont ordinairement les machines simples et particulierement celles qu'on employe dans la marine. C'est aussi pour cela que j'ai parlé seulement des machines dont le mouvement change par degrés insensibles car lorsque le but est de faire produire a une machine le plus grand effet possible il y a un avantage décidé a ce qu'il n'y ait jamais aucun choc ni changement brusque dans le mouvement: qu'un homme par exemple se propose d'élever le poids p a la hauteur H et supposons que pendant le mouvement necessaire pour cela il arrive un choc quelconque nommons q la quantité d'action qu'auroit eu a produire cet homme s'il n'y avoit pas eu de choc Q celle qu'il aura a produire effectivement h la hauteur ou sera le corps p au moment du choc X la somme des forces vives du systeme immédiatement avant ce choc et Y la somme des forces vives immédiatement apres cela posé on a (148) $q = PH$ de plus la quantité d'action que l'homme aura a produire jusqu'au moment du choc sera (148) $ph + \frac{X}{2}$ et celle qu'il aura a produire apres le choc sera (148) $p(H-h) - \frac{Y}{2}$ ainsi en tout il aura a produire la quantité d'action $PH + \frac{X-Y}{2}$ c'est adire qu'on a $Q = PH + \frac{X-Y}{2}$ or (121) X est toujours plus grande que Y donc $Q > PH$ ou $Q > q$ donc de quelque maniere que se fasse le choc il y a necessairement une perte de quantité d'action, perte si réele que sans ce choc l'homme avec le meme travail c'est adire en employant a chaque instant la meme force et la meme vitesse que dans le premier cas auroit pu elever dans le meme tems le poids p a une hauteur H' plus grande que H car on a (148) $Q = PH'$ et $q = PH$ or nous venons de trouver $Q > q$ donc $H' > H$.

quant a la maniere d'estimer les forces appliquées aux machines chacun peut en choisir une a son gré peu importe aufond pourvu qu'on raisonne toujours conséquemment a ses definitions cependant si l'on veut se determiner pour le parti qui paroit le plus simple on ne peut gueres faire autrement que d'estimer comme on fait ordinairement la force d'un corps par le produit de sa masse et de sa vitesse ou bien comme le faisoint m^{rs} de lebnitz et bernouilli par le produit de sa masse et du carré de sa vitesse; ces deux methodes sont toutes les deux tres simples et tres naturelles, en effet lorsque deux corps de masse differentes venant

en sens opposés a la rencontre l'un de l'autre se font mutuellement équilibre
on ne peut gueres se defendre d'appeller égales les forces de ces corps et par
consequent de regarder la force d'un corps comme le produit de la masse
par la vitesse. d'un autre coté si l'on imagine qu'un corps ayant frappé un
ressort ait été obligé depuiser tout son mouvement et par conséquent toute
sa force pour le fermer d'une certaine quantité et qu'ayant remis le ressort
dans son premier etat un nouveau corps de masse differente vienne a son tour
le frapper et soit de meme obligé depuiser aussi tout son mouvement pour
le fermer de la meme quantité que le premier on ne pourra gueres
s'empecher d'appeller égales les forces qui ont produit en se consommant deux
effets si semblables c'est a dire de regarder la force du premier corps qui s'est
epuisée pour fermer le ressort comme égale a celle du second qui s'est aussi
epuisée pour produire le meme effet or si l'on adopte cette methode d'estimer
les forces il faudra les regarder comme le produit des masses par les carrés
des vitesses car ce qu'on vient de dire ne seroit point arrivé si les produits
des masses de chacun de nos corps par le carré de sa vitesse n'avoint été égaux
il est donc tres naturel aussi d'estimer la force d'un corps en mouvement
par le produit de sa masse et du carré de sa vitesse. en un mot chacun
est maitre d'appeller force d'un corps ce qui lui plait il ne s'agit que de
raisonner consequemment a sa définition pour ne pas se tromper. mais tout
le monde ne sçait pas se preserver de l'erreur comme mr bernouilli et c'est
précisément parceque ces deux manieres d'estimer les forces sont si naturelles
et si simples toutes les deux qu'on les confond souvent et qu'il est meme tre
difficile d'eviter cet écueil lorsqu'on n'a que des idées vagues sur la mechanique
comme la plus part des machinistes. j'ai eu occasion de voir plusieurs chercheurs
de mouvement perpetuel et particulierement un auquel on ne peut refuser
une grande sagacité car c'est un paisan qui n'ayant absolument aucune
connoissance en mathématique a trouvé par ses propres reflections presque
toute la theorie des machines. cet homme m'a fait voir les desseins de
plusieurs machines de son invention presque toutes extremement compliquées
et dont l'objet est le mouvement perpetuel quoiqu'il soit presque impossible de comprendre
ce qu'il veut dire il m'a paru que tous ses raisonnements et le fondement de toutes
ses machines revenoint a ceci. imaginés deux corps attachés a une meme machine
que l'un m ne puisse etre mu que dans un plan horisontal et qu'au contraire
l'autre M plus grand que le premier se meuve dans un plan vertical je suppose
les deux corps dans un meme plan horisontal au premier instant et deplus
qu'il ne puisse y avoir de choc cela posé j'imprime un certain mouvement a M
lequel par le moyen de la machine le communique a M et le fait monter
or a mesure que celuici monte l'autre perd son mouvement et il arrive enfin
un moment ou il n'en a plus aucun et ou la machine reviendroit d'ellememe
a sa premiere position si les corps y restoint attachés mais je suppose que le

poids M soit lâché par le moyen d'une détente il tombera sur le plan
horisontal ou je le suppose amené par une courbe pour empecher qu'il n'y ait choc
cela posé je dis que lors qu'il sera arrivé sur ce plan horisontal ou est m
il aura une force ou quantité de mouvement plus grande que celle que j'avois
imprimée a celuici or m avec la quantité de mouvement que je lui avois
imprimée a fait monter M a la hauteur H donc suivant cet homme
M avec une quantité de mouvement plus grande sera capable de faire
remonter un poids egal a une plus grande hauteur que M n'y est monté
la premiere fois et par conséquent si je me contente de l'elever a la meme
hauteur il me restera de la force pour vaincre les frottements force que je
pourrai meme augmenter si elle n'est pas suffisante en diminuant le rapport
de m a M. cet homme a raison jusqu'a un certain point car m etant
moindre que M si l'on appelle u la vitesse que je lui imprime et V la vitesse
que M aura aquise quand il sera retombé dans le plan horisontal d'ou il est
parti on aura effectivement comme il le dit MV > mu car on a (142)
$mu^2 = 2MH$ et par la théorie de la chute des graves $MV^2 = 2MH$
donc $mu^2 = MV^2$ or M > m donc MV > mu et partant m aura
communiqué a M plus de force qu'il n'en avoit reçu luimeme et cet homme
s'imaginoit qu'en vertu de cet exes de force il etoit en etat de faire remonter
le corps M plus haut que la premiere fois ou du moins de vaincre les frottement
et autres résistances qui l'empecheroint de monter a la meme hauteur. or cela
est évidemment faux pour qui connoit les loix du mouvement mais n'est
point du tout aisé a faire comprendre a quelqu'un qui n'a aucune idée de théorie
l'erreur vient visiblement de ce qu'il confond les deux manieres d'estimer
les forces; tantot il regarde la force de m comme en raison composée de m et
de u et tantot en raison composée de M et H parce que veritablement avec
mu de force il eleve M a la hauteur H.

voici donc en deux mots a quoi reviennent les differentes opignons
sur l'estimation des forces. concevons qu'un corps p tombé de la hauteur H
ait la vitesse V cela posé les uns appellent pV force du corps et les autres
donnent ce nom a pH or cela est certainement tres indifferent pourvu qu'on
n'attache pas la meme idée au mot force dans les deux cas. tenons nous en donc
aux idées reçues donnons simplement puisque c'est l'usage le nom de force a
pV et reservons a pH ou pV^2 celui de force vive souvenons nous que
les forces vives et les forces ordinaires ont des proprietes toutes differentes
mais toutes également dependantes des memes loix immuables de la nature
et d'apres cela nous ne pourrons plus nous tromper la force de pression d'un poids
d'un homme d'un ressort &c pourra s'estimer facilement par la quantité
de mouvement quelle pourroit faire naitre dans un mobile pendant un tems
donné. et ce que nous avons appellé (131) la quantité d'action de cette force ou effet
s'estimera par la quantité de force vive quelle fera naitre dans le meme mobile

pendant le meme tems pris pour unité.

je vais terminer cette premiere Section par une observation sur les résistances auxquelles les machines sont sujetes et par la solution du probleme général dont les applications particulieres doivent faire le sujet de la Seconde et derniere Section

158. les formules que nous avons trouvées pour le frottement s'accordent toutes a prouver que ce frottement est Sensiblement proportionel a la pression car les tables d'ou elles sont tirées font voir qu'une augmentation de plusieurs centaines de livres dans la pression sur une base médiocre font une diminution a peine Sensible dans le rapport du frottement a la pression

il suit de la qu'il est inutile de connoitre bien exactement la pression pour avoir Son rapport au frottement aussi approché qu'on peut le desirer dans la pratique et par conséquent lorsque nous aurons une valeur passablement exacte de cette pression nous pourrons l'employer Sans craindre aucune erreur Sensible pour trouver la valeur du rapport du frottement a cette pression rapport qui Sans cela ne pourroit souvent s'obtenir que par des calculs immenses. voici maintenant le probleme annoncé dans l'article précedent.

probleme.

159. ayant l'etat actuel de repos ou de mouvement d'une machine donnée déterminer l'etat de repos ou de mouvement ou doit se trouver cette machine un instant apres en ayant égard au frottement et a la roideur des cordes.

Sol. puisque l'etat actuel de la machine est donné il est visible que par la théorie developée dans le cours de cette section on pourroit trouver l'etat de la machine pour l'instant suivant et les pressions qui s'exercent entre Ses differentes parties si le frottement quelles eprouvent étoit connu ainsi que la roideur des cordes or on a l'expression de ces résistances par les formules trouvées dans la premiere partie de ce mémoire on obtiendra donc facilement toutes les équations necessaires a la solution de chaque cas particulier

Corollaire

160. on voit par la qu'il est aisé de trouver dans chaque cas particulier les équations necessaires a la solution du probleme mais il s'en faut beaucoup qu'il soit toujours facile d'en tirer la valeur des quantités inconnues qu'on cherche; il y a au contraire tres peu de cas dont on puisse avoir la solution complete sans avoir a faire des calculs rebutants ce n'est donc avoir rien fait pour la pratique que d'avoir indiqué une solution générale dont l'application est si difficile qu'on doit la regarder dans l'usage ordinaire des machines comme absolument impraticable; et le Seul parti ce me semble qu'on

ait a prendre est de chercher parmis le nombre infini de questions que renferme le probleme géneral celles de ces questions qui peuvent avoir une utilité réele de s'y borner et d'en trouver s'il est possible une solution particuliere qui puisse etre mise en pratique voyons donc d'abort quelles peuvent etre les questions les plus utiles qu'on ait a proposer sur les machines.

pour cela il faut remarquer qu'en employant une machine a quelque objet quelconque d'utilité on n'a pas simplement le but vague de mettre cette machine en mouvement mais on a celui de la mettre en mouvement de telle ou telle maniere affin de produire tel ou tel effet le mouvement qu'il convient de faire prendre a cette machine doit donc etre regardé comme determiné par l'objet qu'on se propose de remplir et la question est d'appliquer a la machine des forces capables de produire le mouvement or ces forces sont de deux sortes les unes sont destinées a produire l'effet proposé et le produiroint en effet sans les frottements et roideurs de corde qui tendent continuellement a alterer le mouvement qu'on a jugé necessaire pour cela les autres sont destinées a rendre a la machine les degrés de mouvement que lui otent a chaque instant ces resistances les premieres de ces forces se determinent par le calcul mathematique de la machine c'est a dire en faisant abstraction des frottements et roideurs de corde or on a la dessus toutes les regles necessaires: la difficulté qui nous regarde actuellement se réduit donc a la question suivante.

le mouvement que doit prendre une machine etant donné et sachant les forces qui seroint necessaires a chaque instant pour le produire s'il n'y avoit ni frottement ni roideur de corde déterminer ce qu'il faut ajouter aux forces sollicitantes pour que le mouvement soit en effet tel qu'on le veut malgré les résistances.

la partie vraiment utile du probleme general etant réduite a cela voyons s'il ne seroit pas possible de la ramener a quelquechose de plus simple encore sans rien diminuer de sa generalité: or cela est tres facile comme on va voir.

les forces cherchées n'ayant d'autre objet que de reparer les degres de mouvement absorbés continuellement par les frottements et roideurs de corde il est évident quelles doivent etre simplement capables de faire equilibre a chaque instant a l'action reunie de ces resistances; cela posé le mouvement réel etant donné par hypotese ainsi que celui qui auroit pris chaque mobile s'il eut été libre il est clair que les mouvements detruits a chaque instant sont aussi connus: de plus si ces mouvements detruits etoint les seules

forces du systeme il y auroit équilibre donc la question
se réduit à trouver la loi d'équilibre entre ces forces connues en ayant
égard au frottement et a la roideur des cordes c'est adire quil s'agit
d'ajouter au systeme de nouvelles forces qui seroint capables de
mettre la machine au point de se mouvoir si elle n'etoit animée
que de ces forces connues et des frottements et roideurs de cordes
qui resultent des pressions et tensions exercées par toutes les forces
du systeme donc le probleme auquel nous avons réduit la
partie utile du probleme géneral se reduit a ceci.

les forces appliquées a une machine données etant connues
et telles que sans le frottement et la roideur des cordes il y auroit
équilibre déterminer ce quil faut ajouter aux forces sollicitantes
pour mettre la machine au point de se mouvoir de sorte que
si l'on augmentoit davantage ces memes forces le mouvement
naitroit aussitot.

or quoique la question ainsi réduite soit le cas le plus simple
du probleme géneral la solution seroit encore extremement
compliquée si l'on vouloit proceder en toute rigueur ainsi nous
sommes obligés de renoncer a une précision géometrique et de
nous borner a une exactitude qui puisse satisfaire tout homme
raisonnable. la methode que nous allons proposer et que nous
suivrons dans la section suivante remplit je crois cet objet. elle
consiste a réduire la difficulté au meme que si le frottement
etoit réellement proportionnel a la pression par ce moyen si l'on
veut se contenter d'une approximation telle qu'on en a besoin
dans l'usage ordinaire des machines on se bornera a un
premier calcul qui n'aura pas plus de difficulté que si
le rapport du frottement a la pression etoit constant. si
l'on a besoin d'une tres grande exactitude on aura encore
a faire un nouveau calcul tout semblable au premier et
en continuant ainsi de faire des calculs toujours semblables
on approchera tant qu'on voudra de la solution cherchée
voici cette methode

on commencera par determiner les pressions qui
s'exerceront entre les differentes parties du systeme s'il
n'y avoit point de frottement alors par les formules de la
premiere partie on cherchera quel seroit le rapport du
frottement a cette pression sur chaque point si cette pression
etoit réellement celle qu'on a trouvée et l'on achevera la
solution comme si ce rapport etoit exact. cette supposition

étant fausse il en résultera une erreur mais 1° cette erreur sera très petite car les pressions réeles ne peuvent différer que peu de celles qui auroint lieu s'il n'y avoit pas de frottement parceque le frottement étant petit relativement a la pression il ne faut en général qu'une petite augmentation dans l'une des forces du système pour vaincre ce frottement cette petite augmentation est la cause qui fait que la pression vraie diffère de ce qu'elle seroit sans le frottement donc ces deux pressions diferent peu l'une de l'autre Donc (158) on pourra sans inconvénient prendre l'une pour l'autre dans la formule qui exprime le rapport du frottement a la pression: cela une fois admis on aura toutes les forces agissantes et resistantes exercées dans le système et la difficulté sera réduite au calcul mathématique d'une machine a laquelle sont appliquées des forces données problème facile a résoudre par la théorie de cette section. 2° de plus après ce premier calcul on aura une solution approchée qui donnant a peu près la valeur de l'augmentation cherchée fera connoitre en meme tems a peu près la petite quantité qui avoit ete négligée dans le rapport du frottement a la pression on n'aura donc qu'a resoudre de nouveau le problème comme si l'augmentation cherchée etoit précisément celle qu'a donnée le premier calcul ce second calcul tout semblable au premier fera trouver un nouveau resultat plus exact qu'on pourra employer de meme pour en avoir un troisieme plus exact encore et ainsi de suite de sorte qu'on approchera tant qu'on voudra de la solution cherchée par des calculs toujours semblables et les memes que si le frottement étoit réelement proportionnel a la pression.

en un mot j'adopte en tout les methodes ordinaires toute la difference consiste en ce que je commence par déterminer plus exactement qu'on ne le fait ordinairement le rapport du frottement qui s'exerce sur chaque partie du système a la pression qu'elle souffre et j'y parviens en employant pour trouver ce rapport non la pression réele comme l'indiquent les formules de la premiere partie mais celle qui auroit lieu sans le frottement c'est adire que je néglige quelquechose dans la pression qu'il faudroit employer pour avoir exactement son rapport au frottement ce qui facilite infiniment les calculs et ne peut (158) qu'influer extremement peu sur l'exactitude du résultat qu'on peut meme rectifier tant qu'on veut sans difficulté.

FOOTNOTES TRANSCRIBED FROM
CARNOT'S MEMOIRS

1778 MEMOIR (APPENDIX B) § 27, second page
The reference follows the second line, thus:
... vitesse qu'il a perdue[c] estimée dans le sens de celle ...
And reads:

 [c] Si l'on decompose la vitesse V en deux dont l'une soit u l'autre est ce que j'entends par la vitesse perdue et cette vitesse estimée dans le sens de u est evidement $V \cos y - u$.

§ 58
The reference follows the first line of the paragraph (Cor 3), thus:
58 Si l'on considere les moments[d] de toutes les forces appliquées ...
And reads:

 [d] par moment d'une force relativement a un axe j'entends la projection de cette force sur un plan perpendiculaire a cet axe multipliée par la distance de sa direction a cet axe.

1780 MEMOIR (APPENDIX C) § 113, second page
The reference follows the second to last line of text, thus:
... sera nommé mouvement géometrique.[a] 2° il est clair qu'en ...

 [a] pour distinguer par un exemple tres simple les mouvements que j'appelle géometriques de ceux qui ne le sont pas imaginons deux globes parfaitement egaux agissants l'un sur l'autre par un contact immediat et du reste libres et dégagés de tout obstacle // imprimons a ces globes des vitesses egales et dirigées dans le meme sens suivant la ligne des centres // ce mouvement est géometrique parceque les corps pourroint de meme etre mus en sens contraire avec la meme vitesse comme il est evident // mais supposons qu'on imprime a ces corps des mouvements egaux si l'on veut et meme dirigés dans la ligne des centres mais qui aulieu d'etre comme précedemment dirigés dans le meme sens tendent au contraire a les eloigner l'un de l'autre // ces mouvements quoique possibles ne sont pas ce que j'entends par mouvements géometriques parceque si l'on vouloit faire prendre a chacun de ces mobiles une vitesse egale et contraire a celle qu'il reçoit dans ce premier mouvement on en seroit empeché par l'impénétrabilité des corps.

de meme si deux corps sont attachés aux extremités d'un fil inextensible et qu'on fasse prendre au systeme un mouvement arbitraire mais tel que la distance des deux corps soit constement égale a la longeur du fil ce mouvement sera géometrique parceque les corps peuvent prendre un pareil mouvement dans un sens tout contraire // mais si ces mobiles se rapprochent l'un de l'autre le mouvement n'est point géometrique parcequils ne pourroint prendre un mouvement égal et contraire sans s'eloigner l'un de l'autre ce qui est impossible à cause de l'inextensibilité du fil.

en général quelque soit la figure du systeme et le nombre des corps, s'il est possible de lui faire prendre un mouvement tel qu'il n'en resulte aucun changement dans la position respective des corps ce mouvement sera géometrique, comme il est évident apres tout ce qui vient d'etre dit: il est clair en effet qu'il ne s'exerce aucune pression ni action quelconque entre les differentes partie du systeme et que par conséquent la connoissance des loix de la communication des mouvements est inutile pour les determiner; la determination de ces mouvements est donc une affaire de pure géometrie et c'est pour cette raison que je les appelle géometriques.

il ne s'ensuit pourtant pas dela que le mouvement d'un systeme de corps ne puisse etre géometrique sans que toutes ses parties conservent entre elles la meme position respective; c'est ce qui sera facile a comprendre par quelques exemples. // imaginons un treuil a la roue et au cilindre duquel soient attachés des poids suspendus par des cordes // si l'on fait tourner la machine de maniere que le poids attaché a la roue descende d'une hauteur égale a sa circonference tandis que celui du cilindre montera d'une quantité égale a la sienne ce mouvement sera géometrique parcequil est egalement possible de faire descendre le poids attaché au cilindre d'une hauteur égale a sa circonference tandis que le poids de la roue monteroit d'une hauteur égale a la sienne.

de meme si plusieurs corps sont attachés aux extremités de differents fils reunis par les autres extremités a un meme noeud et qu'on fasse prendre au systeme un mouvement tel que chacun des corps reste constement eloigné du noeud d'une meme quantité égale a la longeur du fil auquel il est attaché le mouvement sera géometrique quand meme ces differents corps se rapprocheroint ou s'eloigneroint les uns des autres.

si deux corps sont attachés aux extrémités d'un fil dans lequel soit enfilé un grain mobile il suffira pour que le mouvement soit géometrique que la somme des distances du grain mobile a chacun des deux autres corps soit constement egale a la longeur du fil desorte que si ces autres corps étoint fixes le grain ne sortiroit pas d'une courbe elliptique.

si un corps se meut sur une surface courbe par exemple sur la convexité d'une calote spherique le mouvement sera géometrique tant que le corps sera mu tangentiellement a la surface mais s'il s'en ecarte le mouvement n'est plus géometrique parceque le mouvement directement opposé est visiblement impossible.

si deux corps viennent a se choquer le mouvement du systeme apres le choc sera toujours géometrique et partant si l'on décompose le mouvement virtuel en deux dont l'un soit tel que s'il etoit seul il y auroit équilibre et dont l'autre soit géometrique celui ci sera le vrai mouvement du systeme apres le choc // mais il ne suffiroit pas pour l'assurer de prouver que ce mouvement est possible c'est adire compatible avec l'impénetrabilité ou que les corps puissent le prendre sans se gener mutuellement // il faut de plus que le mouvement egal et contraire soit possible c'est adire que ce mouvement doit etre géometrique.

il resulte visiblement de tout cela que si a la place du mouvement réel d'un systeme on en substitue un autre tel qu'il n'en résulte aucun changement dans l'action réciproque des corps ce mouvement sera ce que j'entends par géometrique mais j'ai preferé une définition suivant laquelle il fut clair par l'ennoncé meme que ces mouvements sont réelement géometriques c'est adire absolument indépendants des regles de la dinamique.

1780 Memoir (Appendix C)

The reference is in the last line of § 116, thus:

. . . aux indeterminées u differents rapports[b] ou differentes directions . . .

And reads:

[b] je dis qu'il faut attribuer aux indeterminées u differents rapports ou differentes directions parceque si l'on se contentoit de leur attribuer differentes valeurs sans que les rapports ni les directions fussent changés on auroit des équations qui seroint toutes justes a la vérité mais qui deviendroint identiques en les multipliant par différentes constantes comme on le verra aisément en jettant les yeux sur l'équation (AA).

§ 153, second page

The reference is in the 15th line from the top of the page, thus:

plus avantageuse, le travail de la force mouvante[c] doit s'estimer par la

[c] il faut prendre garde de confondre le travail absolu ou la peine de l'agent avec ce que j'appelle ici le travail de la force mouvante ou sollicitante car celleci n'est qu'une partie de l'autre // ce n'est dis je que la partie de l'effort absolu de l'agent qui est transmise aux forces résistantes par l'entremise de la machine. // mais l'agent est en outre obligé comme on l'a deja dit de faire un autre effort pour se mouvoir luimeme et l'on doit avoir égard a cet effort aussi bien qu'a celui qui est transmis a la machine et qui est proprement la force mouvante pour juger de son travail absolu. // nous disons ici que la force mouvante doit produire autant d'action quelle en consomme mais ce n'est point la seule condition a remplir pour tirer d'un agent dont le travail absolu est déterminé tout l'effet qu'on a droit d'en attendre. // il faut encore que la partie de ce travail employée a mouvoir l'agent luimeme soit la moindre possible affin que la partie transmise soit la plus grande et c'est a quoi l'on ne peut parvenir que par des experiences multipliées sur la force des animaux comme on verra dans l'article suivant.

BIBLIOGRAPHICAL NOTE

BIBLIOGRAPHICAL NOTE[1]

A. BIOGRAPHIES AND STUDIES OF CARNOT

The two principal biographies of Carnot are discussed in the preface: by his son Hippolyte, *Mémoires sur Carnot* (2 vols., 1861-63), and by Marcel Reinhard, *Le grand Carnot* (2 vols., 1950-52). In addition there may be mentioned Maurice Dreyfous, *Les trois Carnot* (1888), which deals in a semipopular way with the political careers of Carnot himself and of his son Hippolyte and grandson Sadi. It was published on the occasion of the death of the former of these descendants and the election of the latter to the presidency of the Republic. There is one biography to recommend in English, S. J. Watson, *Carnot* (London, 1954). Carnot's official letters and papers during his rise to political and military prominence and his service on the Committee of Public Safety were edited by E. Charavay, *Correspondance générale de Carnot* (4 vols., 1892-1907). In addition a volume of memorabilia was published by "la Sabretache" in 1923, *Centenaire de Lazare Carnot, 1753-1823, notes et documents inédits.* Two studies exist of his contributions to the art of fortification and to military thought: an essay by Ernest Marsh Lloyd in *Vauban, Montalambert, Carnot: engineer studies* (London, 1887); and R. Warschauer, *Studien zur Entwicklung der Gedanken Lazare Carnots über der Kriegsführung* (Berlin, 1937). Reinhard's biography indicates the archival location of manuscript materials bearing on Carnot's public life (II, 349), and lists the titles of his "Principaux Ouvrages" (II, 347). Since, however, Reinhard does not include his scientific writings in this category, it may be useful to set out the bibliographical detail in the chronological summary that follows.

B. PRINCIPAL SCIENTIFIC WRITINGS OF LAZARE CARNOT

1. Mémoire sur la théorie des machines pour concourir au prix de 1779 proposé par l'Académie Royale des Sciences de Paris. The manuscript is dated 28 March 1778. It is conserved in the *Archives de l'Académie des sciences, Institut de France,* and consists of 85 sections in 63 folios. The diagrams to which the text refers are missing. Sections 27-60 are reproduced in Appendix B of the present work.

2. Mémoire sur la théorie des machines pour concourir au prix que l'Académie Royale des Sciences de Paris doit adjuger en 1781. The manuscript is dated from Béthune 15 July 1780. It is conserved in the *Archives de l'Académie des sciences, Institut de France,* and consists of 191 sections in 106 folios. Sections 101-160 are reproduced in Appendix C of the present work.

3. *Essai sur les machines en général* (Dijon, 1783). This first printing appeared anonymously, the title page stating only that it is by "Un Officier du Corps Royal du Génie." It is very rare. The *Bibliothèque nationale* has no copy, the only one I have seen being in the family archives at Nolay. A second, identical printing was pub-

[1] Unless otherwise indicated, all French titles were published in Paris.

lished in 1786 (A Dijon, de l'Imprimerie de Defay. Et se vend à Paris, chez Nyon l'aîné, Libraire, rue de Jardinet). It carries the identification, "Par M. Carnot, Capitaine au Corps Royal du Génie, de l'Académie des Sciences, Arts, et Belles-Lettres de Dijon, Correspondant du Musée de Paris."

4. *Lettre sur les aérostats*. A manuscript memoir addressed to the *Académie royale des sciences* on 17 January 1784, and preserved in its archives.

5. Dissertation sur la théorie de l'infini mathématique, ouvrage destiné à concourir au prix qu'a proposé L'académie Royale des Sciences, arts et belles-lettres de Berlin, pour l'année 1786. The manuscript is dated from Arras 8 September 1785. It is conserved in the Archives of the *Deutsche Akademie der Wissenschaften zu Berlin*, and consists of 100 paragraphs in 90 folios. It is reproduced in its entirety as Appendix A of the present work.

6. *Réflexions sur la métaphysique du calcul infinitésimal* (Chez Duprat, an V-1797). Carnot published a second, much enlarged edition in 1813, chez veuve Courcier, the text of which was reissued by Bachelier in 1839, Mallet-Bachelier in 1860, Gauthier-Villars in 1881, and Gauthier-Villars again in 1921 (Collection "Les Maitres de la pensée scientifique"). For translations, see the essay in the present work by A. P. Youschkevitch, following Chapter V.

7. *Oeuvres mathématiques du citoyen Carnot* (Basel: J. Decker, 1797). Consists of the unchanged texts of (3) and (6) above, published with a portrait of Carnot.

8. "Lettre du citoyen Carnot au citoyen Bossut, contenant quelques vues nouvelles sur la trigonométrie," in Charles Bossut, *Cours de mathématiques, géométrie et application de l'algèbre à la géométrie* (Nouvelle édition, T. II, 1800), pp. 401-421.

9. *De la corrélation des figures de géométrie* (Duprat, 1801).

10. *Géométrie de position* (Duprat, 1803).

11. *Principes fondamentaux de l'équilibre et du mouvement* (Deterville, 1803).

12. Rapport (with Lacroix) fait à la première classe de l'Institut de France, 22 Nivôse an 11 (1804), "Mémoire sur une nouvelle notation d'algèbre descriptive," par Samson Michel, Institut de France, Académie des sciences, *Procès-verbaux des séances de l'Académie tenues depuis la fondation de l'Institut jusqu'au mois d'août 1835*, II (Hendaye, 1912), pp. 611-613.

13. *Mémoire sur la relation qui existe entre les distances respectives de cinq points quelconques pris dans l'espace, suivi d'un essai sur la théorie des transversales* (Courcier, 1806).

14. Rapport (with Berthollet) . . . sur la machine appelée Pyréolophore, 15 December 1806, Institut de France, *Procès-verbaux* III (1913), 465-467.

15. Rapport (with Charles, Montgolfier, and Prony) . . . sur la nouvelle machine à feu de M. Cagnard, 8 May 1809, *ibid.*, IV (1913), 200-202.

16. Rapport (with Monge and Prony) . . . sur la nouvelle machine hydraulique de M. Lingois, 13 July 1812, *ibid.*, V (1914), 72-75.

17. Rapport (with Périer and Prony) . . . sur les machines hydrauliques de l'invention de M. Mannoury, 28 December 1812, *ibid.*, V, 133-137.

18. Rapport (with Périer and Prony) . . . sur les *Moulins* de Mannoury d'Ectot, 21 June 1813, *ibid.*, V, 221-224.

19. Rapport (with Sané and Poinsot) . . . "Mémoire sur la *Stabilité des corps flottants*" par M. Dupin, 30 August 1814, *ibid.*, V, 392-395.

Carnot's less important reports, together with those of other commissions on which he served as a member but not as "rapporteur," may be located by means of the index published in each volume of the *Procès-verbaux*.

C. SADI CARNOT

The only work that Sadi published himself was *Réflexions sur la puissance motrice du feu et sur les machines propres à développer cette puissance* (Bachelier, 1824). The treatise was reprinted in *Annales scientifiques de l'Ecole normale supérieure*, 2ᵉ série, I (1872) and republished in 1878 (Gauthier-Villars) accompanied by a biographical memoir by Hippolyte Carnot, a letter presenting the original manuscript to the *Académie des sciences*, and a somewhat random selection from notes on mathematics, physics, and the nature of heat that Sadi had left in manuscript at his death. In 1927 Emile Picard published a facsimile edition of the complete notes, *Sadi Carnot, biographie et manuscrit* (Gauthier-Villars). The text of the *Réflexions* together with a complete and ordered printing of these notes was reissued in 1953 (Blanchard).

A German translation, *Betrachtungen über die bewegende Kraft des Feuers . . .* , appeared in Ostwalds "Klassiker der exacten Wissenschaften," No. 37 (Leipzig, 1892) accompanied by an introduction and commentary. There is an English translation in W. F. Magie, ed., *The Second Law of Thermodynamics* (New York: Harper, 1899), which also contains papers by Clausius and William Thomson, Lord Kelvin. Another English translation was published by R. H. Thurston (New York: The American Society of Mechanical Engineers, 1943). The most recent edition in English is by E. Mendoza, *Reflections on the motive power of fire, by Sadi Carnot; and other papers on the second law of thermodynamics by E. Clapeyron and R. Clausius* (New York: Dover, 1960).

The only other known work by Sadi, the memoir entitled "Recherche d'une formule propre à représenter la puissance motrice de la vapeur d'eau," has been published by W. A. Gabbey and J. W. Herivel, "Un manuscrit inédit de Sadi Carnot," *Revue d'histoire des sciences*, 19 (1966), 151-166.

There is a considerable secondary literature dealing with Sadi's work, its context and its importance for the development of the science of thermodynamics: Emmanuel Ariès,

L'oeuvre scientifique de Sadi Carnot; introduction à l'étude de la thermodynamique (Payot, 1921); Charles Brunold, *L'entropie, son role dans le développement historique de la thermodynamique* (Masson, 1930); Paul Renaud, *Analogies entre les principes de Carnot, Mayer et Curie* (Hermann, 1937); L. Rosenfeld, "La genèse des principes de la thermodynamique," *Bulletin de la Société Royale de Liège*, 10 (1941), 197-212; Milton Kerker, "Sadi Carnot," *The Scientific Monthly*, 85 (1957), 143-149; T. S. Kuhn, "Carnot's version of Carnot's cycle," *American Journal of Physics*, 23 (1955), 91-94; and "Engineering precedent for the work of Sadi Carnot," *Actes du IXᵉ Congrès International d'Histoire des Sciences* (Barcelona, 1959), pp. 530-535; E. Mendoza, "Contributions to the study of Sadi Carnot and his work," *Archives internationales de l'histoire des sciences*, 12 (1959), 377-396; C. C. Gillispie, *The Edge of Objectivity* (Princeton, 1960), Chapter IX; T. S. Kuhn, "Sadi Carnot and the Cagnard Engine," *Isis*, 52 (1961), 567-574; E. Mendoza, "Sadi Carnot and the Cagnard Engine," *Isis*, 54 (1963), 262-263; J. Payen, "Une source de la pensée de Sadi Carnot," *Archives internationales de l'histoire des sciences*, 21 (1968), 18-32; Robert Fox, "Watt's Expansive Principle in the Work of Sadi Carnot and Nicolas Clément," *Notes and Records of the Royal Society of London*, 24 (1970), 233-253; and "The intellectual environment of Sadi Carnot: a new look," *Actes du XIIᵉ Congrès International d'Histoire des Sciences* (Paris: in press); James Challey, "Sadi Carnot," *Dictionary of Scientific Biography*, Vol. III (New York: 1971), pp. 79-84.

D. SCIENTIFIC AND HISTORICAL CONTEXT

It would surpass the scope of this monograph to attempt a full bibliography of the history of mechanics, mathematics, technology, or politics during Carnot's lifetime. Relevant items in the contemporary scientific literature are cited in the notes and may be located through reference to the authors in the index. There is a wide-ranging bibliography of the original literature exhibiting the interactions of mechanics with what may be called rational technology in the issue of *Thalès* (12, 1966) recording the conclusions of a seminar devoted to that subject in the *Institut d'histoire des sciences et des techniques de l'Université de Paris*, under the direction of Georges Canguilheim. In the secondary literature the two writers who have contributed most signally to our knowledge of the history of mechanics in recent years are C. A. Truesdell at the level of rational mechanics and D.S.L. Cardwell at the level of technology. Truesdell has recently gathered his papers on particular topics into a book, *Essays in the History of Mechanics* (New York and Berlin: Springer, 1968), in which I have found most generally helpful, "A program toward rediscovering the rational mechanics of the age of reason," pp. 85-137; and most specifically helpful, "Whence the law of moment of momentum?" pp. 239-304. The papers by Cardwell that touch closely on the sector in which Carnot worked are "Power technologies and the advance of science, 1700-1825," *Technology and Culture*, 6 (1965), 188-207; and "Some factors in the early development of the concepts of power, work and energy," *British Journal for the History of Science*, 3 (1967), 209-224.

Other valuable studies of particular topics are C. S. Gillmor, *Coulomb and the Evolution of physics and engineering in eighteenth-century France* (Princeton, in press); Roger Hahn, "The chair of hydrodynamics in Paris, 1775-1791," *Actes du Xe Congrès international d'histoire des sciences* (Ithaca, 1962), pp. 751-754; and "L'hydrodynamique au 18e siècle" (Paris: Conférence du Palais de la Découverte, 1964); Thomas L. Hankins, "Eighteenth-century attempts to resolve the *vis-viva* controversy," *Isis*, 56 (1965), 281-297; Erwin Hiebert, *Historical roots of the principle of conservation of energy* (Madison, Wis.: State Historical Society, 1962); J. R. Ravetz, "The representation of physical quantities in eighteenth-century mathematical physics," *Isis*, 52 (1961), 7-20; Wilson L. Scott, "The significance of 'hard' bodies in the history of scientific thought," *Isis*, 50 (1959), 199-210. Also to be mentioned are René Dugas, *Histoire de la mécanique* (Neuchâtel, 1950); and Emile Jouguet, *Lectures de la mécanique*, 2nd ed., 2 vols. (Gauthier, 1924).

In the history of mathematics, René Taton, *L'oeuvre scientifique de Monge* (Presses Universitaires de France, 1951) is the most important work for the revival of geometry and for the mathematical milieu of late eighteenth-century France in general. Very helpful also are Pierre-Léon Boutroux, *Les principes de l'analyse mathématique; exposé historique et critique*, 2 vols. (Hermann, 1914-1919); Felix Klein, *Vorlesungen über die Entwicklung der Mathematik im 19 Jahrhundert*, ed. R. Courant and O. Neugebauer, 2 vols. (Berlin: Springer, 1926-1927); and Carl B. Boyer, *The Concepts of the Calculus; a critical and historical discussion of the derivative and the integral*, 2nd ed. (New York: Hafner, 1949). Professor Boyer is one of the few modern historians of mathematics who have noticed Carnot's work—"Carnot and the concept of deviation," *American Mathematical Monthly*, 61 (August-September 1954), 459-463; and "The great Carnot," *The Mathematics Teacher*, 49 (January 1956), 7-14. A work that has the merit of near contemporaneity as well as greater fullness than the histories just mentioned is Michel Chasles, *Aperçu historique sur l'origine et le développement des méthodes en géométrie, particulièrement de celles qui se rapportent à la géométrie moderne* (Bruxelles: Hayez, 1837). A second edition appeared in Paris in 1875.

Among special historical studies, the most pertinent are Georges Bouchard, *Prieur de la Côte-d'Or* (Clavreuil, 1946); Maurice Crosland, *The Society of Arcueil* (Cambridge, Mass.: Harvard University Press, 1967); Robert R. Palmer, *Twelve who Ruled: the Committee of Public Safety during the Terror* (Princeton: Princeton University Press, 1941); Robert S. Quimby, *The background of Napoleonic warfare: the theory of military tactics in 18th-century France* (New York: Columbia University Press, 1941); and René Taton, ed., *Enseignement et diffusion des sciences en France au 18e siècle* (Hermann, 1964). Maurice Crosland has recently published a most useful edition of the account of French scientific institutions in the 1790's by a contemporary Danish visiting astronomer: *Science in France in the revolutionary era, described by Thomas Bugge* (Cambridge, Mass.: the M.I.T. Press, 1969).

INDEX

Index

accelerative force, *see* force

action and reaction, 39, 40, 51, 67, 74, 133

action, quantity of, 77, 78, 80, 84, 115. *See also* moment-of-activity

d'Alembert, Jean le Rond, 7, 10, 27, 31, 63n., 120, 143, 154-155, 157-158, 163n; principle of, 39, 48, 71, 114, 118-119, 120

algebra, 27, 31, 42, 109, 125, 128, 129n., 155, 156, 162

Ampère, André-Marie, 117

analysis (in general), 4, 5, 31, 71, 92, 105, 121, 124, 128, 149, 151, 162, 166, 168; compared to synthesis, 121, 142, 143-145; finite, 134, 143, 154, 163n.; infinitesimal, 134, 136, 143, 149-168. *See also* calculus

angular momentum, *see* motion

Apollonius, 153

approximation, method of, 135, 138, 141, 158, 160, 167

Arago, François, 13, 117

Arbogast, Louis-François-Antoine, 164

Archimedes, 14, 37

d'Arcy, Patrick, 89

Aristotle, 22

artillery, 6, 15, 16, 102

d'Aumont, Duc, 6

Babbage, Charles, 141

Babeuf, François-Emile, 22

Bailleul, J.-Ch., 23n.

Bailly, Jean-Sylvain, 156n.

Barère, Bertrand, 21

Barras, Paul-Jean, 23

Barré de Saint-Venant, 101, 104

Beguelin, Nicolas, 157n.

Bélidor, Bernard Forest de, 10, 102, 106, 111, 112, 114, 116, 117

Bellavitis, Giusto, 122

Bénézech de Saint-Honoré, 8

Bérard, J. E., 109

Berkeley, George, 151-155, 160

Berlin, Academy of Sciences, 11, 24, 122, 150, 155-156, 157n., 158, 168

Bernoulli, Daniel, 10, 31, 34, 89-90, 112-113, 141

Bernoulli, Jakob, 31, 141

Bernoulli, Johann I, 31, 141, 164

Bernoulli, Johann III, 157n.

Berthaud, 7n.

Berthollet, Claude-Louis, 25, 28n., 29

Bézout, Etienne, 63n.

Billaud-Varenne, Jean-Nicolas, 21

Biron, Duc de, 19

Blanchard, François, 92n.

Bolzano, Bernard, 149, 168

Borda, J.-C. de, 112-113

Bossut, Abbé Charles, 7, 8, 10, 26-27, 63n., 105, 106, 128n.

Bouffet, 7n.

Bougainville, Louis-Antoine, 29

Bouillet, Chevalier de, 9

Bouillet, Ursule, 9

Boyer, Carl B., 151n.

Bréguet, 25

Brunold, Charles, 6on., 82n., 91n.

Bureaux de Pusy, 8

Cagniard de Latour, 28

Cajori, Florian, 151n.

calculus, Carnot on, 121-145, 149-168; differential, 116, 140, 152-154, 155, 158, 159-160, 164-168; Euler on (calculus of zeros), 154, 161, 164-165; foundations of, 5, 27, 82, 92, 116, 122, 142, 144, 149-168; history of, 141, 149, 151, 153; integral, 116, 138, 158, 161; Leibniz on, 141-142, 152, 158, 160-162, 164, 167; method of exhaustion, 141, 142, 151, 157, 162; method of first and last ratios, 142-143, 160, 164; method of indivisibles, 141, 161, 162; method of infinitesimals, 134-143, 149-168; method of limits, 27, 141-143, 151, 154, 157-158, 160, 162, 163-168. *See also* continuity, principle of; Newton on (fluxions), 141, 151-152, 160, 162, 165

caloric, 28, 60, 93-99

Camus, Charles-Etienne-Louis, 10

Cantor, M., 149-150n.

Cardwell, D.S.L., 82n., 110n.

Carnot, Claude, 5-6

Carnot cycle, 4, 61, 97-100. *See also* continuity, principle of

Carnot de Feulint, Claude-Marie, 6, 7, 11, 16-17, 18, 23

Carnot, Hippolyte, 6, 7n., 8, 9, 10n., 11, 12n., 13, 15n., 21, 23, 24, 29, 30, 92, 95, 140n., 159n.

Carnot, Joseph, 9

CARNOT, LAZARE

1) general character of his work, 4-5, 13-14, 23-24, 31-33, 35, 59, 66-67, 71, 79, 105, 123, 131-132, 134, 139, 141, 144-145n.; principle of, 4, 56-58, 70n., 77, 80, 90, 108, 119-120; reputation, 5, 17-18, 23, 31, 82, 101, 106-107, 113-115, 117, 121-123, 131, 149, 168.

2) Convention, 18-20; death, 30, 92; education, 6-7; *Eloge de Vauban* (1784), *10-13*; exile, 3, 23, 30, 91; family, 5-6, 24, 26, 29; Institut de France, technological reports to, 26-29, 101, 121; Legislative Assembly, 16-18; love affair and imprisonment, 9; member of the Committee of Public Safety, 3, 19-22; member of the Directory, 3, 22-23; Minister of the Interior, 3, 30; Minister of War, 3, 24; professional life, 9-13, 14-16, 23, 25-29, 101; relations with Napoleon, 3, 24, 26, 29-30; representative on mission, 18-19.

3) *Mémoire sur la théorie des machines* (1778 Prize Essay), 11, *62-72*, App. B; *Mémoire sur la théorie des machines* (1780 Prize Essay), 11, *72-81*, 145n., App. C; *Essai sur les machines en général* (1783), 4, 5, 11, 14, 25, *31-61*, 62, 66, 67, 68, 70, 71, 72, 73, 74, 75, 77, 78, 79, 80, 81, 82, 83, 84, 86, 87, 88, 89, 90, 92, 101, 113, 116, 133; *Lettre sur les aérostats* (1784), 11-12; *Dissertation sur la théorie de l'infini mathématique* (1785 Berlin Prize Essay), 11, 122, 135, *149-168*, App. A; *Réflexions sur la métaphysique du calcul infinitésimal* (1797 & 1813), 3, 11, 24, 82, 121-122,

CARNOT, LAZARE—*Réflexions (cont.)*
134-143, 149-150, 159, 160, 162,
166, 167-168; *Lettre . . . au
citoyen Bossut . . . sur la
trigonométrie* (1800), 24, *128n.*;
*De la corrélation des figures de
géométrie* (1801), 24, 82,
123-125, 128, 144; *Géométrie
de position* (1803), 3, 24,
81, 86, 115, 121-122, 125,
128-133, 141, 162; *Principes
fondamentaux de l'équilibre
et du mouvement* (1803), 3,
24, 31, 35-36n., 43, 44, 62, 68,
75, *81-90*, 91, 95n., 101, 106,
113, 116, 128, 133, 140, 143;
*Essai sur la théorie des
transversales* (1806), 24, 125n.,
131; "Digression sur les
quantités dites négatives"
(1806), *125-127*, 144n.
Carnot, Sadi, 3, 4, 5, 24-25, 28, 29,
30, 50, 62, 101, 107, 110-111;
work compared to Lazare's, 31,
60-61, 86, *90-100*, 144-145
Carnot, Sadi (grandson), 3
Carroll, Lewis, 139
Carruccio, Ettore, 122n.
Castillon, Jean-François, 157n.
Cauchy, Augustin, 27, 118, 119-120,
134, 149, 166, 168
Cavalieri, Bonaventure, 141
center of gravity, principle of,
36-38, 48, 54-56, 69-70, 74, 79,
85, 105, 132
Challey, James, 95n.
Clapeyron, Benoît-Paul-Emile,
100, 101
Charles, Jacques-Alexandre-César,
26
Chasles, Michel, 101, 121, 141
Clément, Nicolas, 91, 95n.
collision theory, 33-34, 35, 39-50,
56-57, 60, 67-68, 76, 80, 81, 85,
86, 87-89, 105, 106, 112-113,
118-120, 133, 144
Committee of Public Safety, 3, 12,
19-20, 22, 24, 25
Condé, Prince de, 11
Condorcet, Antoine-Nicolas
Caritat, Marquis de, 63n.
conservation of moment of the
quantity-of-motion, *see* motion
Constantine, Grand Duke, 30
continuity, principle of, 4, 5, 14,
34, 47, 54, 70, 96, 108, 124, 141,

142, 144. *See also* Carnot cycle;
Carnot, principle of; calculus,
method of limits
Coriolis, G.-G. de, 5, 70, 101,
104, 115-119
Corps of Engineers, *see*
engineering, profession of
corpuscular theory of matter, 33,
39, 40-41, 44, 51, 58, 66, 67, 69,
134, 136
correlative systems, *see* geometry,
correlation of figures
Coulomb, Charles-Augustin, 10,
11, 26, 63, 73, 89, 106, 113
Coustard, Anne-Pierre, 19
Couthon, Georges, 21
cyclic processes, 98-100, 144

Daguerre, Louis-Jacques-Mandé, 27
Dalton, John, 91
Darboux, Gaston, 12n.
dead force, *see* force
Desaguliers, Jean-Théophile, 112
Desargues, Girard, 121
Descartes, René, 14, 47, 48, 102,
141; principle of, 36, 38, 47, 54,
69, 74, 102
Didot, Firmin, 102
Dietrich, Philippe-Frédéric, Baron
de, 19
differentials, *see* calculus
Dijon, Academy of, 10, 11, 12n.
dimensional glossary, 35-36
dirigeable, 12
displacements, virtual, 42, 75, 86,
120. *See also* geometric motion
double refraction, 27
Drappier, Pierre-Thomas, 117
Duesme, Chevalier de, 9
Dugas, René, 82n.
Dulong, Pierre-Louis, 107
Dumouriez, General, 19
Dupin, Charles, 29, 95n., 101, 117
Dupont de Nemours,
Pierre-Samuel, 29
Duthuron, G., 12n.
dynamics, 10, 34, 39, 83, 85, 86, 105

Earle, Edward Mead, 15n.
Ecole Centrale, 104
*Ecole Centrale des Travaux
Publics. See Ecole Polytechnique*
Ecole d'Application, 104
Ecole des Ponts et Chaussées, 103,
106, 116
Ecole Polytechnique, 7n., 24-25, 82,

105, 115, 116, 140; graduates of,
30, 91, 102, 103, 104, 106, 117.
See also engineering, education in
elastic bodies, 33, 34, 35, 42, 44, 50,
69, 74, 76, 81, 85, 86, 88, 89, 118,
119. *See also* collision theory
elasticity, theory of, 33, 47, 49, 103,
104, 108-109
energy (in general), 4, 5, 34, 39,
43, 47, 56, 61, 67, 84, 87-88, 89,
94, 99, 110-111; kinetic, 4, 34,
35, 70, 84, 88, 111, 116; potential,
35, 84, 88. *See also* live force
engine: air, 28-29, 92, 96, 99, 110,
121; Cagniard, 28-29, 92, 96n.,
121; column of water, 108-110;
heat, 4, 12, 28-29, 50, 60, 90-100,
107, 110, 144; Niepce, 27-29, 92,
110, 121; steam, 12, 28-29, 91-100,
103, 106, 110
engineering, education in, 6-8, 10,
37, 82, 91, 115, 116. *See also*
Ecole Polytechnique; *Mézières,
Ecole Royale du Génie de*;
mechanics in, 4, 26n., 34, 101-105,
111, 116, 121; profession of, 5,
6, 8, 10, 11, 13, 15, 16, 23-24, 25,
26, 32, 34, 50, 59, 64, 73, 81, 82,
101-102, 105-106, 107, 112, 121,
134, 139, 144
equations, theory of, 138-139,
144-145, 159, 160-161, 164,
165-166, 167, 168
equilibrium and motion, principles
of, 4, 29, 32, 34-40, 44, 47-60,
65-71, 73-90, 93-94, 99, 102,
105-106, 107-109, 111, 113, 114,
118-119, 132-133, 141, 143, 145n.
See also center of gravity,
principle of; Descartes, principle
of; machines, science of; motion,
quantity of
equipollencies, 122
error, compensation of, 134-143,
145n., 149, 150, 152-155, 158-159,
160, 161, 164
Euclid, 132
Euler, Leonhard, 10, 27, 31, 88,
89n., 106, 112, 153-154, 161
exhaustion, method of, *see* calculus,
method of exhaustion

Fermat, Pierre de, 131
Feulint, *see* Carnot de Feulint,
Claude-Marie
first and last ratios, *see* calculus

flight, 11-12
fluxions, *see* calculus
force (in general), 5, 35-39, 42,
    45-48, 50-59, 67-72, 76, 80-81,
    83-84, 88-90, 99, 105, 107,
    111-112, 113-114, 115, 132, 145n.;
    accelerative, 36, 37, 67-68, 83,
    105; dead, 36, 58, 76, 81;
    impelling and resisting, 38, 40,
    52-59, 67, 72, 77-80, 113, 145n.;
    motive, 28, 36, 48-50, 52-53, 67,
    72, 75, 78-79, 84, 91, 96, 106,
    108-110, 158. *See also* live force;
    quantity-of-motion; power;
    weight
Formey, J.-A.-S., 150
fortification, 13, 15-16, 29, 102;
    of Calais, 9; of Cherbourg, 9;
    perpendicular, 14-15
Fourcroy, Antoine, 15
Fourcroy de Ramecourt, C.-F., 15
Fourier analysis, 103
Fox, Robert, 95n.
Frederick the Great, 15
friction, 51, 62-66, 72-73, 79, 97,
    105, 106, 108, 112, 118, 145n.
Fructidor, coup d'état of, 3, 23,
    24, 26
Fulton, Robert, 27

Gabbey, W. A., 91n., 94n.
Galilei, Galileo, 34, 37, 85,
    111-112, 131
Gauss, Charles-Frederic, 122, 149,
    168
Gauthey, E.-Marie, 103
geometric motion (in general), 5,
    120, 132-133; in the *Essai sur les
    machines*, 39, 40-45, 47, 49, 50,
    52-54, 57, 61, 67, 68, 74-75, 77,
    79, 86-89, 94, 132-133; in the
    1778 *Mémoire*, 67, 68-69, 75;
    in the 1780 *Mémoire*, 74-76,
    78-79, 81, 86; in the *Principes
    Fondamentaux*, 43, 44, 68, 75,
    86-89, 133; in the *Géométrie de
    Position*, 86, 133; in relation to
    reversible processes, 29n., 43, 86,
    97, 99, 133
geometry (in general), 5, 10, 12, 31,
    38-39, 82, 83, 86, 105, 121-122,
    125, 128-129, 131-133, 144, 151,
    155-156, 158; Carnot's definition
    of, 132; correlation of figures,
    123-125, 127, 128n., 129-131;
    descriptive, 10, 121, 131;

geometric analysis, 123-133;
    history of, 131; influence of
    Carnot's, 115; of position, 123,
    128-133; projective, 121;
    topology, 27, 132
Gillmor, C. Stewart, 10n., 26n.,
    63n., 89n.
Girard, P. de, 29, 117n.
Gosselin, 104
Guenyveau, André, 106, 117
Guibert, Comte de, 15
Guillaume, James, 25n.
Guyton de Morveau,
    Louis-Bernard, 26-27

Hachette, J.-N.-P., 105, 106-107,
    110n.
Hahn, Roger, 6n.
Halley, Edmond, 131
hard bodies, 33-35, 39, 40, 42, 44,
    46-50, 58, 66-67, 74, 76, 81, 85-89,
    113, 118-119. *See also* collision
    theory
Hauff, J.-K.-F., 122, 150
heat, 12, 50, 86, 91-97, 101,
    109-111; mechanical equivalent
    of, 100. *See also* caloric
Henderson, L. J., 103
Henry, Prince of Prussia, 11
Hérault de Séchelles, 21
d'Herbois, Collot, 21
Herivel, J. W., 91n., 94n.
Hofmann, J. E., 150n.
L'Hôpital, Marquis de, 134, 141,
    152
L'Huilier, Simon, 156-158, 163,
    164, 166n.
Huygens, Christiaan, 112, 131
hydrodynamics, 5, 10, 34, 86, 106,
    112-113, 128, 133, 144.
    See also machines, hydraulic

ideal systems, *see* machines,
    idealized
impact, laws of, *see* collision theory
impenetrability, 33
indeterminate coefficients, method
    of, 162, 167
indivisibles, *see* calculus
inelastic bodies, *see* collision theory;
    hard bodies
inertia, 32, 34, 55, 72, 74, 79-80, 105
infinitesimals, *see* calculus
Institut de France, 5, 25-29, 82, 121
integration, *see* calculus
inventions, 12, 26-29, 101, 121

Jacobins, 8, 20, 25, 30
Janvier, Aristide, 25
Jouguet, Emile, 35n., 82n.
Joule, J., 100

Kästner, Hofräth, 122
Klein, Felix, 122, 151
Kolmogorov, A. N., 165
Kretz, X., 104
Kuhn, Thomas S., 28n., 96

La Caille, N. L. de, 155
Laclos, Choderlos de, 14-15
Lacroix, Sylvestre-François, 27, 29,
    122, 125, 128n., 168
Lagrange, Joseph-Louis, 27, 31, 37,
    82, 85-86, 87, 106, 113, 139-140,
    154-160
Lakhtin, L. K., 149
Landen, John, 141
Laplace, Pierre-Simon, 25, 27, 29,
    106, 116
Laroche, François de, 109
Lauzun, Duc de, 19
Least Action, Principle of, 39, 48,
    57n., 86, 87-89
Legendre, Adrien-Marie, 27
Leibniz, G. W., 27, 56, 132, 141,
    142, 152, 158, 160-162, 164, 167
light, polarization of, 27n.
limit theory, *see* calculus
Lindet, Robert, 21
Lingois, Abbé, 29
Liouville, J., 103
live force, 34-35, 47, 48, 56-57, 67,
    70, 76, 78-79, 81, 84-85, 87-90, 99,
    107-112, 114-120; conservation of,
    4, 5, 34-35, 39, 42, 44, 48-50, 58,
    69, 71, 76, 78, 85, 87, 96,
    111-114, 117; history of, 111-114;
    latent, 84-85, 88, 90
Longpré, Louis-Siméon de, 7
Louis XVI, 19, 23, 30
Louis XVIII, 29
Lückner, General, 19
Lucretius, 62, 72

machines (in general), 4, 26, 63, 66,
    70, 85, 87, 89-90, 94, 101, 112, 118;
    conception and definition of,
    32, 50, 51, 72, 74; funicular, 71,
    72; hydraulic, 4, 29, 34, 40, 58,
    59-60, 91, 95, 108-110, 111n.,
    112-113. *See also* hydrodynamics;
    idealized, 37, 60, 95, 96n., 99, 144;
    operation of, 4, 40, 56, 59, 61,
    69, 88, 99, 110; science of, 11,

machines—science of *(cont.)*
28, 31-62, 67, 71-83, 91-92,
101-107, 111-117, 121, 134, 141,
145; simple, 32, 50, 62, 65, 71, 73,
80, 93, 105; spring-driven, 90;
weight-driven, 36, 56-57, 59,
78, 79
Maclaurin, Colin, 131, 141, 151, 154
magnitude, geometric, 123, 126,
128, 161
Mallet, Charles-François, 117
Malus, Etienne-Louis, 27n.
Mannoury d'Ectot, 29
Marie, Abbé Joseph-François, 155
mass, 32, 36, 37, 41, 46, 48, 50, 51,
54, 56-58, 66, 69, 70, 72, 74, 84,
88, 108, 109
mass-points, 37, 114
mathematics, 3, 5, 8, 10, 42, 52,
105, 121-123, 125, 128-129,
131, 134, 139-140, 143, 151, 155,
167; British, 141-142. *See also*
algebra; analysis; calculus;
geometry; trigonometry
matter, states of, 33; structure of,
see corpuscular theory
Maupertuis, P.-L. Moreau de, 33, 77
maxima and minima, 37, 48, 57, 59
Mayer, Julius Robert, 100
mechanical advantage, 40, 59, 71,
80, 89
mechanics (in general), 3, 4, 5, 7,
8, 10, 11, 13, 24, 31-37, 40, 42, 46,
47, 50-53, 60, 62-63, 66-67, 68,
73-75, 81, 83, 85-87, 101, 103, 110,
113-116, 119, 121, 128, 132-134,
139, 140, 144, 155, 158; applied,
101-107, 115, 117; principles of,
4, 40, 51, 66, 71-72, 76, 83, 89,
93, 96, 114; rational, 31, 38, 40,
42, 82, 85-87, 102, 104, 107, 115,
116. *See also* action and reaction;
d'Alembert, principle of;
Descartes, principle of;
equilibrium and motion; force;
Least Action; live force;
moment-of-activity; motion; work
mechanics, engineering, *see*
engineering, mechanics in
Merz, J. T., 103
metaphysics, 35, 135, 140, 151, 154,
155, 162, 168
Metz, *Ecole d'Artillerie et du
Génie,* 116, 117
Meusnier, de la Place,
Jean-Baptiste, 8, 10, 12

*Mézières, Ecole Royale du Génie
de,* 6-8, 10, 11, 24, 105
Michel, Samson, 27
moment-of-activity, 5, 34, 35, 36n.,
39, 40, 43, 47, 51-61, 67, 69-70,
75, 76-77, 81, 84-85, 88, 89, 90,
93-94, 96, 99, 104, 111, 115-116.
See also work, concept of
moment-of-momentum, *see* motion
moment of the quantity-of-motion,
see motion
momentum, *see* motion
Monge, Gaspard, 7-8, 24, 27, 29,
105, 121, 131
Montalembert, General the
Marquis de, 14-15, 16
Montgolfier brothers, 12, 110
Motion, conservation of moment of
quantity-of-motion, 47, 49, 53,
57, 67, 75, 77, 89; geometric, *see*
geometric motion; lost and
gained, 46, 51, 53, 67, 75, 119;
moment of quantity-of-motion
(moment of momentum), 39, 40,
45-51, 53, 57, 66-67, 71, 75-76,
78, 81, 89; quantity of motion
(momentum), 34, 36, 40-42, 47,
51, 66, 68-69, 77, 84, 87, 118-119;
virtual, 32, 40, 44, 74, 87, 120.
See also equilibrium and motion
motive force, *see* force
motor, definition of, 106-108;
internal combustion, 27
Mulcrone, Thomas F., 8n.

Napier, J., 131
Napoleon Bonaparte, 3, 6, 22, 23,
24, 26, 29-30, 105, 123
Napoleon II, 30
Navier, C.-L.-M.-H., 101, 102-104,
111-114, 117-118, 119n.
Newton, Isaac, 39, 131, 141-143,
151-152, 154-155, 160, 165
Niepce, Claude, 27, 92, 110, 121
Niepce, Joseph-Nicéphore, 27, 92,
110, 121
Nollet, Abbé Jean-Antoine, 8

Palissot de Beauvois, 29
Palmer, R. R., 15n., 19n., 21n.
Parent, Antoine, 112
Paris, Academy of Sciences, 11, 25,
26, 31, 62-63, 72-74, 81, 95n., 102,
106, 114, 115, 116, 117, 119n.,
145n.
Pascal, Blaise, 121

Peacock, George, 141
percussion, 4, 40, 47, 56-57, 60, 80,
89, 112-113, 118
perpetual motion, 37, 40, 59, 96, 99
Petit, A.-T., 101, 107-111, 114,
117, 118
photography, 27
physics, 5, 10, 24, 34, 37, 39, 52-53,
60-61, 67, 93, 96, 97, 98-99,
111, 151
Podgorny, M. E., 142
Poinsot, Louis, 27
Poisson, Siméon, 107, 118
Poncelet, J.-V., 101, 103, 116,
117-121
porisms, 131
position, geometric, *see* geometry
power (in general), 5, 34, 35, 36n.,
40, 50-56, 59-60, 67-71, 75-77, 81,
89-92, 99, 101-102, 105, 112,
115-116; hydraulic, 40. *See also*
machines, hydraulic; motive,
28-29, 75, 91-99, 110; technology
of, 26-29, 90-100, 105, 107-111;
transmission of, 4, 40, 47, 51-52,
54-55
Prieur de la Côte-d'Or,
Claude-Antoine, 8, 19, 20, 21,
24-25, 29, 140
Prieur de la Marne, 21
priority, questions of, Carnot and
Lagrange, 139-140, 158-160;
Leibniz and Newton, 160, 167;
Poncelet and Coriolis, 116-117
prize contests, *see* Berlin, Academy
of Sciences; Dijon, Academy of;
Paris, Academy of Sciences
Prony, Gaspard, 26, 103, 106, 107,
116, 117n.
Prussian Academy of Sciences, *see*
Berlin, Academy of Sciences

quantity, arbitrary, 138-139,
154; auxiliary, 111, 139, 142,
143-144, 159; designated and
non-designated, 138-139, 159,
160-161, 164; direct and inverse,
124, 127-128, 130, 143, 145; finite,
138, 141, 154, 165, 166;
imaginary, 129, 130, 134,
143-144, 161; indefinitely small,
163-164, 165; infinite, 138, 151,
155, 159, 163; infinitesimal,
138-139, 140-141, 149-168;
negative, 121, 123-134, 143, 159,
161, 162; potential infinitesimal,

158; projection of, 41n., 83-84,
114-115, 131-132; vanishing, 141,
142, 153-154, 155, 161, 164-165,
167
Quimby, Robert S., 14n.

Reinhard, Marcel, 7n., 8, 9n., 10n.,
13-14, 16, 17, 18, 19n., 20n., 22n.
resistances, 37, 50, 51, 55, 58, 72,
106, 107-110, 118-119
reversibility, 4, 5, 29n., 35, 43, 61,
86, 93, 94-95, 96n., 98-99, 110,
133, 144
Revolution, French, 3, 4, 5, 6, 8,
12, 14, 16, 17, 19-20, 22-24, 25,
26, 29, 103, 128n.
Robespierre, M., 3, 9, 21
Rosati, society of, 9
rotation, moment of, 89
Rousseau, Jean-Jacques, 12n.
Royal Corps of Engineers, see
engineering, profession of

Saint-André, Jeanbon, 21
Saint-Just, A., 3, 21
Saint-Simonianism, 30
Schatounova, E. S., 157n.
Schumacher, H. G., 122, 125
Scott, Wilson L., 33n.
Serret, J. A., 113n.
siegecraft, 13, 102
signs, algebraic, 124-128, 130, 144,
157
Simpson, Robert, 141
soft bodies, see collision theory
Solovine, Maurice, 134

springs, 113-114
statics, 10, 32, 83, 85, 105, 133
steam engine, see engine, steam
Stevin, Simon, 37
Stewart, Matthew, 141
strategy, military, 12-13, 14-16, 29.
See also engineering;
fortification; siegecraft
submarine, specific gravity of, 12
synthesis, 121, 136, 143-145.
See also analysis

Taton, René, 6n., 7n., 8n., 10n.,
105n.
Taylor, B., 155
Terror, Jacobin, 3, 18, 20, 21, 22, 25
thermodynamics, science of, 3, 4,
5, 31, 61, 90-100
Thomson, William (Lord
Kelvin), 100
Thouin, A., 29
Thouin, J.-A., 29
topology, see geometry
torque, 39, 47, 54, 66
Torricelli, E., 85
transversals, theory of, 125, 129n.,
131
trigonometry, 5, 24, 31, 41n.-42n.,
46, 48, 53, 59, 83-84, 114,
126-132
Trudaine de Montigny, 63n.
Truesdell, Clifford A., 46-47
turbulence, 4, 40, 58, 60, 113
Turreau, Louis, 21

units, 116

Vandermonde, Alexandre, 25
Vauban, Sébastien, 10, 12, 14-15
vector analysis, 5, 35, 38, 68, 71, 83,
86, 114. See also geometric
motion; quantity, projection of
velocity (in general), 5, 34, 36,
40-42, 44, 45-46, 48, 49, 52-55,
59, 60, 69-71, 76, 77-80, 83-84,
88, 115, 118-120; geometric, 43n.,
44, 51. See also geometric
motion; lost and gained, 41n.,
44, 48, 49n., 66-67, 84-85,
118-120. See also motion, lost
and gained; relative, 119-120;
virtual velocities, principle of,
38-39, 41, 66, 85-86, 114, 120
Viète, François, 131
virtual displacements, see
displacements, virtual; geometric
motion
vis viva, see live force
Vivanti, G., 149-150

Wallis, John, 33
warfare, theory of, see strategy,
military
waterwheel, see machines,
hydraulic
weight, 36-38, 43, 52, 54, 56-58,
63-65, 69-70, 72, 74, 76, 79, 81, 90,
108, 109, 111-112, 113, 115
work, concept of, 58, 69, 70, 78, 79,
87-88, 89, 90, 110

Youschkevitch, A. P., 24, 134, 142,
143, 152n., 154n.